Green Energy Harvesting

Green Energy Harvesting

Materials for Hydrogen Generation and Carbon Dioxide Reduction

Edited by

Pooja Devi
CSIR, Delhi
India

WILEY

This edition first published 2023

Registered Offices
John Wiley & Sons, Inc., 111 River Street, Hoboken, NJ 07030, USA
John Wiley & Sons Ltd, The Atrium, Southern Gate, Chichester, West Sussex, PO19 8SQ, UK

Editorial Office
Boschstr. 12, 69469 Weinheim, Germany

For details of our global editorial offices, customer services, and more information about Wiley products visit us at www.wiley.com.

Wiley also publishes its books in a variety of electronic formats and by print-on-demand. Some content that appears in standard print versions of this book may not be available in other formats.

A catalogue record for this book is available from the Library of Congress

Paperback ISBN: 9781119776055; ePub ISBN: 9781119776079; ePDF ISBN: 978111977606;
oBook ISBN: 9781119776086

Cover image: © Elnur/Shutterstock
Cover design by Wiley

Set in 9.5/12.5pt STIXTwoText by Integra Software Services Pvt. Ltd, Pondicherry, India

Contents

List of Contributors

Chandraraj Alex
Centre for Nano and Soft Matter Sciences,
Shivanapura, Bengaluru,
Karnataka, India

Soumen Basu
School of Chemistry and Biochemistry,
Thapar Institute of Engineering and
Technology, Patiala, Punjab, India
soumen.basu@thapar.edu

Laxmidhar Besra
Materials Chemistry Department, CSIR-
Institute of Minerals and Materials
Technology, Bhubaneswar, Odisha,
India

Rajani Kumar Borah
Inorganic and Physical Chemistry
Laboratory (IPCL), Council of Scientific
and Industrial Research (CSIR)-Central
Leather Research Institute (CLRI),
Chennai, India

Tribani Boruah
Institute of Nano Science and Technology
(INST), Mohali, Punjab, India

Rapaka S. Chandra Bose
Centre for Materials for Electronics
Technology, Thrissur, Kerala,
India

Sriparna Chatterjee
Materials Chemistry Department, CSIR-
Institute of Minerals and Materials
Technology, Bhubaneswar,
Odisha, India
sriparna251@gmail.com

Sampath Chinnam
Department of Chemistry, M.S Ramaiah
Institute of Technology (Affiliated to
Visvesvaraya Technological University,
Belgaum), Bengaluru,
Karnakata, India
sampathchinnam@gmail.com

Tulsi Satyavir Dabodiya
Department of Chemical and Materials
Engineering, University of Alberta,
Alberta, Canada
er.tulsikumar@gmail.com

Pooja Devi
Materials Science and
Sensor Application,
Central Scientific Instruments Organisation,
Chandigarh, Punjab, India
poojaiitr@csio.res.in

Ramendra Sundar Dey
Institute of Nano Science and Technology
(INST), Mohali, Punjab, India
rsdey.kgp@gmail.com

Adarsh P. Fatrekar
Inorganic and Physical Chemistry Laboratory (IPCL), Council of Scientific and Industrial Research (CSIR)-Central Leather Research Institute (CLRI), Chennai, India

Twinkle George
Centre for Nanoscience and Technology, Madanjeet School of Green Energy Technologies, Pondicherry University Kalapet, Puducherry, India

Demudu Babu Gorle
Materials Research Centre, Indian Institute of Science, Bangalore, Karnataka, India

Neena S. John
Centre for Nano and Soft Matter Sciences, Shivanapura, Bengaluru, Karnataka, India
jsneena@cens.res.in

Kamlesh
Academy of Scientific & Innovative Research (AcSIR), Ghaziabad, Uttar Pradesh, India

Maryam Sadat Kiai
Nano-Science and Nano-Engineering Program, Graduate School of Science, Engineering and Technology, Istanbul Technical University, Istanbul, Turkey

N. Usha Kiran
Materials Chemistry Department, CSIR-Institute of Minerals and Materials Technology, Bhubaneswar, Odisha, India

Indu Kumari
Department of Biotechnology, Chandigarh College of Technology, Chandigarh Group of Colleges, Landran, Mohali, Punjab, India

A. Lakshman Kumar
CSIR-Central Electrochemical Research Institute, Karaikudi, Tamil Nādu, India

Rameez Ahmad Mir
Department of Materials Science Engineering, University of Toronto, Canada

Manish Mudgal
Academy of Scientific & Innovative Research (AcSIR), Ghaziabad, Uttar Pradesh, India

Annapurna Nowduri
Department of Engineering Chemistry, Andhra University College of Engineering (A), Andhra University, Visakhapatnam, India
dr.nannapurna@andhrauniversity.edu.in

Swapna Pahra
Materials Science and Sensor Application, Central Scientific Instruments Organisation, Chandigarh, Punjab, India

Panchami R.
Inorganic and Physical Chemistry Laboratory (IPCL), Council of Scientific and Industrial Research (CSIR)-Central Leather Research Institute (CLRI), Chennai, India

O.P. Pandey
Center of Excellence for Emerging Materials (CEEMS)-Virginia Tech (VT), TIET, Patiala, Punjab, India
oppandey@thapar.edu

Srikanth Ponnada
Sustainable Materials and Catalysis Research Laboratory (SMCRL), Department of Chemistry, Indian Institute of Technology Jodhpur, Karwad, Jodhpur, India

Satya Prakash
Academy of Scientific & Innovative Research (AcSIR), Ghaziabad, India

Pankaj Raizda
School of Chemistry, Shoolini University, Himachal Pradesh, India

Rakesh K. Sharma
Sustainable Materials and Catalysis Research Laboratory (SMCRL), Department of Chemistry, Indian Institute of Technology Jodhpur, Jodhpur, India
rks@iitj.ac.in

Sweta Sharma
Academy of Scientific & Innovative Research (AcSIR), Ghaziabad, India

Archana Singh
CSIR – Advanced Material and Processes Research Institute, Bhopal, India
archanasingh@ampri.res.in

Kapil Dev Singh
Department of Material Science and Engineering, National Institute of Technology, Hamirpur, Himachal Pradesh, India

Pardeep Singh
School of Chemistry, Shoolini University, Himachal Pradesh, India

Shelly Singla
School of Chemistry and Biochemistry, Thapar Institute of Engineering and Technology, Patiala, India

A.K. Srivastava
CSIR – Advanced Material and Processes Research Institute, Bhopal, India

Deepika Tanwar
Academy of Scientific & Innovative Research (AcSIR), Ghaziabad, India

Anupma Thakur
Discipline of Chemical Engineering, Indian Institute of Technology Gandhinagar, Gujarat, India
anupmathakur92@gmail.com

Sanjay Upadhyay
School of Physics and Materials Science (SPMS), Thapar Institute of Engineering and Technology, Patiala, India

Amit A. Vernekar
Inorganic and Physical Chemistry Laboratory (IPCL), Council of Scientific and Industrial Research (CSIR)-Central Leather Research Institute (CLRI), Chennai, India
amitvernekar@clri.res.in

Arumugam Vadivel Murugan
Centre for Nanoscience and Technology, Madanjeet School of Green Energy Technologies, Pondicherry University Kalapet, Puducherry, India

Preface

This book, *Green Energy Harvesting: Materials for Hydrogen Generation and Carbon Dioxide Reduction*, concisely summarises the possible ways to harvest hydrogen from water and also reduce CO_2 into various hydrocarbons. A special emphasis is given to the figure-of-merits for the currently developed system/materials for hydrogen generation and CO_2 reduction. We further have summarised the trends in materials innovation and the corresponding state of the art to achieve the desired efficiency and stability, while also considering the cost of production. Finally, the future prospects of this sustainable alternative fuel is summarized for the possible future strategy in adopting these sustainable solutions at the commercial level.

This book can be used to develop an understanding in this field in terms of fundamentals, materials advances, and devices deployment. The students and researchers from energy, environment, materials, chemistry, electrochemistry, and similar backgrounds will find it useful in their respective fields.

Acknowledgements

The kind permission of the Director of CSIO to execute this book project is highly acknowledged. All the reviewers who have reviewed the chapters in this book and suggested necessary improvements are also acknowledged.

Abbreviations

AB	ammonia borane
ABPE	applied bias photon to current efficiency
ac	aberration-corrected
AC	activated carbon
AEL	alkaline electrolysis
AEM	alkaline exchange membrane
AFC	alkaline fuel cell
Ag	silver
Al	aluminum
ALD	atomic layer deposition
APCE	absorbed photon-to-current efficiency
Au	gold
B	boron
BASF	Baden Aniline and Soda Factory
BC_7N	borocarbonitride
BDC	benzenedicarboxylic acid
BHT	benzene-1,2,3,4,5,6-hexathiol
C_3N_4	carbon nitride
C	carbon
CA	California
CB	conduction band
CBE	conduction band edge
C_{dl}	double layer capacitance
CdS	cadmium sulfide
CH_4	methane
CNT	carbon nanotube
Co	cobalt
CO	carbon monoxide
CO_2	carbon dioxide
COD	Chemical Oxygen Demand
COOH	carboxyl intermediate
CoPC	Co phthalocyanine
CoPS	Co-phosphosulphides

CoP|S	Co-phosphosulfate nanoparticles
CO_2RR	CO_2 reduction reaction
COVID-19	Coronavirus disease 2019
CS	catalytic selectivity
C_s	specific capacitance
CTF	covalent triazine framework
Cuf	copper foam
CUMS	coordinatively unsaturated metal sites
CV	cyclic voltammetry
CVD	chemical vapor deposition
1D	one-dimensional
2D	two-dimensional
3D	three-dimensional
DBD	dielectric barrier discharge
DFT	density functional theory
DMSO	dimethyl sulfoxide
DOE	Department of Energy
DOS	density of states
DRIFTS	CO-diffuse reflectance infrared Fourier transform spectroscopy
DTM	double transition-metal
EC	electrocatalyst
EC	electrochemical
ECSA	electrochemical active surface area
EELS	electron energy-loss spectroscopy
EF	energy efficiency
EG	ethylene glycol
EH_{vac}	H-vacancy energy
EIA	Energy Information Administration
EIS	electrochemical impedance spectroscopy
ENE-FARM	energy and farm
EV	electrovolts
EXAFS	extended X-ray absorption fine structure
EY	Ernst & Young Global Ltd
FA	formic acid
fcc	face-centred cubic
FCH JU	Fuel Cells and H_2 Joint Undertaking
FE	faradic efficiency
Fe	iron
FeO_x	iron oxide
FTO	conductive surface
ΔG	Gibbs's free energy
GCE	glassy carbon electrode
GDL	gas diffusion layer
GGNR	graphene/graphene nanoribbon
GO	graphene oxide

Δh	enthalpy
Δs	entropy
2H	hydrogen
H_2	hydrogen
H_{ads}	hydrogen adsorption
HAADF	high-angle annular dark-field
h-BN	hexagonal boron nitride
HCN	heptazine-based crystalline carbon nitride
HCOOH	formic acid
hcp	hexagonal close packing
HDH	hetero-dimensional hybrid architecture
HDS	hydrodesulfurization
HEP	H_2 evolution photocatalyst
HER	hydrogen evolution reaction
HES	hydrogen energy storage
HF	hydrofluoric acid
HRTEM	high resolution transmission electron microscope
HSSA	high specific surface area
IC	ion chromatograph
ICP-AES	inductively coupled plasma atomic emission spectroscopy
IL	ionic liquid
i-MAX	in-plane MAX
IPCC	Intergovernmental Panel on Climate Change
IPCE	Incident Photon-to-Current Efficiency
IPHE	International Partnership for H_2 and Fuel Cells in the Economy
i-PrA	isopropylamine
IQE	internal quantum efficiency
iR drop	ohmic potential drop
IrO_2	iridium oxide
j_o	exchange current density
KOH	potassium hydroxide
LB	Langmuir Blodgett
LBL	layer by layer
LDH	layered double hydroxide
LM Wind	Lunderskov Møbelfabrik
LOHC	liquid organic hydrogen carrier
LSV	linear sweep voltammetry
M	metal
MCFC	molten carbonate fuel cell
MD	molecular dynamic
MEA	membrane electrode assembly
MILD	minimally intensive layer delamination
Mo	molybdenum
Mo_2C	molybdenum carbide
MOF	metal-organic framework

MoP	molybdenum phosphide
MoS_2	molybdenum disulfide
$MoSe_2$	molybdenum diselenide
MWCNT	multi-walled carbon nanotubes
MX	metal complex
N	nitrogen
N_2O	nitrous oxide
NASA	National Aeronautics and Space Administration
Nb	niobium
ND	nanodisc
ND	nano-dots
NEXAFS	near edge X-ray absorption fine structure
NF	nanoflake
NG	N-doped graphene
NGO	N-doped graphene oxide
NH_4HF_2	ammonium bifluoride
NHE	normal hydrogen electrode
Ni	nickel
NiCo-UMOFN	Ni-Co MOF nanosheet
Ni-G	Ni-graphene
NP	nanoparticle
O_2	oxygen
OEP	O^2 evolution photocatalyst
OER	oxygen evolution reaction
OH	hydroxyl
o-MAX	out-of-plane MAX
OPEC	organic photoelectrochemical
ORR	oxygen reduction reaction
Os	osmium
Ov	oxygen vacancy
QD	quantum dots
P	phosphorus
PAFC	phosphoric acid fuel cell
PC	photocatalytic
PCE	photo-chemical-efficiency
PCG	porous conductive graphene
Pd	palladium
PDMS	polypyrrole, polydimethyl siloxane
PEC	photoelectrocatalyst
PEC	photoelectrochemical
PEC-HER	photoelectrochemical-hydrogen evolution reaction
PEM	(polymer) electrolyte membrane
PEM	proton exchange membrane
PEMEL	Proton Exchange Membrane Electrolysis
PEMFC	proton exchange fuel cell

PH_3	phosphine gas
PL	photoluminescence
PLD	pulsed laser deposition
POM	polyoxometalate
POMOF	polyoxometalate-based metal-organic framework
Pt	platinum
PV	photovoltaic
PVEC	photovoltaic electrocatalyst
PXRD	powder X-ray diffraction
QE	quantum efficiency
R_{ct}	charge transfer
R&D	Research and Development
RDS	rate determining step
RECAI	Renewable Energy Country Attractiveness Index
RES	renewable energy resources
rGO	reduced graphene oxide
RHE	reversible hydrogen electrode
Ru	ruthenium
S	sulfur
SA	surface area
SAA	single-atom alloy
SAC	single-atom catalyst
SCE	saturated calomel electrode
SCWG	supercritical water gasification
Se	selenium
SEM	scanning electron microscope
SFE	Solar-to-Fuel efficiency
SMR	steam methane reforming
SOEL	high-temperature solid oxide water electroysis
SOFC	solid oxide fuel cell
SPR	van der Waals
SSA	specific surface area
STEM	scanning transmission electron microscopy
STH	solar-to-hydrogen
Ta	tantalum
TA	terminal alkyne
TaS_2	tantalum disulfide
TBA	tetrabutylammonium
TEOA	triethanolamine
TNAOH	tetrabutylammonium hydroxide
TDOS	total density of states
TEM	transmission electron microscope
THT	triphenylene-2,3,6,7,10,11-hexathiolate (THT)
TM	transition metal
TMAOH	tetramethylammonium hydroxide

TMC	transition metal carbide
TMD	transition metal dichalcogenide
TMN	transition metal nitrides
TMO	transition metal oxide
TMP	transition metal phosphide
TMPS	TM-phosphosulphides
TOF	turnover frequency
TON	turnover number
TOP	trioctylphosphine
TV	television
TW	terawatt
UCLA	University of California, Los Angeles
UPS	UV photoelectron spectroscopy
USEPA	United States Environmental Protection Agency
UV	ultraviolet
VB	valance band
vdW	van der Waals
VO_2	vanadium dioxide
VS_2	vanadium sulfide
VSe_2	vanadium selenide
VS_2s	vanadium sulfides
$W_{use.out}$	useful work output
$W_{rev.out}$	reversible work output
WC	tungsten carbide
WCHN	WSe_2/CNTs hybrid network
WG	waved graphene
WGSR	water-gas shift reaction
WHO	World Health Organisation
φ_M	work function
φ_M	semiconductor work function
WS_2	tungsten disulfide
WSe_2	tungsten diselenide
XPS	X-ray photoelectron spectroscopy
XRD	X-ray diffraction
XANES	X-ray absorption near-edge spectroscopy
YSZ	yttria-stabilized zirconia
ZIF	zeolite imidazolate framework

1

Renewable Energy

Introduction, Current Status, and Future Prospects

Srikanth Ponnada[1], Indu Kumari[2], Sampath Chinnam[3], Maryam Sadat Kiai[4], A. Lakshman Kumar[5], Rapaka S. Chandra Bose[6], Demudu Babu Gorle[7], Annapurna Nowduri[8,], and Rakesh K. Sharma[1,*]*

[1] *Sustainable Materials and Catalysis Research Laboratory (SMCRL, Department of Chemistry, Indian Institute of Technology Jodhpur, Karwad, Jodhpur 342037, India*
[2] *Department of Biotechnology, Chandigarh College of Technology, Chandigarh Group of Colleges, Landran, Mohali, Punjab 140307, India*
[3] *Department of Chemistry, M.S Ramaiah Institute of Technology (Affiliated to Visvesvaraya Technological University, Belgaum), Bengaluru, Karnataka 560054, India*
[4] *Nano-Science and Nano-Engineering Program, Graduate School of Science, Engineering and Technology, Istanbul Technical University, Istanbul 34469, Turkey*
[5] *CSIR-Central Electrochemical Research Institute, Karaikudi 630003, Tamil Nādu, India*
[6] *Centre for Materials for Electronics Technology, Thrissur 680581, Kerala, India*
[7] *Materials Research Centre, Indian Institute of Science, Bangalore 560012, India*
[8] *Department of Engineering Chemistry, Andhra University College of Engineering (A), Andhra University, Visakhapatnam 530003, India*
[*] *Corresponding Author*

1.1 Introduction

Continuous large-scale exploitation of our valuable natural resources, i.e., water, energy, and land has resulted in a drastic change in average global temperature [1]. While considering the world's future needs, mitigating climate change without misusing these resources becomes the prime challenge of human civilization today. However, based on our former scrutiny of energy resources, it is possible to sustain and broaden a prosperous civilization by improving air quality, energy access, and energy security [2]. Energy resources mainly consist of three groups, i.e., fossil fuels, renewable resources, and nuclear resources [3]. Since the recovery of non-renewable resources (i.e., fossil fuels and nuclear resources) is not possible after their depletion, the demand of renewable energy resources (RES) increases.

Renewable energy is the form of sustainable energy that can be derived directly or indirectly from the environment and sources that are persistently replenished by nature. The main advantages of RES include no wastage, low maintenance cost, are economical, and no depletion. Renewable energy plays a major role in energy security and reducing greenhouse gas emissions. In general, roughly 8 billion metric tons of carbon are being consumed and dumped into the atmosphere each year; deforestation contributes to 1.5 billion, with 6.5 billion tons from fossil fuels [4]. The great consumption of fossil fuels has caused serious damage to the environment and disrupted the whole ecological cycle. According to the experts, nonrenewable resources

Green Energy Harvesting: Materials for Hydrogen Generation and Carbon Dioxide Reduction, First Edition. Edited by Pooja Devi.

will become depleted within 53 to 110 years and therefore are not sufficient to fulfill the world's energy needs [5]. In addition, the burning of fossil fuels has led to poor air quality and global warming. According to the World Health Organisation (WHO), around 7 million deaths were recorded globally in 2016 due to household and ambient air pollution. In this data, around 94% of deaths were from low- and middle-income countries [6].

Thus, many countries have turned to renewable resources to meet their rising energy demands and to reduce air pollution. However, at present, RES provides only 14% of the total energy world energy demands [7], though several efforts have been taken up by countries worldwide. For instance, the binding target of 27% (by year 2030) has been adjusted by the European Union, that was earlier decided in 2014 to reach 32% by June 2018. According to this new target, by 2023, countries are going to discuss an even higher target [8]. The Government of India has also set an ambitious renewable energy target of 175 GW to be completed by 2022, consisting of 60 GW of wind and 100 GW of solar energy, and 10 GW of bio-power and 5 GW from small hydro-power [9]. In 2019, India was ranked fifth in wind power and solar power and fourth in renewable power installed capacity. The Government of India is planning to achieve 227 GW of renewable energy capacity by 2022, that includes 114 GW of solar capacity and 67 GW of wind power capacity, i.e., more than its 175 GW target [10]. Since July 2021, India holds 25.2% of the overall installed capacity of hydro projects and provides great options for green data centers' development. The Government of India's target is to establish a renewable energy capacity of 523 GW by 2030, including 73 GW from hydropower and about 280 GW expected from solar power. Throughout 2023, around 5000 compressed biogas plants are planned to be set up across India [11].

China, the largest energy producer and consumer, has a pivotal role in the global energy transition. China has also set targets to reduce carbon emissions per unit of gross domestic product by 60–65% from 2005 to 2030 [12]. In 2017, more than half of all global solar photovoltaic (PV) capacity additions of 94 GW were contributed by China. Also, solar PV deployment quotas were introduced by the Government of China in 2018 [13]. By the end of 2021, China and U.S. aimed to produce 600T Wh and 400 TWh, respectively, i.e., jointly representing more than half of the global wind power capacity. Figure 1.1 represents the geographical breakdown of the renewable power generation capacity additions, wherein China accounts for over one-third, followed by the United States, India, and the European Union [10].

In 2021, the U.S. Energy Information Administration's (EIA), with the recent invention of electricity generators, enabled power plant owners to generate 39.7 GW of new electricity capacity to start commercial operation [14], wherein solar accounts for the largest share of new capacity at 39% and wind accounts for 31% [14]. The U.S. primary energy consumption, in terms of energy source, is represented in Figure 1.2. According to the EIA, the tendency of large-scale battery storage more than quadrupled by late 2021. In Florida, the world's largest solar powered battery was construction and scheduled to be operational by the end of 2021 [14].

The main advantage of RES is its distribution over a wide range of geographical areas. The most common types of renewable resources include hydropower, biomass energy, geothermal power, wind energy, solar energy, and tidal energy (Scheme 1.1). These forms of energy are interconnected to each other in various ways. For instance, the Sun's heat drives the winds, and wind turbines capture its energy. Then, the Sun's heat and wind collectively lead to the evaporation of water that converts into rain or snow and finally flows downhill forming rivers

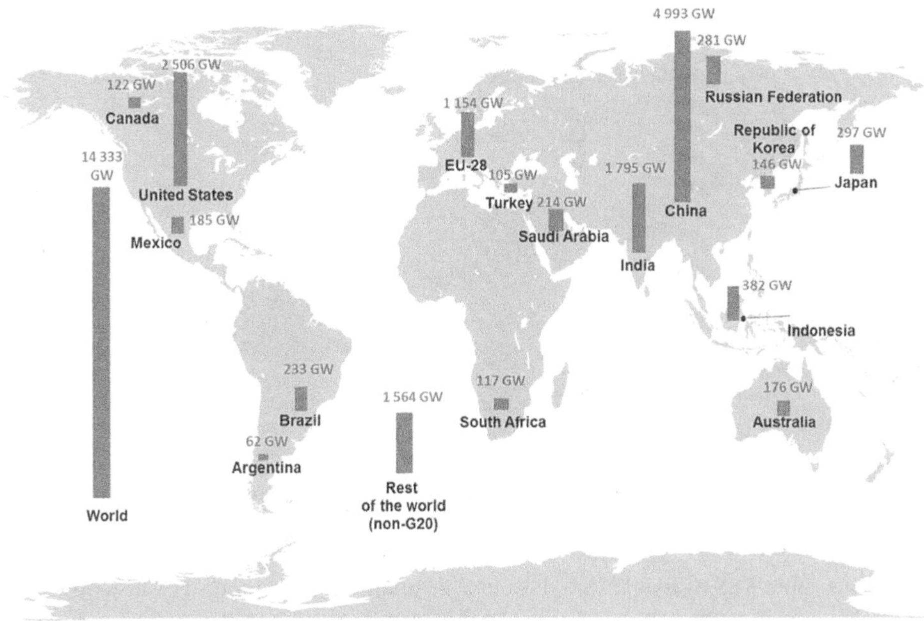

Figure 1.1 Geographical breakdown of renewable power generation capacity additions, 2018–2050. Reproduced from [10] / With permission of Elsevier.

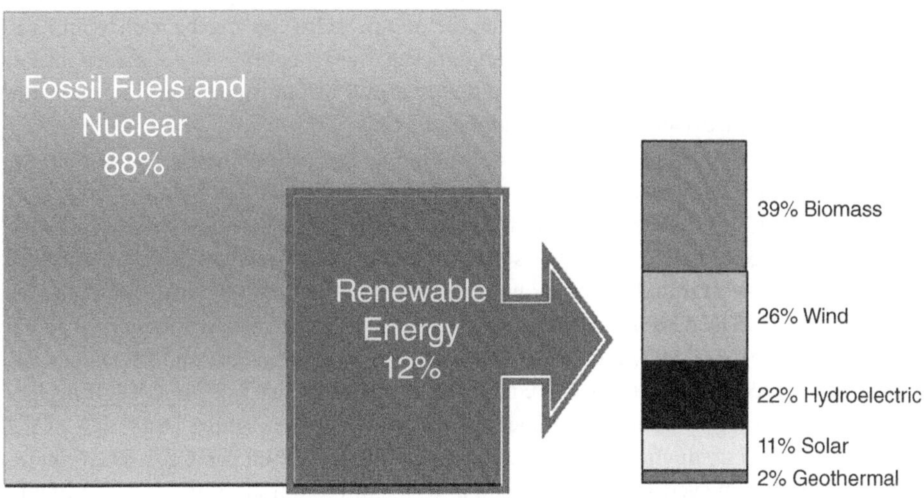

Figure 1.2 U.S. primary energy consumption by energy source, 2020.

or streams. Their energy can be utilized by hydroelectric power. In addition to rain and snow, sunlight is also responsible for the growth of plants and vegetation. The organic matter made by plants is the biomass that can be used for various purposes, such as transportation, fuel, electricity, or chemicals that lead to the generation of bioenergy. Hydrogen can be burned as a fuel or transformed into electricity. Though it is always found in combination with other elements, it can be used only after its separation from another element.

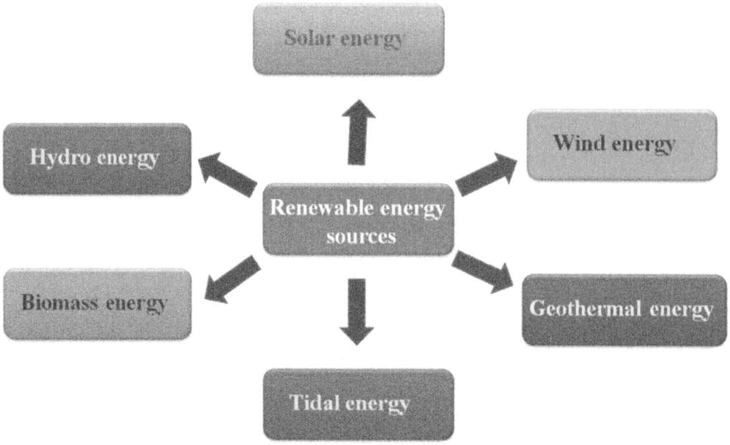

Scheme 1.1 Schematic illustration of different types of renewable energy.

There is some RES available that does not come directly from Sun. For instance, geothermal energy uses the heat present inside the Earth and can be used in various applications, including electric power production and heating of buildings. Geothermal energy was first used for commercial purposes in 1900s by the Italians [15]. Turkey is known for its rich geothermal energy resources and ranked fifth after China, Japan, USA, and Iceland [16]. Additionally, the energy produced from the oceans' tides can also be used as an RES. There are many sources available that can generate ocean energy. For instance, ocean energy can be generated from the the gravitational pull of the moon and Sun upon the Earth. Also, it can be driven by both the tides and winds [17].

Climate change and local air pollution are among the major factors responsible for energy transition worldwide. Countries such as China and India are greatly impacted by local air pollution. In Europe the rise in harmful health effects have been observed due to air pollution, largely related to energy supply and use. Thus, energy transition needs to lessen emissions substantially, whilst ensuring that sufficient energy is still available for economic growth. The data in Figure 1.3 shows that the CO_2 emissions intensity of global economic activity needs to be reduced by 85% between 2015 and 2050, and CO_2 emissions need to be lowered by more than 70% compared to the Reference Case in 2050. It is clear that renewable energy and energy efficiency measures can successfully attain 94% of the necessary emissions reductions by 2050, as compared to the Reference Case. The remaining 6% would be achieved via other options in terms of reduction of activities leading to CO_2 emissions, i.e., fossil fuel switching, continued use of nuclear energy, and carbon capture and storage [10] (Figure 1.3).

Renewable energy and sustainable development are very much related to each other. The development of renewable energy with reduced CO_2 emissions has generated new interest in storage, thus it has become a chief component of sustainable development. Energy storage can improve the system flexibility, mitigate power variations, and enable the storage and transmission the electricity produced by different RES, including solar and wind energy. The various storage technologies are used in electric power systems such as

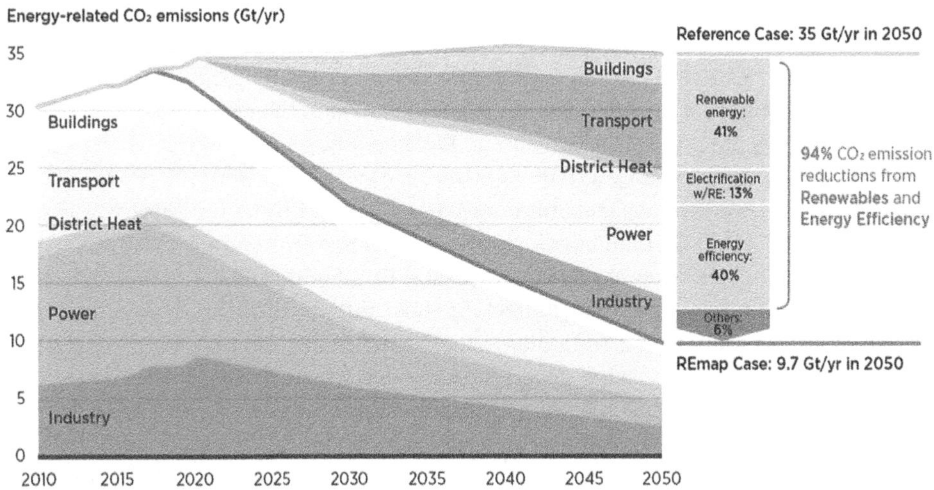

Figure 1.3 CO_2 emission reduction potential by technology in the Reference Case and REmap, 2010–2050. Note: the figure shows the breakdown of energy-related CO_2 emissions by technology in the REmap Case compared to the Reference Case. The figure excludes emissions from non-energy use (feedstocks). Reproduced from [10] / With permission of Elsevier.

chemical or electrochemical, mechanical, thermal, or electromagnetic storage [18]. For electrochemical storage, different batteries are available, including lithium-sulfur, nickel-cadmium, nickel-zinc, lead-acid, ZnO, etc. [18, 19a–c]. These batteries have remarkable properties; for example, high charge/discharge efficiency, long life, and low self-discharge. For hydrogen energy storage (HES), the energy is stored in the form of hydrogen where it is retransformed to electricity by a fuel cell to energize the power plants. Hydrogen can store energy for a long time by using various HES models such as compressed, liquefied, metal hydride, etc. [18]. Mechanical energy includes flywheel energy storage, pumped HES, and compressed air energy storage [18]. In thermal energy storage, the energy is stored by varying the temperature of the material such as by heating or cooling [18]. In India, it is predicted that about 49% of the total electricity will be produced by renewable energy owing to the more efficient batteries for the storage of electricity, which will further cut the solar energy by 66% as compared to present costs [18, 19].

Based on the above discussion, the renewable concept has been accepted worldwide and is now a central energy policy unit. Though the RES has numerous advantages, the concept might even be hazardous toward the efforts taken to combat climate change or power sustainable development [20]. This is because of the dependency of these solutions on geographical sites and climatic conditions. The careful planning, measures, and location selection can help to eliminate these limitation of RES. Among the different types of RES, many organizations have discussed the exceptional role of biomass combustion in various renewable energy strategies and scenarios [21, 22]. However, it has few environmental issues, such as biomass energy being an insufficient source of energy when compared with fossil fuels, growing and harvesting biomass, transportation to the power plant, and combustion, all of which can add to global warming emissions.

In the case of hydropower, major disadvantages include high costs of facilities, changes in stream regimens (where it can affect plants, fish, and wildlife by changing stream levels, flow patterns, and temperature), dependence on precipitation, deluge of land and wildlife habitat, and dislocation of people living in the vicinity of the reservoir [23]. Among RES, solar power is considered the true renewable resource and the most abundant renewable resource on the Earth. However, solar sources provide basic power, out of which humans consume only 0.04% due to the high cost of PV panels, which are more expensive than burning fossil fuels. Apart from its advantages, such as no wastage and no emission of greenhouse gases, its main disadvantages are the costs involved, and dependency on sunshine [15]. Like other RES, wind energy also has some limitations; for example, high maintenance and transmission costs, the irregular and unpredictable nature of wind power, noise pollution, interruption of TV and radio signals, killing of migratory birds, and requirement of large geographical areas for the setups [15, 24]. Similarly, the drawbacks of geothermal energy resources include finding a suitable location for the setup; safety issues, such as volcanoes concentrated near geothermal energy sources and earthquakes at these points being more frequent; relatively lesser energy than other RES; and the steam can include toxic materials such as mercury, ammonia, arsenic, etc. [15, 25]. Thus, more attention has to be paid toward these issues, and energy policies are much needed that should focus on solving such disadvantages of RES.

1.2 Impact of COVID-19 on Renewable Energy Resources

The ongoing COVID-19 pandemic is having a major impact on the renewable power sector around the world. During the pandemic, the full-lockdown measures ordered by governments worldwide resulted in depressed electricity demand (~15–30%) in many countries with the generation of an oversupply of existing power capacity.

As the crisis hit, a huge drop in global energy investment became apparent with spending plunging in each main sector in 2020 [26]. For instance, a wind power plant in North Dakota was closed due to the spread of the pandemic [27]. In Spain, LM Wind and Siemens Gamesa, top competitors in the wind energy market when the government announced a nationwide lockdown, stopped their wind turbine blade plant production [28]. The same effects have been observed in the solar industry; for example, delays in the supply chain and difficulties in tax stock markets [29]. India, the world's fourth largest in the wind sector, was also affected by the outbreak of the pandemic. Its chief aims of generating 60 GW of energy by the end of 2022 and 450 GW by 2030, both were affected by these unforeseen situations [30]. Reports show that around 600 MW of new wind power addition is expected to overcome 2.60 GW of loss in the coming few years. In 2019, nations such as China, the U.S., India, the UK, and Spain had accounted for 70% of new wind power additions; however, at present they are among the countries most affected by the pandemic [31]. Additionally, many thermal plants were closed during the lockdown period [32]. Thus, the RES has faced various obstacles due to the pandemic; however, followed by new capacity additions, the energy sector has disregarded the pandemic and sustained its growth.

1.3 Green Hydrogen as Promising RES

Among different types of RES, hydrogen energy is one of the very versatile forms of energy that can be used in liquid or gaseous form. Hydrogen exists in abundant amounts and its supply is almost unlimited. Hydrogen can be produced or transported anywhere and can store large amounts of electricity for extensive periods of time. Every year, around 70 million metric tons of hydrogen is manufactured globally that is used in different areas; for example, food processing, steel manufacturing, ammonia production, chemical and fertilizer production, metallurgy, etc. It is predicted that in the universe, around 90% of all atoms are hydrogen, more than any other element. However, hydrogen atoms are not present in nature by themselves. Thus, hydrogen atoms need to be decoupled from other elements or molecules with which they occur to produce hydrogen. The sustainability of hydrogen energy depends on the method of decoupling used.

Hydrogen energy can be transformed into electricity or fuel and various methods are available for its production. However, hydrogen can be generated at very low cost from entirely carbon-free sources by means of wind and solar energy. Based on the process and source of production, H_2 is classified into four different categories (Figure 1.4) [33, 34].

1.3.1 Types of Hydrogen

Grey Hydrogen: H_2 produced from fossil fuels (i.e., hydrogen produced from methane using steam methane reforming (SMR) or coal gasification) is categorized as "greyH_2." Production of grey H_2 results in CO_2 emission. The majority of H_2 produced globally is grey H_2.

Blue Hydrogen: H_2 produced from fossil fuels, where the generated carbon emissions are captured or utilized, is considered "blue H_2." Hydrogen produced from nuclear energy is also considered as blue H_2 due to the small amount of carbon emissions.

Turquoise Hydrogen: H_2 that makes use of natural gas as a feedstock while emitting no CO_2. The carbon in methane is converted to solid carbon black by the pyrolysis process. Since there is already a market for carbon black, this provides an extra revenue source. Carbon black can be stored more easily than CO_2. Production of Turquoise hydrogen is still in the pilot stage.

Green Hydrogen: H_2 generated from hydrocarbon-free renewable resources or excess process heat via a non-fossil process such as electrolysis of water is "green H_2," with very low carbon emissions (illustrated in Figure 1.4).

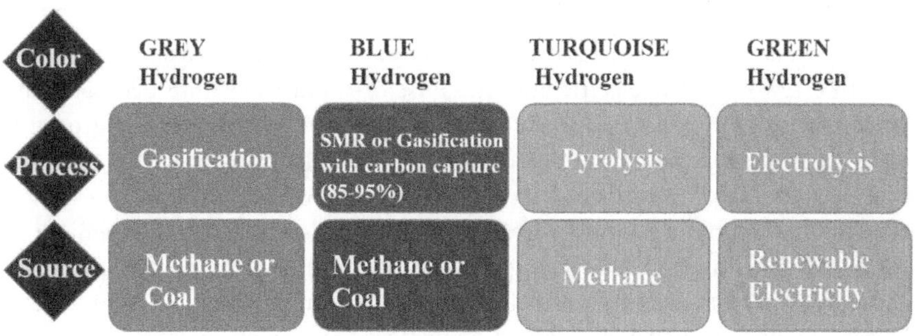

Figure 1.4 Illustration of types of hydrogen and its sources.

1.3.2 Need for Green Hydrogen Production?

As already discussed in previous sections, global warming is a major challenge for the entire world. A growing number of countries have pledged to achieve net-zero carbon dioxide (CO_2) emissions by the middle of this century (2050), with the objective of keeping global warming to 1.5°C. This necessitates a significant change in electricity generation from fossil fuels to renewable sources such as solar and wind energy. Nature offers various renewable sources such as solar energy, wind energy, tidal energy, biomass energy, etc. (Scheme 1.1). However, such energy sources suffer from discontinuous availability due to regional or seasonal factors [35]. As a result, in conjunction with the exploration of renewable energy sources for large-scale use, an efficient energy conversion and storage system is also required [36]. This requirement is the primary driving force behind numerous innovations in energy conversion and storage systems. Hydrogen production from electrolysis of water, fuel cells for converting hydrogen to electricity, and lithium-ion or metal-air batteries for energy storage have all received a lot of attention in recent decades [37]. For the battery-based energy storage systems, it is increasingly difficult to store excess electricity generated from a large-scale production facility, which is very expensive and also needs a large facility area. Hence, large-scale solar or wind-generated electricity require alternate energy storage pathways. Green hydrogen generation using electricity-driven water splitting has emerged as a promising approach for converting huge amounts of excess electrical energy from renewable energy sources into clean fuel hydrogen. When this is used as a fuel in the hydrogen fuel cell, it not only converts energy efficiently but also creates no pollution because it only emits water as a by-product. As a result, the development of green hydrogen production from renewable sources has become a global push toward a future power package that is both sustainable and affordable. This advancement is paving the way for many of the difficult issues encountered during conversion and storage of renewable energy.

In addition, approximately 4 billion tonnes of hydrogen is required annually, with 95% of hydrogen production derived from fossil fuel [38]. Around 830 million tonnes of CO_2 are emitted annually when hydrogen gas is produced using fossil fuels. Hence, swapping to production of green hydrogen utilizing renewable energy sources will reduce the CO_2 emissions to a greater extent in the next few decades and will become independent of fossil energy carriers.

1.3.3 Uses and Limitations of Green Hydrogen

Generally, hydrogen can be generated using the electrolysis of water releasing oxygen as a by-product. In electrolysis, the electric current is used to split water into hydrogen and oxygen in an electrolyzer. Among different types of hydrogen energy, green hydrogen is generated by electrolysis, wherein the electricity is generated by using renewable sources; for example, solar or wind. Here, electricity is fed to an electrolyzer which requires water and electricity for the production of hydrogen and oxygen, with zero carbon emissions (Scheme 1.2) [33]. The main advantage of green hydrogen is that it only needs water and electricity to produce more electricity or heat. It can be used in industry and can be transported in gas pipelines to power household appliances. The green hydrogen produced

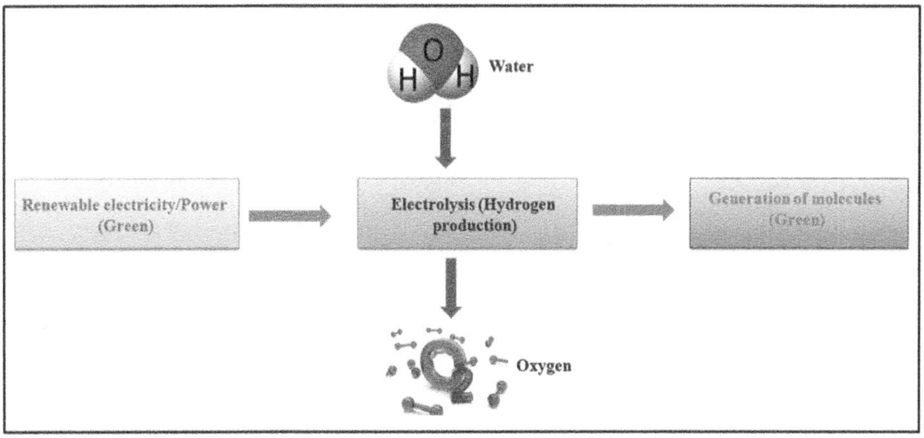

Scheme 1.2 Production of green hydrogen.

could be directly blended and added to natural gas networks up to a definite percentage. This results in less consumption of natural gas as compared to the case of no green hydrogen. Additionally, synthetic methane can be produced via steam methane reforming process and can be directly added to gas networks. This is a proficient method for the reduction of carbon dioxide emission. Green hydrogen can be stored and used in aviation, marine, and other transportation systems via the hydrogen supply chain. Figure 1.5 illustrates the production of green hydrogen, its conversion into numerous beneficial compounds, transport, and multiple end uses across the energy system [34]. The total cost of hydrogen generation changed from $6/kg in 2015 to an estimated figure of $2/kg by 2025 by using cheap renewable energy. This fast decline in cost of renewable energy is one of the chief reasons for the growing interest in green hydrogen worldwide. The current decade is critical for green hydrogen technology development as one of the most promising options for the long duration storage of electricity. By this, the aim of 40% share of electricity in the worldwide energy portfolio in 2050 would be reached and therefore the Paris Agreement regarding the decarbonized energy will likely be accomplished [39]. Green hydrogen is basically considered as an alternative fuel produced with clean energy and thus identified as the clean energy source that could meet the world's future energy demands and transform the world with net-zero emissions. However, the economics of green hydrogen are challenging today due to the underlying costs and that the availability of renewable energy sources vary widely [40].

Although green hydrogen is gaining popularity across industries, it still faces the future power systems with numerous challenges in the planning and operational phases. Several factors such as market, public, demand uncertainty, and environmental constraints may impose further pressures on the network. There is less knowledge on optimum demand and return on investment, therefore limited bankability. In order to fulfil market demands, organizations have to scale up and advance their green hydrogen plant designs. However, optimizing plant designs and green hydrogen systems can be expensive and complex on the basis of limited market demand. Though green hydrogen will generate numerous new opportunities, so many individuals still need the essential training and skills to support the

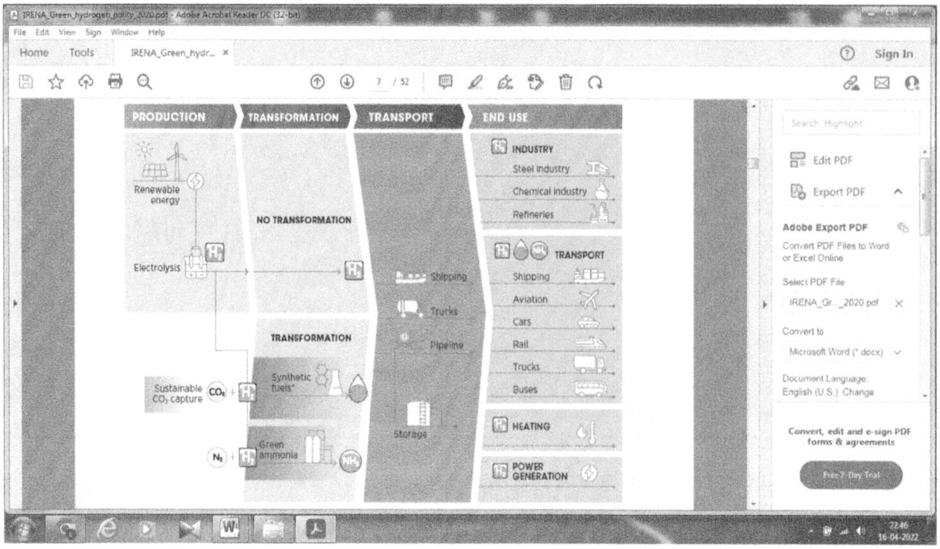

Figure 1.5 Illustration showing the production of green hydrogen, its transformation, transport, and end uses across the energy system. Image from [34]. https://www.irena.org/publications/2020/Nov/Green-hydrogen.

hydrogen economy. The best way forward seems to incorporate hydrogen generation to dedicated solar or wind power plants that can reach suitable annual load factors in chosen locations. Moreover, green hydrogen is expensive to store and transport, thus requiring high operational costs in specialized pipelines and carriers [41]. In addition to this, high energy loss at every point in the supply chain of green hydrogen is also a major concern. Around 30–35% of the energy utilized for the generation of hydrogen is lost during the electrolysis process, liquefying, or transforming hydrogen to other carriers; for example, ammonia, and this results in a 13–25% energy loss. Around 10–12% of the extra energy is required in the transporting of hydrogen [42]. Such inefficiencies will need significant renewable energy deployment to nourish green hydrogen electrolyzers that can compete with electrification. Apart from these challenges, another major challenge is the way to monetize green hydrogen. The condition of geographical area for green hydrogen creates a requirement for dedicated pipelines with all linked lead times and costs.

Transition to green hydrogen is one of the key requirements to reduce emissions, especially in the hard-to-abate areas. The Government of India has set a target of production of 5 million tonnes of green hydrogen before 2030. Thus, they have considered different policy measures to assist transition from fossil fuels to green hydrogen, both as energy carriers and chemical feedstock for different sectors [43]. The U.S. hydrogen economy could generate $140 billion and support 700,000 jobs. There are numerous green energy projects in the U.S. and around the world attempting to deal with these challenges and support hydrogen adoption. California is planning to invest $230 million on hydrogen projects before 2023. In Lancaster, CA, the world's largest green hydrogen project is located. This plant uses waste gasification, combusting 42,000 tonnes of recycled paper waste annually to generate green hydrogen. European countries including Germany, Spain, and France announced the installation of 4, 5, and 6.5 GW of green hydrogen by 2030, respectively [44]. Green hydrogen national targets of France, Portugal, Germany, Netherlands, and Spain contributed to more than 50% of the European Union's targeted 40 GW of installed electrolyzer capacity in 2030.

1.3.4 Green Hydrogen Production Pathways from Renewable Energy Sources and Their Current Level of Maturity

Various technology choices are available for creating hydrogen from renewable energy sources [39]. Water electrolysis is the most well-established technology choice for creating green hydrogen from RES. Biomass gasification and pyrolysis, thermochemical water splitting, photocatalysis, biomass supercritical water gasification, and coupled dark fermentation and anaerobic digestion are less developed routes. In this chapter, we restrict our discussion to the production of green hydrogen through electrolysis of water using renewable energy resources.

Currently, there are three main types of electrolysis technologies: (1) proton exchange membrane electrolysis (PEMEL); (2) alkaline electrolysis (AEL); and (3) high-temperature solid oxide water electrolysis (SOEL). While the low-temperature technologies, AEL and PEM, both provide high-technology readiness levels, the high-temperature SOEL technology is still in the development stage [38].

Alkaline water electrolysis uses concentrated lye as an electrolyte, and a gas-impermeable separator is required to keep the resultant gases from mixing. Non-noble metals, such as nickel, are used as electrodes with an electrocatalytic coating. The electrolyte in PEMEL is a humidified polymer membrane, and the electrocatalysts are noble metals like platinum and iridium oxide. Both systems can function at temperatures ranging from 50 to 80°C and at pressures up to 30 bar. Both technologies have a nominal stack efficiency of roughly 70% [45, 46]. SOEL is also known as high-temperature or steam electrolysis. Here gaseous water is transformed into hydrogen and oxygen at temperatures between 700 and 900°C. Due to beneficial thermodynamic effects on power usage at higher temperatures, stack efficiencies of 100% are theoretically possible. However, for cost-effective operation, the increased thermal demand needs a sufficient waste heat supply from the chemical, metallurgical, or thermal power generation industries. Moreover, the corrosive environment demands further material development [46, 47]. As a result, compared to 6 MW for AEL and 2 MW for PEMEL, SOEL only offers tiny stack capacities below 10 kW [46].

Generally, the overall water electrolysis reaction can be divided into two half-cell reactions: hydrogen evolution reaction (HER) and oxygen evolution reaction (OER). During HER, water is reduced at the cathode to produce H_2, and during OER, water is oxidized at the anode to produce O_2. One of the critical barriers that keep water splitting from being of practical use is the sluggish reaction kinetics of OER and HER due to high overpotentials [48], a measure of the kinetic energy barriers. A broad range of highly effective catalysts are developed to minimize the overpotentials for OER and HER toward efficient H_2 and O_2 production. Platinum (Pt) is the most advanced catalyst for HER and OER at this time, and noble-metal-based catalysts continue to be the most efficient catalysts for HER and OER [49–52]. The creation of earth-abundant catalysts with high activity, as a result, becomes one of the most important tasks in the development of cost-effective and efficient water electrolysis systems. There have been numerous reports of earth-abundant catalysts with significant catalytic activity toward OER and, in particular, HER [53–60]. A great deal has been done in the field of HER research on transition metal dichalcogenides (TMDs) [61–66], transition metal phosphides (TMPs) [67–70], carbides [71–73], and nitrides [74, 75]. Several heterostructured catalysts have recently emerged from the crowd, demonstrating superior catalytic performance in electrochemical water splitting compared to their conventional equivalents [76–80]. By depositing MoS_2 on the surface of $CoSe_2$, for example, Gao et al. developed a $MoS_2/CoSe_2$ heterostructure catalyst that was effective. When tested in 0.5 m H_2SO_4, the $MoS_2/CoSe_2$ heterostructure had excellent HER performance. It displayed a high overpotential of 68 mV at 10 mA cm^{-2}, a Tafel slope of 36 mV dec^{-1}, and good performance durability [81]. In one study, Chen et al. synthesized a 3D core/shell catalyst composed of metallic Co cores and amorphous Co_3O_4 shells, and the Co/Co_3O_4 heterostructure delivered 10 mA cm^{-2} at a low overpotential of only 90 mV in 1 m potassium hydroxide (KOH) [82]. Most heterostructured catalysts, including active/active and active/nonactive types of heterostructures, exhibit higher HER activities than their single counterparts [83].

Since the first study of photochemical H_2 evolution from water splitting on TiO_2, single crystal electrodes utilizing photoelectrochemistry was published in 1972 [84], and photochemical H_2 evolution from water splitting has remained a hot topic for both academic and industrial researchers. It is possible to divide photochemical H_2 evolution processes into two categories based on the reaction mechanism they use: (1) photoelectrocatalytic H_2 evolution, and (2) photocatalytic H_2 evolution. Photoelectrocatalytic methods employ photocatalysts as electrodes in addition to light irradiation, and necessitate the application of an additional bias voltage in order to prevent recombination of the photo-generated carriers. Because of its great efficiency, photoelectrocatalytic H_2 evolution is frequently used in industry. However, it requires additional energy, and hence will not be discussed in detail in this chapter. Photocatalysis, on the other hand, directly utilizes the abundant solar energy to split water into H_2 through a four-electron or two-electron process, which can successfully avoid environmental contamination as well as the consumption of additional energy. It is the quickest and most straightforward method of water splitting, and it produces H_2 at a low cost and on a huge scale. It can easily be seen that the number of studies focusing on the creation of photocatalysts that do not contain noble metals (also known as non-noble metal photocatalysts) has expanded dramatically in recent years. However, they are still insignificant in comparison to the photocatalysts containing noble metals, which have a far higher number of active sites (named as noble-metal photocatalysts). Noble metals have the potential to be used as effective redox co-catalysts in general because of their good physicochemical and electrical properties, as well as their high catalytic activity [85, 86]. But due to the low abundance of noble metal elements in noble-metal photocatalysts, they are expensive and have only a few practical uses in the field of water splitting. However, because of their low cost and high efficiency, non-noble-metal photocatalysts are an attractive candidate for water-splitting applications due to their low toxicity. Non-noble-metal photocatalysts, on the other hand, have excellent stability and do not suffer from deactivation under particular conditions, in contrast to Pt-based photocatalysts (noble-metal photocatalysts), which suffer from Pt deactivation in the presence of halide ions. It follows that photocatalysts made of non-noble metals are acceptable for the conversion of wastewater to H_2. All of the lanthanides, as well as the other elements in the s, p, and d regions of the periodic table, are stated to be capable of generating H_2 [87]. The photocatalytic H_2 evolution process can be classified into three steps that are followed by each other: (1) when a semiconductor absorbs high-energy photons with a wavelength greater than the bandgap, electrons in the valence band (VB) are excited and transmitted to the conduction band (CB), resulting in the generation of holes in the VB; (2) the induced electron–hole pairs separate and transfer to the surface of the material; and (3) the electrons in the CB reduce the adsorbed H^+ to H_2, and the holes in the VB oxidize water to oxygen. However, a prerequisite must be met for H_2 production according to the following redox reactions [88]:

$$\textit{Oxidation}: 2H_2O \rightarrow O_2 + 4H^+ + 4e^- \;\; E_0 = 1.23\,eV \tag{1.1}$$

$$\textit{Reduction}: 2H^+ + 2e^- \rightarrow H_2 \;\; E_0 = 0.00\,eV \tag{1.2}$$

$$\textit{Overall}: 2H_2O \rightarrow O_2 + 2H_2 \;\; E_0 = -1.23\,eV \tag{1.3}$$

A lower than 0 V ($E(H^+/H_2)$) CB energy level under the Normal Hydrogen Electrode (NHE) is required in order for the H_2 evolution reaction to proceed under the normal hydrogen electrode (HER). In addition, the VB energy level should be more than 1.23 V ($E(O_2/H_2O)$) in order for the reduction process of H_2O to proceed [89]. The production of H_2 on non-noble-metal photocatalysts is hampered primarily by their weak visible light sensitivity, rapid recombination of the photo-generated carriers, low surface reaction rate, and high thermodynamic potential barriers, among other characteristics. There have been a variety of ways taken to circumvent the restrictions mentioned above. These include energy band engineering, heterojunction building [90], and reactive activity improvement [91]. It is possible to immobilize the co-catalyst on semiconductor photocatalysts, which is a potential strategy for overcoming the constraints outlined above. Co-catalysts are categorized into three categories based on the type of material they are made of: metallic, non-metallic, and semiconducting. Metallic co-catalysts can be further divided into two categories: precious metal catalysts and non-noble (base) metal catalysts, which are distinguished by the cost of the metal used [92]. The use of noble metals as efficient co-catalysts for hydrogen production has increased dramatically in recent decades, with the most common being Ru [93], Rh [94], Pd [95], Ag [96], Pt [97], and Au [98]. With the inclusion of precious metal co-catalysts, the activity of the photocatalyst can be dramatically increased. The usage of noble-metal co-catalysts, on the other hand, is not suited for large-scale applications due to the high cost and restricted storage space associated with them. So, the development of high-performance photocatalytic materials including non-precious metals is a promising strategy for the sustainable and large-scale production of hydrogen from water splitting.

1.4 Current Scenario of RES in India

Over the past few years, India has developed a sustainable path for its energy supply and emerged as one of the top leaders in the world's most attractive renewable energy sectors. India is the world's third largest producer and consumer of electricity with 38% of total installed energy capacity in 2020 from RES [99]. India occupies the third position after the U.S. and China as per the Ernst & Young Global Ltd. (EY) 2021 Renewable Energy Country Attractiveness Index (RECAI) [40].

To meet the future nation's energy needs, the Government of India has taken several initiatives. For instance, from FY 2016/17 to FY 2020/21, wind energy capacity in India has been augmented by 2.2 times. In March 2021, solar power capacity has increased by more than five times in the last five years from 6.7 to 40 GW [100]. On 16 June 2021, the installed renewable energy capacity was raised by over two and half times and stands at more than 141 GW, which is almost equal to 37% of the country's total capacity [100]. At the same time, the installed solar energy capacity has augmented by over 15 times and stands at 41.09 GW. By 31 June 2021, the total installed energy capacity for renewable was 96.95 GW [100].

From 2020 to 2021, a decrease in the utility power generation by 0.8% has been observed along with the decline in power generation from fossil fuels by 1%. During the year 2020/2021, India exported more electricity than it imported from bordering countries. India's grid-connection electricity generation capacity reached 100 GW from non-conventional renewable technologies and 46.21 GW from conventional renewable power or major hydroelectric power plants in 2021 [101].

The EY's RECAI ranking (in July 2021) [99] in terms of installed capacity and investment in renewable energy is as shown in Table 1.1.

The technology-specific RECAI scores (and rank) [102, 103] for 2021 are tabulated in Table 1.2.

By 12 August 2021, India achieved the target of installing 100 GW of renewable energy capacity according to the Union Ministry of New and Renewable Energy. This data does not involve the large hydroelectricity capacities installed in the country. India has set a

Table 1.1 RECAI scores and rank in July 2021.

Country	Score	RECAI Rank
U.S.	70.1	1
China	68.7	2
India	66.2	3

Table 1.2 Technology-specific RECAI scores (and rank).

Technology	U.S.	China	India
Solar PV	57.6	60.3	62.7 (1)
Onshore wind power	58.1	55.7	54.2 (6)
Offshore wind power	55.6	60.6	28.6 (29)
Biofuels	45.3	52.8	47.4 (10)
Hydroelectricity	57.6	60.3	46.4 (3)
Solar CSP power plants	46.2	54.3	09.2 (4)
Geothermal power	46.0	31.7	23.2 (16)

Table 1.3 Installed capacity of non-conventional non-renewable power.

Wind Power	39,870.45
Solar Power – Ground Mounted	39,347.92
Solar Power – Roof Top	5574.12
SPV Systems (Off-grid)	1353.10
Small Hydro Power	4809.81
Biomass (Bagasse) Cogeneration)	9403.56
Biomass (non-Bagasse) Cogeneration)/Captive Power	772.05
Waste to Power	168.64
Waste to Energy (Off-grid)	233.20
Total	101,532.85

target of 459 GW renewable energy capacity to be installed by 2030. This value of installed renewable energy capacity can be increased by 146 GW if large hydroelectricity capacity is included. The world's largest renewable energy park with 30 GW capacity solar wind hybrid project is in the pipeline in Gujarat [104]. The installed capacity of non-conventional non-renewable power [105] is tabulated in Table 1.3.

1.5 Future Prospects and Summary

Renewable energy is produced from renewable resources, which are naturally recaptured in the timespan of humans, including biomass, wood and its waste, municipal solid waste, landfill gas and biogas, ethanol, biodiesel, sunlight, wind, rain, tides, waves, and geothermal heat. The sector of renewable energy has experienced an unrivaled boom over the past few years. The research on renewable energy has shown the important roles that it plays such as supply with security, addressing climate change causes, meeting rising power demand, promoting economic growth and development, and industrial policy issues. This chapter discusses the different types of renewable energy resources, including biomass, hydropower, geothermal, wind, solar, and ocean. The advantages of renewable energy that have been known for a long time include cleaner air and water, zero carbon emissions, improved public health, conservation of natural resources, recycling of waste products, and long-term sustainable savings. The disadvantages and drawbacks of renewable energy are discussed in terms of cost, storage capabilities, geographical limitations, unpredictable nature, noise pollutions, and safety issues. The COVID-19 pandemic has shown a major impact on the renewable energy sector worldwide, curbing future energy investments and threatening to slow down the expansion plans of key renewable energy technologies. In addition, the current scenario of renewable energy in India is discussed with supported facts and figures. India is one of the top leaders in most attractive renewable energy sources because of the development of a sustainable path for energy supply, and also the third largest country to produce and consume 38% electricity from total installed energy capacity in the world. Data of installed capacity of renewable and non-renewable energy is shown in Table 1.3. Some key points from this chapter, which will be useful, can be drawn here:

- Renewable energy is derived from naturally abounding resources and plays an important role in energy security and minimizing greenhouse gas emissions. Nearly 8 billion metric tonnes of carbon have been released into the atmosphere from deforestation and fossil fuels each year.
- The Government of India set an ambitious renewable energy capacity target of 175 GW to be completed by 2022, but are planning to achieve more than that target (227 GW) and to establish a renewable energy capacity of 523 GW by 2030.
- Various energy storage systems are used in electric power technologies. Electrochemical energy storage systems such as batteries are one of the potential storage systems and contribute to the implementation of sustainable energy. In the batteries, during charging, electrical energy is stored from an external electrical source and can be used to supply the energy to an external load during discharging.
- India, the U.S., China, the UK, and Spain are the most affected countries by the COVID-19 pandemic, which produced 70% wind power in 2019 worldwide.
- According to RECAI-May 2021 released by EY, India has been ranked third out of 40 countries in terms of installed capacity and investment in the renewable energy sector, whereas the U.S. topped the index and China took the second place.
- Among different types of hydrogen energy, green hydrogen, i.e., the hydrogen produced from RES, significantly contributes to successful energy transition. This promising energy carrier is well-capable to efficiently linking various energy sectors.

Competing Interest

Authors declare no competing interest.

Acknowledgements

All authors would like to thank the Indian Institute of Technology Jodhpur-India, Chandigarh College of Technology, Chandigarh Group of Colleges; IISC-Bengaluru; Andhra University College of Engineering, Andhra University-India, Istanbul Technical University-Turkey, and CMET-India, for resource and technical support. This work was supported as part of the Sustainable Materials and Catalysis Research Laboratory (SMCRL), Department of Chemistry, Indian Institute of Technology Jodhpur with SERB-CRG grant number CRG/2020/002163.

References

1 Panwar, N.L., Kaushik, S.C., and Kothari, S. (2011). Role of renewable energy sources in environmental protection: A review. *Renew. Sustain. Energy Rev.* 15(3): 1513–1524.

2 Harjanne, A. and Korhonen, J.M. (2019). Abandoning the concept of renewable energy. *Energy Policy* 127: 330–340.

3 Demirbas, A. (2000). Recent advances in biomass conversion technologies. *Energy Educ. Sci. Technol.* 6: 19–41.

4 Malkivia-Pyh, I. and Pyh, Y. (2002). *Sustainable Energy Resources, Technology and Planning*. pp. 24–29. Boston: WIT Press.

5 Ellabban, O., Abu-Rub, H., and Blaabjerg, F. (2014). Renewable energy resources: Current status, future prospects and their enabling technology. *Renew. Sustain. Energy Rev.* 39: 748–764.

6 Chriscaden, K. and Osseiran, N. (2016). WHO releases country estimates on air pollution exposure and health impact. *News release GENEVA*, 27 September 2016. https://www.who.int/news/item/27-09-2016-who-releases-country-estimates-on-air-pollution-exposure-and-health-impact

7 *World Energy Assessment 2000 – Energy and the Challenge of Sustainability* (2000). New York: UNDP (ISBN 9211261260).

8 Renewable energy directive, *European Commission*. https://energy.ec.europa.eu/topics/renewable-energy/renewable-energy-directive-targets-and-rules/renewable-energy-directive_en.

9 *Tentative State-wise break-up of Renewable Power target to be achieved by the year 2022 so that cumulative achievement is 175,000 MW* (n.d.). Delhi: MNRE. http://164.100.94.214/sites/default/files/uploads/Tentative-State-wise-break-up-of-Renewable-Power-by-2022.pdf

10 Gielen, D., Boshell, F., Saygin, D. et al. (2019). The role of renewable energy in the global energy transformation. *Energy Strategy Rev.* 24: 38–50.

11 *Renewable Energy Industry in India, India Brand Equity Foundation (IBEF)* (February 2022). https://www.ibef.org/industry/renewable-energy

12 The 13th five-year plan for energy development (2016). *Beijing: NDRC*. https://policy.asiapacificenergy.org/node/2918

13 *Market Report Series: Renewables 2018* (2018). Paris: OECD/IEA. https://www.iea.org/reports/renewables-2018

14 Renewables account for most new U.S. electricity generating capacity in 2021, *Today in Energy*, 11 January 2021. https://www.eia.gov/todayinenergy/detail.php?id=46416

15 Mohtasham, J. (2015). International conference on technologies and materials for renewable energy, Environment and Sustainability, TMREES15. *Energy Procedia* 74: 1289–1297.

16 Kılıc, E.F. and Kılıc, M. (2013). *Engineer & the Machinery Magazine* 639: 45–56.

17 Alrikabi, N.K.M.A. (2014). Renewable energy types. *J. Clean Energy Technol.* 2(1): 61–64.

18 Amrouche, S.O., Rekioua, D., Rekioua, T. et al. (2016). Overview of energy storage in renewable energy systems. *Int. J. Hydrogen Energy* 41(45): 20914–20927.

19 (a) Ponnada, S., Kiai, M.S., Gorle, D.B. et al. (2021). History and recent developments in divergent electrolytes towards high-efficiency lithium–sulfur batteries: A review. *Mater. Adv.* 2(13): 4115–4139; (b) Ponnada, S., Kiai, M.S., Gorle, D.B. et al. (2021). Insight into lithium–sulfur batteries with novel modified separators: Recent progress and perspectives. *Energy & Fuels* 35(14): 11089–11117; (c) Ponnada, S., Kiai, M.S., Gorle, D.B. et al. (2021). Improved performance of lithium–sulfur batteries by employing a sulfonated carbon nanoparticle-modified glass fiber separator. *Nanoscale Advances* 3(15): 4492–4501.

20 Maradin, D. (2021). Advantages and disadvantages of renewable energy sources utilization. *Int. J. Energy Econom. Policy* 11(3): 176.

21 Bilgen, S., Kaygusuz, K., and Sari, A. (2004). Renewable energy for a clean and sustainable future. *Energy Sources* 26(12): 1119–129.

22 Demirbaş, A. (2006). Global renewable energy resources. *Energy Sources* 28(8): 779–792.
23 Liu, J., Zuo, J., Sun, Z. et al. (2013). Sustainability in hydropower development: A case study. *Renew. Sustain. Energy Rev.* 19: 230–237.
24 Fetter, S., Flynn, M., Rothstein, L. (2000). Energy 2050. *Bull. At. Sci.* 56(4): 28–38.
25 *Clean Energy for a Healthy Planet, Geothermal Engineering Limited*. http://geothermalengineering.co.uk
26 *COVID-19: The Impact on Renewable Energy, Energy and Economic Growth, Applied Research Program*, 15 June 2020. https://www.energyeconomicgrowth.org/blog/covid-19-impact-renewable-energy
27 Anmar, F. (2020). *Coronavirus at wind power plant in North Dakota shuts down production*. https://www.cnbc.com/2020/04/20/coronavirus-at-wind-power-plant-in-north-dakota-shuts-down-production.html
28 McPhee, D. (2020). *Moray east of shore wind project "on track" despite COVID-19 delays, Director claims. Energy Voice*. https://www.energyvoice.com/renewables-energy-transition/235798/moray-east-offshore-wind-project-on-track-despite-covid-19-delays-director-claims
29 Eroğlu, H. (2021). Effects of Covid-19 outbreak on environment and renewable energy sector environment. *Environment, Development and Sustainability.* 23(4): 4782–4790.
30 Ghose, D., Naskar, S., and Uddin, S. (2019). Q-GIS-MCDA based approach to identify suitable biomass facility location in Sikkim, India. *Proceedings of the Second International Conference on Advanced Computational and Communication Paradigms*, IEEE, Sikkim, India. Proceedings published in IEEE Xplor.
31 Hosseini, S. (2020). An outlook on the global development of renewable and sustainable energy at the time of COVID-19. *Energy Res. Soc. Sci.* 68: 101633–101636.
32 Pradhan, S., Ghose, D., and Shabbiruddin (2020). *Present and future impact of COVID-19 in the renewable energy sector: A case study on India. Energy Sources. Part A: Recovery, Utilization, and Environmental Effects*, pp. 1–11.
33 Eljack, F. and Kazi, M.K. (2021). Prospects and challenges of green hydrogen economy via multi-sector global symbiosis in Qatar. *Front. Sustain.* 1: 612762.
34 *Green Hydrogen: A Guide to Policy Making. Tech. rep.* International Renewable Energy Agency (IRENA), November 2020, ISBN:978-92-9260-286-4. https://www.irena.org/publications/2020/Nov/Green-hydrogen
35 Song, J., Wei, C., Huang, Z-F. et al. (2020). A review on fundamentals for designing oxygen evolution electrocatalysts. *Chem. Soc. Rev.* 49(7): 2196–2214.
36 Khan, M.A., Zhao, H., Zou, W. et al. (2018). Recent progresses in electrocatalysts for water electrolysis. *Electrochem. Energy Rev.* 1(4): 483–530.
37 Suen, N-T., Hung, S-F., Quan, Q. et al. (2017). Electrocatalysis for the oxygen evolution reaction: Recent development and future perspectives. *Chem. Soc. Rev.* 46(2): 337–365.
38 Suleman, F., Dincer, I., and Agelin-Chaab, M. (2015). Environmental impact assessment and comparison of some hydrogen production options. *Int. J. Hydrogen Energy* 40: 6976–6987.
39 Hydrogen from Renewable Power: Technology Outlook for the Energy Transition. *Tech. rep.*, International Renewable Energy Agency (IRENA), September 2018. ISBN: 978-92-9260-077-8. https://www.irena.org/%20publications/2018/Sep/Hydrogen-from-renewable-power
40 Rabiee, A., Keane, A., and Soroudi, A. (2021). Green hydrogen: A new flexibility source for security constrained scheduling of power systems with renewable energies. *Int. J. Hydrogen Energy* 46(37): 19270–19284.

41 Dempsey, J.L., Winkler, J.R., and Gray, H.B. (2010). Kinetics of electron transfer reactions of H_2-evolving cobalt diglyoxime catalysts. *J. Am. Chem. Soc.* 132(3): 1060–1065.

42 Ouziel, S. and Avelar, L. (2021). *4 technologies that are accelerating the green hydrogen revolution*. 29 Jun 2021. https://www.weforum.org/agenda/2021/06/4-technologies-accelerating-green-hydrogen-revolution

43 Baruah, R. and Bhaskar, U. (2022). *Govt unveils hydrogen policy as India takes aim at fossil fuel*, Mint, 18 February 2022. https://www.livemint.com/industry/energy/concessional-green-power-manufacturing-zones-among-raft-of-incentives-under-india-s-green-hydrogen-policy-11645105443544.html

44 National Hydrogen Strategies, KPMG. https://home.kpmg/xx/en/home/insights/2021/08/national-hydrogen-strategies.html

45 Carmo, M., Fritz, D.L., Mergel, J. et al. (2013). A comprehensive review on PEM water electrolysis. *Int. J. Hydrogen Energy* 38: 4901–4934.

46 Buttler, A. and Spliethoff, H. (2018). Current status of water electrolysis for energy storage, grid balancing and sector coupling via power-to-gas and power-to-liquids: A review. *Renew. Sustain. Energy Rev.* 82: 2440–2454.

47 Wang, M., Wang, Z., Gong, X. et al. (2014). The intensification technologies to water electrolysis for hydrogen production: A review. *Renew. Sustain. Energy Rev.* 29: 573–588.

48 Santos, D.M.F., Sequeira, C.A.C., and Figueiredo, J.L. (2013). Hydrogen production by alkaline water electrolysis. *Química Nova* 36: 1176–1193.

49 Wang, P., Zhang, X., Zhang, J. et. al. (2017). Precise tuning in platinum-nickel/nickel sulfide interface nanowires for synergistic hydrogen evolution catalysis. *Nat. Commun.* 8: 14580.

50 Lim, J., Park, D., Jeon, S.S. et al. (2018). Ultrathin IrO_2 nanoneedles for electrochemical water oxidation. *Adv. Funct. Mater.* 28: 1704796.

51 Ying, J., Jiang, G., Cano, Z.P. et al. (2017). Nitrogen-doped hollow porous carbon polyhedrons embedded with highly dispersed Pt nanoparticles as a highly efficient and stable hydrogen evolution electrocatalyst. *Nano Energy* 40: 88.

52 Zhang, L., Han, L., Liu, H. et al. (2017). Potential-cycling synthesis of single platinum atoms for efficient hydrogen evolution in neutral media. *Angew. Chem., Int. Ed.* 56: 13694.

53 Zhang, Y., Zhou, Q., Zhu, J. et al. (2017). Nanostructured metal chalcogenides for energy storage and electrocatalysis. *Adv. Funct. Mater.* 27: 1702317.

54 Fan, H., Yu, H., Zhang, Y. et al. (2017). Fe-doped Ni_3C nanodots in N-doped carbon nanosheets for efficient hydrogen-evolution and oxygen-evolution electrocatalysis. *Angew. Chem.* 129: 12740.

55 Feng, J.X., Ding, L.X., Ye, S.H. et al. (2015). $Co(OH)_2$@PANI hybrid nanosheets with 3D networks as high-performance electrocatalysts for hydrogen evolution reaction. *Adv. Mater.* 27: 7051.

56 Cheng, H., Su, C.-Y., Tan, Z.-Y. et al. (2017). Interacting $ZnCo_2O_4$ and Au nanodots on carbon nanotubes as highly efficient water oxidation electrocatalyst. *J. Power Sources* 357: 1.

57 Chen, G.F., Ma, T.Y., Liu, Z.Q. et al. (2016). Efficient and stable bifunctional electrocatalysts Ni/Ni_xM_y (M = P, S) for overall water splitting. *Adv. Funct. Mater.* 26: 3314.

58 Cheng, H., Su, Y.-Z., Kuang, P.-Y. et al. (2015). Hierarchical $NiCo_2O4$ nanosheet-decorated carbon nanotubes towards highly efficient electrocatalyst for water oxidation. *J. Mater. Chem. A* 3: 19314.

59 Yan, D., Li, Y., Huo, J. et al. (2017). Defect chemistry of nonprecious-metal electrocatalysts for oxygen reactions. *Adv. Mater.* 29: 1606459.

60 Wang, A.-L., Xu, H., and Li G.-R. (2016). NiCoFe layered triple hydroxides with porous structures as high-performance electrocatalysts for overall water splitting. *ACS Energy Lett.* 1: 445.

61 Voiry, D., Yamaguchi, H., Li, J. et al. (2013). Enhanced catalytic activity in strained chemically exfoliated WS_2 nanosheets for hydrogen evolution. *Nat. Mater.* 12: 850.

62 Kim, Y., Jackson, D.H., Lee, D. et al. (2017). *In situ* electrochemical activation of atomic layer deposition coated MoS_2 basal planes for efficient hydrogen evolution reaction. *Adv. Funct. Mater.* 27: 1701825.

63 Xu, C., Peng, S., Tan, C. et al. (2014). Ultrathin S-doped $MoSe_2$ nanosheets for efficient hydrogen evolution. *J. Mater. Chem. A* 2: 5597.

64 Ouyang, C., Wang, X., and Wang, S. (2015). Phosphorus-doped CoS_2 nanosheet arrays as ultra-efficient electrocatalysts for the hydrogen evolution reaction. *Chem. Commun.* 51: 14160.

65 Tao, L., Duan, X., Wang, C. et al. (2015). Plasma-engineered MoS_2 thin-film as an efficient electrocatalyst for hydrogen evolution reaction. *Chem. Commun.* 51: 7470.

66 Feng, J.-X., Wu, J.-Q., Tong, Y.-X. et al. (2018). Efficient hydrogen evolution on Cu nanodots-decorated Ni_3S_2 nanotubes by optimizing atomic hydrogen adsorption and desorption. *J. Am. Chem. Soc.* 140: 610.

67 Chung, D.Y., Jun, S.W., Yoon, G. et al. (2017). Large-scale synthesis of carbon-shell-coated FeP nanoparticles for robust hydrogen evolution reaction electrocatalyst. *J. Am. Chem. Soc.* 139: 6669.

68 Zhang, J., Wang, T., Liu, P. et al. (2016). Engineering water dissociation sites in MoS_2 nanosheets for accelerated electrocatalytic hydrogen production. *Energy Environ. Sci.* 9: 2789.

69 Zhang, Z., Lu, B., Hao, J. et al. (2014). FeP nanoparticles grown on graphene sheets as highly active non-precious-metal electrocatalysts for hydrogen evolution reaction. *Chem. Commun.* 50: 11554.

70 Wang, A.-L., Lin, J., Xu, H. et al. (2016). Ni_2P–CoP hybrid nanosheet arrays supported on carbon cloth as an efficient flexible cathode for hydrogen evolution. *J. Mater. Chem. A* 4: 16992.

71 Wan, C., Regmi, Y.N., Leonard, B.M. et al. (2014). Multiple phases of molybdenum carbide as electrocatalysts for the hydrogen evolution reaction. *Angew. Chem.* 126: 6525.

72 Ang, H., Tan, H.T., Luo, Z.M. et al. (2015). Hydrophilic nitrogen and sulfur Co-doped molybdenum carbide nanosheets for electrochemical hydrogen evolution. *Small* 11: 6278.

73 Ang, H., Wang, H., Li, B. et al. (2016). 3D hierarchical porous Mo_2C for efficient hydrogen evolution. *Small* 12: 2859.

74 Chen, W.F., Sasaki, K., Ma, C. et al. (2012). Hydrogen-evolution catalysts based on non-noble metal nickel–molybdenum nitride nanosheets. *Angew. Chem., Int. Ed.* 51: 6131.

75 Yan, H., Tian, C., Wang, L. et al. (2015). Phosphorus-modified tungsten nitride/reduced graphene oxide as a high-performance, non-noble-metal electrocatalyst for the hydrogen evolution reaction. *Angew. Chem., Int. Ed.* 54: 6325.

76 Lin, H., Shi, Z., He, S. et al. (2016). Heteronanowires of MoC–Mo_2C as efficient electrocatalysts for hydrogen evolution reaction. *Chem. Sci.* 7: 3399.

77 Wang, D., Li, Q., Han, C. et al. (2017). When NiO@Ni meets WS_2 nanosheet array: A highly efficient and ultrastable electrocatalyst for overall water splitting. *ACS Cent. Sci.* 4: 112.

78 Rui, K., Zhao, G., Chen, Y. et al. (2018). Hybrid 2D dual-metal–organic frameworks for enhanced water oxidation catalysis. *Adv. Funct. Mater.* 28: 1801554.

79 Dou, S., Wu, J., Tao, L. et al. (2015). Carbon-coated MoS_2 nanosheets as highly efficient electrocatalysts for the hydrogen evolution reaction. *Nanotechnology* 27: 045402.

80 Chen, Y., Zhou, Q., Zhao, G. et al. (2018). Electrochemically inert g-C_3N_4 promotes water oxidation catalysis. *Adv. Funct. Mater.* 28: 1705583.

81 Gao, M.R., Liang, J.X., Zheng, Y.R. et al. (2015). An efficient molybdenum disulfide/cobalt diselenide hybrid catalyst for electrochemical hydrogen generation. *Nat. Commun.* 6: 5982.

82 Yan, X., Tian, L., He, M. et al. (2015). Three-dimensional crystalline/amorphous Co/Co_3O_4 core/shell nanosheets as efficient electrocatalysts for the hydrogen evolution reaction. *Nano Lett.* 15: 6015.

83 Zhao, G.Q., Rui, K., Dou, S.X. et al. (2018). Heterostructures for electrochemical hydrogen evolution reaction: A review. *Adv. Funct. Mater.* 28: 1803291.

84 Fujishima, A. and Honda, K. (1972). Electrochemical photolysis of water at a semiconductor electrode. *Nature* 238(5358): 37–38.

85 Kahng, S., Yoo, H., and Kim, J.H. (2020). Recent advances in earth-abundant photocatalyst materials for solar H_2 production. *Adv. Powder Technol.* 31(1): 11–28.

86 Pareek, V., Bhargava, A., Gupta, R. et al. (2017). Synthesis and applications of noble metal nanoparticles: A review. *Adv. Sci. Eng. Med.* 19(7): 527–544.

87 Xie, Y.P., Wang, G.S., Zhang, E.L. et al. (2017). Photocatalytic hydrogen evolution from water splitting using semiconductors: Advance, challenge and prospects. *Chin. J. Inorg. Chem.* 33(2): 177–209.

88 Acar, C., Dincer, I., and Naterer, G.F. (2016). Review of photocatalytic water-splitting methods for sustainable hydrogen production. *Int. J. Energy Res.* 40(11): 1449–1473.

89 Kudo, A. and Miseki, Y. (2009). Heterogeneous photocatalyst materials for water splitting. *Chem. Soc. Rev.* 38(1): 253–278.

90 Gao, T., Chen, Z., Huang, Q. et al. (2015). A review: Preparation of bismuth ferrite nanoparticles and its applications in visible-light induced photocatalyses. *Rev. Adv. Mater. Sci.* 40(2): 97–109.

91 Rongchen S.J.X., Xiang, Q., Chen, X. et al. (2019). Ni-based photocatalytic H_2-production cocatalysts. *Chin. J. Catal.* 40(3): 240–288.

92 Fang, S.Y. and Hu, Y.H. (2019). Recent progress in photocatalysts for overall water splitting. *Int. J. Energy Res.* 43(3): 1082–1098.

93 Kong, C., Li, Z., and Lu, G.X. (2015). The dual functional roles of Ru as co-catalyst and stabilizer of dye for photocatalytic hydrogen evolution. *Int. J. Hydrogen Energy* 40(17): 5824–5830.

94 Kang, H.W., Lim, S.N., Song, D. et al. (2012). Organic-inorganic composite of g-C_3N_4-$SrTiO_3$: Rh photocatalyst for improved H_2 evolution under visible light irradiation. *Int. J. Hydrogen Energy* 37(16): 11602–11610.

95 Bai, S., Xie, M.L., Kong, Q. et al. (2016). Incorporation of Pd into Pt Co-catalysts toward enhanced photocatalytic water splitting. *Part. Part. Syst. Char.* 33(8): 6.

96 Yang, Q., Jones, W., Wells, P.P. et al. (2016). Exploring the mechanisms of metal co-catalysts in photocatalytic reduction reactions: Is Ag a good candidate? *Appl. Catal. a-Gen.* 518: 213–220.

97 Li, X.G., Bi, W.T., Zhang, L. et al. (2016). Single-atom Pt as Co-catalyst for enhanced photocatalytic H_2 evolution. *Adv. Mater.* 28(12): 2427–2431.

98 Shen, P.C., Zhao, S., Su, D. et al. (2012). Outstanding activity of sub-nm Au clusters for photocatalytic hydrogen production. *Appl. Catal. B Environ.* 126: 153–160.

99 *India ranked 3rd, US 1st in renewable energy country attractiveness index*. Released by EY, AC Team, 20 May 2021. https://affairscloud.com/india-ranked-3rd-us-1st-in-renewable-energy-country-attractiveness-index-released-by-ey

100 Manohar, A. *Renewable Energy Sector, Invest India, National Investment Promotion and Facilitation Agency*, Government of India. https://www.investindia.gov.in/sector/renewable-energy

101 Power Industry Report, India Brand Equity Foundation, February 2021. https://www.ibef.org/download/Power-March-2021.pdf

102 *Renewable Energy Country Attractiveness Index (PDF)*. (Accessed 24 August 2020.) https://www.ey.com/en_in/recai

103 Renewable Energy Country Attractiveness Index (RECAI), Ernst & Young, 2021. https://currentaffairs.adda247.com/india-ranks-third-in-2021-ey-index/#:~:text=India%20has%20retained%20the%20third,Ernst%20%26%20Young%20(EY)

104 Power Industry Report, India Brand Equity Foundation, February 2022. https://www.ibef.org/industry/power-sector-india

105 *Programme/Scheme wise Cumulative Physical Progress as on March 2022*, Ministry of New & Renewable Energy, Government of India. https://mnre.gov.in/the-ministry/physical-progress

2

Hydrogen and Hydrocarbons as Fuel

Chandraraj Alex and Neena S. John

Centre for Nano and Soft Matter Sciences, Shivanapura, Bengaluru 562162, Karnataka, India

2.1 Introduction

Global warming and air pollution are significant challenges to the environment and human health [1]. Fossil fuels emit tremendous amounts of CO_2 that lead to faster climate changes [2]. Thus, there is an urgent need to find cleaner and efficient energy sources. Hydrogen fuel is a zero-carbon fuel, and does not produce any carbon by-products on combustion. The only by-product is water, as given in Equation 2.1.

$$2H_2(g) + O_2(g) \rightarrow 2H_2O(g) + \text{Energy} \tag{2.1}$$

The utilization of hydrogen as a fuel has not been realized to the extent of hydrocarbons due to many reasons such as high manufacturing cost, lack of infrastructure for hydrogen storage, and distribution. Hydrocarbons such as petrol, diesel, and liquefied petroleum gases (LPG) are widely used in vehicles, homes, and industries. The combustion engine consumes a huge number of hydrocarbons for energy production. However, the percentage of heat energy that is transformed into work is low. In most gasoline combustion engines, thermal/energy efficiency is 20%, and is around 40% in the diesel-based engine [3]. The chemical equation for burning simple hydrocarbon propane in a combustion engine is given in Equation 2.2.

$$C_3H_8(g) + 6O_2(g) \rightarrow 3CO_2(g) + 4H_2O(g) + \text{Energy} \tag{2.2}$$

The main drawback of the combustion engine is poor energy/thermal efficiency and CO_2 emission, so we need an alternative method to rectify this issue. Hydrogen fuel is an alternative to conventional fuels. Hydrogen can be produced in many ways such as steam reforming of fossil fuels, methane pyrolysis, biomass gasification, photobiological water splitting, and water electrolysis [4].

In steam reforming, hydrogen is produced from natural gas. It is the currently available cheapest method for industrial hydrogen production and nearly 50% of the world's hydrogen is produced by this method. The natural gas (e.g., CH_4) is heated with water at 700–1100°C in the presence of a nickel catalyst, resulting in the formation of CO and hydrogen. The formed carbon monoxide, passed with steam over an iron oxide catalyst at 360°C, leads to the formation of CO_2 and H_2 (Equations 2.3 and 2.4) [4, 5].

Green Energy Harvesting: Materials for Hydrogen Generation and Carbon Dioxide Reduction,
First Edition. Edited by Pooja Devi.

$$CH_4 + H_2O \rightarrow CO + 3\,H_2 \tag{2.3}$$

$$CO + H_2O \rightarrow CO_2 + H_2 \tag{2.4}$$

In methane pyrolysis, methane is heated at very high temperatures of >1000°C on a metal catalyst. The methane decomposes into carbon and hydrogen, as in Equation 2.5.

$$CH_4 \rightarrow C + 2\,H_2 \tag{2.5}$$

In biomass gasification, biomass consists of mainly renewable organic sources such as agricultural crops residues, forest residues, organic municipal solids, and animal wastes. In biomass gasification, it is converted to organics, fossil carbonaceous materials, and hydrogen, with the production of a controlled amount of oxygen and steam by pyrolysis at temperatures of >700°C. An absorber or special membrane is used to separate hydrogen from the products. The formed CO undergoes further reaction with steam to form more hydrogen at a given temperature (Equations 2.6 and 2.7). The produced hydrogen is then separated and purified [6].

$$C_6H_{12}O_6 + O_2 + H_2O \rightarrow CO + CO_2 + H_2 + \text{carbonaceous materials} \tag{2.6}$$

$$CO + H_2O \rightarrow CO_2 + H_2 \tag{2.7}$$

Photobiological water splitting, etc. employs microorganisms such as cyanobacteria or microalgae to produce hydrogen from water using sunlight. The main challenges in this process are the low rate of hydrogen production and poor solar-to-hydrogen production efficiency. Some recent research is focused on improving the activity of hydrogen production enzymes and developing strains for sunlight harvesting in algae. Currently, 10–12 mL of hydrogen is obtained per liter culture in 1 h. Alternatively, the decomposition of water into hydrogen and oxygen using electrical energy is achieved in water electrolysis. About 70–80% efficiency has been achieved by water electrolysis and it is expected to be 82–86% before 2030 [7]. Water electrolysis consists of two half-cell reactions such as anodic oxygen evolution and cathodic hydrogen evolution reactions. The overall reaction for water electrolysis is $H_2O \rightarrow H_2 + ½\,O_2$; the individual electrode reactions are shown in Figure 2.1.

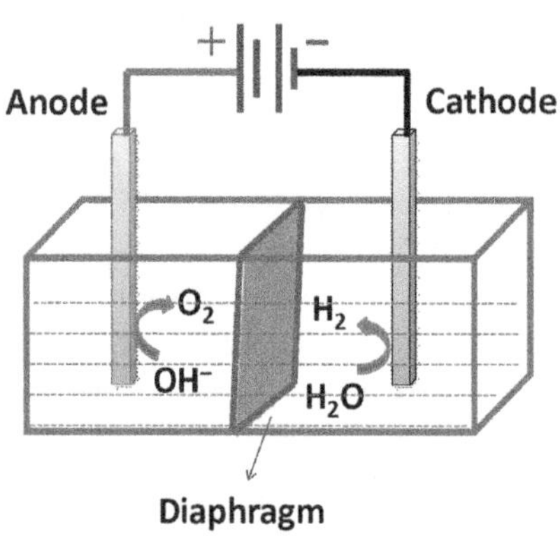

Figure 2.1 Schematic representation of water electrolysis.

2.2 Hydrogen and Hydrocarbons as Fuels

Hydrogen is used as an energy source in many applications such as the hydrogen internal combustion engine, welding gas (oxyhydrogen), synthesis of synthetic fuel, and fuel cells. Among the various applications, hydrogen is mainly used in the fuel cell due to its high efficiency. The energy efficiency of hydrogen fuel cells exceeds 60%. Further, these fuel cells have benefits such as high temperature operation capability and lower or zero CO_2 emissions, making them more eco-friendly [3]. Most of the fuel cells employ hydrogen, hydrocarbons, urea, and alcohols as fuels.

2.2.1 Fuel Cells

Fuel cells are considered as a cleaner source of energy due to their low emissions [8]. The fuel cell is an electrochemical device that produces direct electric current from the chemical energy of fuels [9]. Sir William Robert Grove developed the first fuel cell in 1839, which produces electric energy by combining hydrogen and oxygen [10]. In 1889, Mond and Langer developed porous electrodes and powdered electrocatalysts [11]. Ostwald determined the relationship between different components of fuel cells in 1893 [12]. Bacon developed the first alkaline fuel cells, and the National Aeronautics and Space Administration (NASA) used the alkaline fuel cell for space technology in the 1960s [13]. Thomas Grubb and Leonard Niedrach used the sulfonated polystyrene membrane, and these were replaced when Grot developed Nafion [14]. The experimental membrane was made by the Dow chemical company in 1986, and it produces power four times higher than the proton exchange fuel cell (PEMFC) [15]. Finally, commercialization of the fuel cell started in 2007.

2.2.1.1 Types of Fuel Cells

A fuel cell is an electrochemical cell that converts chemical energy into electricity through redox reactions. Common fuel cells are made up of three components: the anode, cathode, and electrolyte [16]. The anode is the electrode that catalyzes the electrochemical oxidation reaction of a fuel. The fuel may be hydrogen, hydrocarbons, urea, etc. During operation, fuel loses its electron and is oxidized, while the cathode promotes electrochemical reduction reaction of an oxidizing agent. Air, oxygen, oxygen-hydrogen peroxide, and oxygen-carbon dioxide mixtures are some examples, and electrolytes effectively transport the formed ions at the cathode and anode during the redox process. The generated electron passes from anode to cathode through the given external load.

2.2.1.2 Fuels

Hydrogen is a common fuel in all fuel cells. Apart from hydrocarbons such as methane, ethane, propane, butane, etc., and alcohols such as methanol, ethanol, propanol, butanol, etc., urea and carbon are also used as fuels in fuel cells.

The above schematic (Figure 2.2) represents the working principle of the H_2–O_2 fuel cell [17]. In the fuel cell, H_2 (fuel) is oxidized at the anode. It produces H^+ ions with electrons, the formed H^+ ions pass internally through the electrolyte to the cathode, and the electrons travel externally through the load and reach the cathode. The cathode reduces O_2 (oxidizing agent) into OH^- ions which combine with H^+ ions in the electrolyte with resultant formation of H_2O.

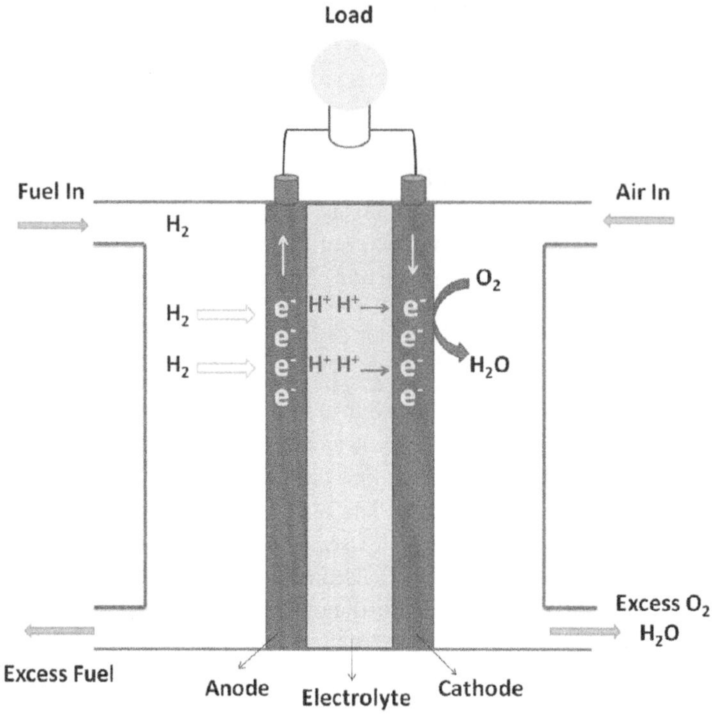

Figure 2.2 Schematic representation of a hydrogen-oxygen fuel cell.

The fuel cells are classified primarily based on the type of electrolyte employed. The five main types of fuel cells are alkaline, molten carbonate, phosphoric acid, solid oxide, and proton exchange membrane fuel cells. Various hydrogen and hydrocarbon fuels can be employed in the above configuration for energy generation. Table 2.1 summarizes details of fuel cells listing the materials used as fuels, catalysts, and other components.

2.2.1.3 Alkaline Fuel Cells (AFCs)

The AFCs were one of the first fuel cell technologies developed. AFCs use an aqueous solution of KOH as an electrolyte. The H_2 and O_2 gases are used as fuel and oxidizers. The electrodes are made from nonprecious materials such as sintered Ni, iron, cobalt, silver, and Raney metal catalysts [18]. The alkaline fuel cells have many advantages: low cost, excellent oxygen reduction kinetics in alkaline media, and high efficiency [19]. One of the major problems in AFCs is the performance loss by CO_2 in the electrolyte. The entrance of carbon dioxide into the electrolyte leads to the formation of alkaline carbonate as a result of lowering of the hydroxide ion concentration, and the porous electrode gets blocked by metal carbonates [20]. To maintain the efficiency of these fuel cells, additional care should be taken to avoid CO_2 entering into the cell.

2.2.1.4 Molten Carbonate Fuel Cells (MCFCs)

MCFCs are high-temperature fuel cells, which operate in the temperature range of 550–700°C with a variety of fuels. The electrolyte is made up of a molten carbonate salt suspended in a porous, chemically stable ceramic matrix composed of alumina-based compounds [21].

Table 2.1 Overall details of fuel cells.

Types of Fuel Cell	AFC	MCFC	PAFC	SOFC	PEM
Electrolyte	Aqueous KOH/NaOH	Molten carbonate salt	H_3PO_4 in silicon carbide	Y_2O_3-stabilized ZrO_2 (YSZ)	Ion exchange membrane
Fuel	H_2	hydrogen, hydrocarbons, bio- and coal fuels	H_2	Ammonia, hydrogen, hydrocarbons, biogas, ethanol, methanol, CO	H_2
Catalyst	Ni, Ag, metal oxides, and noble metals	Copper, cobalt, and nickel with aluminum/chromium	Pt	nickel cermet Ni/YSZ as an anode, and (La, Sr)MnO_3 as a cathode	Platinum carbon
Oxidant	Air/O_2	Air/CO_2	Oxygen	Air/O_2	Air/O_2
Operating temperature	65–220°C	550–700°C	150–220°C	800–1000°C	80–120°C
Reactions	**Anode:** $H_2 + 2OH^- \rightarrow 2H_2O + 2e^-$ **Cathode:** $\frac{1}{2} O_2 + H_2O + 2e^- \rightarrow 2OH^-$	**Anode:** $2H_2 + 2CO_3^{2-} \rightarrow 2H_2O + 2CO_2 + 4e^-$ **Cathode:** $O_2 + 2CO_2 + 4e^- \rightarrow 2CO_3^{2-}$	**Anode:** $H_2 \rightarrow 2H^+ + 2e^-$ **Cathode:** $\frac{1}{2} O_2 + 2H^+ + 2e^- \rightarrow H_2O$	**Anode:** $C_2H_5OH + 3H_2O \rightarrow 2CO_2 + 12H^+ + 12e^-$ **Cathode:** $3O_2 + 12H^+ + 12e^- \rightarrow 6H_2O$	**Anode:** $H_2 \rightarrow 2H^+ + 2e^-$ **Cathode:** $\frac{1}{2} O_2 + 2H^+ + 2e^- \rightarrow H_2O$

(Continued)

Table 2.1 (Continued)

Types of Fuel Cell	AFC	MCFC	PAFC	SOFC	PEM
Advantages	i) Low cost ii) High efficiency iii) Fast electrode kinetics	i) High energy and fuel efficiency ii) Usage of non-noble metal electrodes iii) Is not prone to CO_2 poisoning	i) Long-term stability ii) High CO_2 tolerance iii) Heat and power cogeneration	i) Can work with a variety of fuels ii) Long life span iii) Electrodes made up of non-noble metals	i) High power density ii) Comfortable for portable and stationary applications
Disadvantages	i) CO_2 electrode poisonous	i) Corrosive electrolyte environment and electrode corrosion ii) Not suitable in portable applications	i) Low power density ii) Intolerant to CO	i) Carbon deposition on the electrode when hydrocarbons used as fuel ii) High temperature limits versatile applications	i) Electrodes made by the noble metal catalyst ii) Membrane is sensitivity to water, CO, and operating temperature iii) Require high purity hydrogen

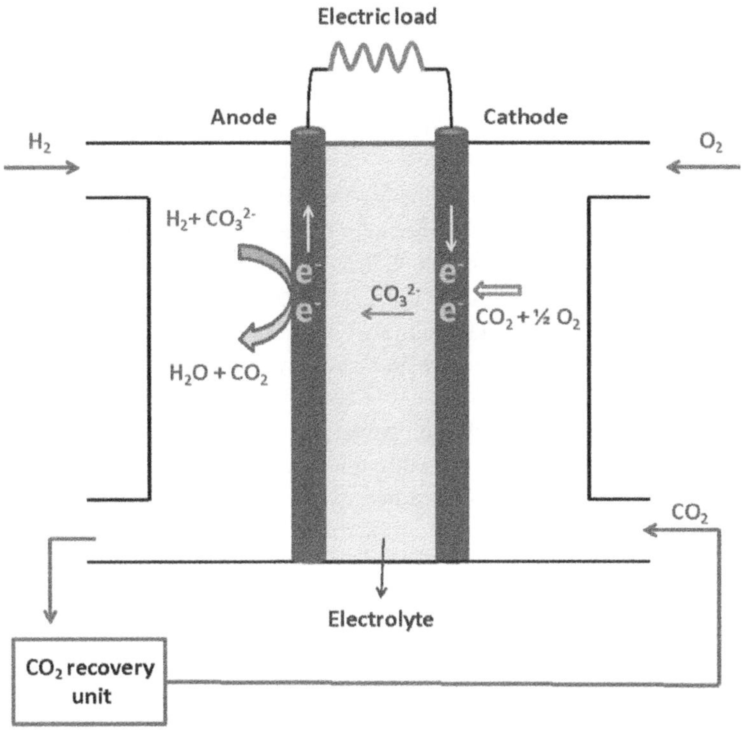

Figure 2.3 Schematic representations of the H_2-O_2 carbonate fuel cell.

The anode and cathode are made up of nickel catalysts. The MCFCs operate in hydrogen, hydrocarbons, bio, and coal fuels [22]. The oxidant is a mixture of carbon dioxide and oxygen. The MCFCs cells have many advantages, such as the following: i) low-cost electrodes; ii) are not prone to poisonous CO_2 and CO; iii) high energy efficiency; and iv) the high operating temperature allows to improve the fuel efficiency by >80% [23, 24]. However, it has some disadvantages such as electrode corrosion, short lifetime, and low power density [25].

In the carbonate fuel cell represented in Figure 2.3, H_2 is used as the fuel. This is introduced into the anodic compartment. The fuel undergoes oxidization and forms CO_2 and H_2O. The CO_2 recovered from outlet gases of the anodic compartment then feeds into the cathodic compartment with the oxygen. The CO_2 feeding induces electrochemical reduction of CO_2 and CO_3^{2-} formation. The carbonate ion moves from cathode to anode through a porous electrolyte.

2.2.1.5 Phosphoric Acid Fuel Cells (PAFCs)

PAFCs work by using 100% phosphoric acid in silicon carbide matrix as the electrolyte. The cathode and anode are made up of Pt catalysts. The cell operation temperature is 150–200°C [26]. Phosphoric acid has many advantages, such as long-term operation and high carbon dioxide tolerance [27]. Some disadvantages include low power density and noble metal catalysts CO intolerance [28].

2.2.1.6 Solid Oxide Fuel Cells (SOFCs)

SOFCs are the most common fuel cells which work at high temperatures of 800–1000°C. They uses a variety of fuels such as ammonia, hydrocarbons, alcohols, hydrogen sulfide, bio-, and synthetic gas [1]. Most of SOFCs use yttria-stabilized zirconia (YSZ) as an electrolyte [29]. The nickel cermet (Ni/YSZ) and (La, Sr)MnO_3 are the commonly used anode and cathode [30]. The fuel is introduced into the anode for oxidation, and oxygen reduction occurs at the cathode. The SOFCs electrodes should have the following advantages: i) good catalytic activity and electrical conductivity; ii) thermal expansion stability; iii) must be durable and fuel flexible at high temperatures; and iv) low coking resistance and cost. The SOFCs produce a high amount of heat that can be utilized in cogeneration or in combined cycle applications. They use non-noble-metal catalysts with a long life span of 40,000–80,000 h [31, 32]. The high operating temperature and electrolyte resistivity restrict their utilization in many applications.

Figure 2.4 represents an SOFC. It has anodic compartments where the feed fuel is oxidized and the electrons pass from anode to cathode through an external load. The oxygen is reduced into oxide ions (O^{2-}) in the cathode compartment. The ions undergo internal transportation via solid electrolytes and reach the anode compartments for further fuel oxidation.

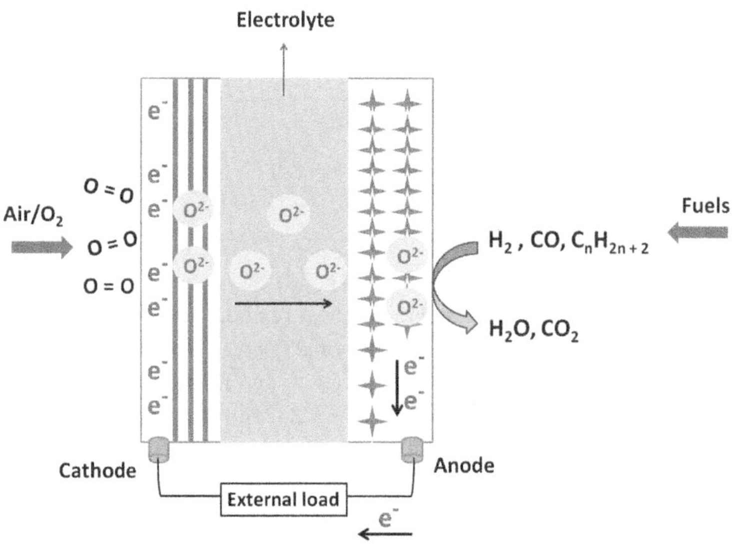

Figure 2.4 Schematic representation of solid oxide fuel cell.

2.2.1.7 Proton Exchange Membrane Fuel Cells (PEMFCs)

PEMFCs are the promising technology in energy conversion. They have significant features such as a lower operating temperature of ~80–120°C, high power density, and are ideal for transport/portable applications [33]. The cathode and anode are made up of Pt/C. The PEMFCs are made by basic components of bipolar plates, diffusion layers, electrodes, and electrolytes [34].

Figure 2.5 shows a membrane electrode assembly (MEA), composed of a proton exchange membrane, catalyst layer, and gas diffusion layer (GDL), sandwiched between

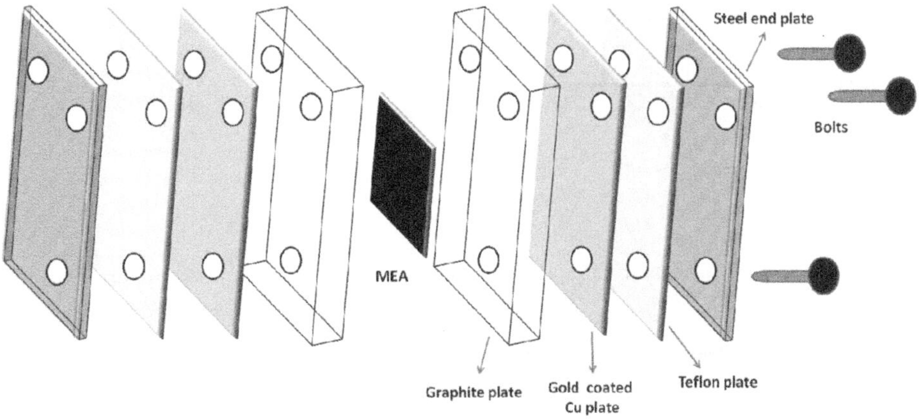

Figure 2.5 Components of a PEM fuel cell (MEA – membrane electrode assembly).

flow field plates. The operation temperature for the PEMFC is around 80°C [35]. The cathode and anode are made up of platinum-supported carbon. Hydrogen and air are used as fuel and oxidants.

2.3 Emerging Technologies in Hydrogen Fuel Cells

The emerging fuel cell technology is based on the varied nature of fuel handling systems rather than the type of electrolytes. Hydrogen is an explosive flammable gas so handling it is a difficult process. In emerging technologies, the hydrogen is stored in other forms such as ammonia, urea, methanol, hydrides, etc. The onsite hydrogen is either generated from the above fuel or the fuel is directly used for fuel cell applications. The technologies based on these fuels for small-scale portable power devices application are at the development stage.

2.3.1 Indirect and Direct Urea Fuel Cells

In indirect fuel cells, the hydrogen is produced from the storage urea through electrolysis and is used as the input in fuel cells. The hydrogen produced by urea electrolysis is introduced into various hydrogen–oxygen fuel cells as the result of normal electrochemical reactions. The direct urea fuel cells consist of two electrodes, such as the anode and cathode. Fuel urea is introduced into the anodic compartment and undergoes electrochemical oxidation with OH^- as a result forming N_2, CO_2 and water. The cathodic compartment reduces O_2 and H_2O into OH^- through the following reaction: $O_2 + 2H_2O + 4e^- \rightarrow 4OH^-$ (Figure 2.6). The formed OH^- is transferred into the anodic compartment by an alkaline exchange membrane electrolyte for urea electro-oxidation. In direct urea fuel cells, the urea is directly used as fuel [36].

The advantages and drawbacks of hydrogen and hydrocarbon fuel cells are discussed below.

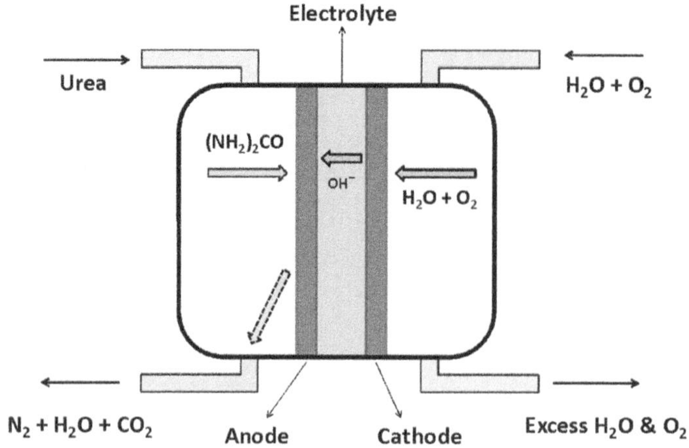

Figure 2.6 Schematic of direct urea fuel cells.

2.4 Advantages and Drawback of Hydrogen and Hydrocarbons as a Fuel

Hydrogen has a high gravimetric energy density, which is higher than hydrocarbons and possesses zero carbon emission. Hydrogen fuel cells offer greater efficiencies with a long operation time. It can be utilized in stationary and mobile applications. Other advantages are low noise pollution compared to other renewable energy sources such as wind turbines and conventional internal combustion engines [37].

One of the main problems with hydrogen fuel is safe storage with a poorer energy storage density than other hydrocarbons. Hydrogen is highly flammable in air at a 4–75% concentration and easily leaks through seals. Currently, most hydrogen is produced through steam reforming of natural gas followed by the water gas shift reaction. These processes emit CO_2 into the atmosphere [38].

The usage of hydrocarbons as fuel can rectify the problems associated with direct hydrogen fuel cells. The direct usage of hydrocarbon as fuel in fuel cells eliminates the process of hydrogen generation from fossil fuels. As a result, there is a reduction in overall energy delivery cost as energy delivery/kWh of energy production. The average emission of CO_2 for hydrogen production through methane decomposition is 7 kg of CO_2/1 kg of H_2. This can be brought down by direct hydrocarbon fuel cells and its higher fuel efficiency [39]. Generally, hydrocarbon fuel cells perform better at high operating temperatures (>200°C) because of the difficulty of breaking the C–C bond for fuel oxidation. They encounter lower electron transfer rates compared to hydrogen for fuel oxidation reaction [40].

2.5 Direct Hydrocarbon Fuels and Their Usage in Various Fuel Cells

2.5.1 Hydrocarbon Fuel in Phosphoric Acid Fuel Cells (PAFCs)

PAFCs operate in the temperature range of 150–200°C. Pratt and Whitney developed the stationary electric power station with 250–400 kW power based on a PAFC [41].

The performance of PAFCs with petrol, diesel, and biodiesel was investigated by Yuanchen Zhu et al. [42]. Their results show performance degradation of PAFCs due to the deposition of carbon-based material on the electrode surface. The operation of the n-hexadecane PAFC was analyzed by Yuanchen Zhu et al. [41] at two different temperatures, 160 and 190°C. Pt/C was used as cathode and anode for reactions. The cathodic and anodic fuel cell reactions are given in Equations 2.8–2.10:

Anodic reaction

$$C_{16}H_{34\,(g)} + 32H_2O_{(g)} \rightarrow 16CO_{2(g)} + 98H^+ + 98e^- \tag{2.8}$$

Cathodic reaction

$$49/2\ O_{2\,(g)} + 98H^+ + 98e^- \rightarrow 49H_2O \tag{2.9}$$

Overall reaction

$$C_{16}H_{34\,(g)} + 49/2\ O_{2\,(g)} \rightarrow 16CO_{2(g)} + 17H_2O \tag{2.10}$$

The results showed a potential decrease over time, related to either formation of CO_2 or carbonaceous deposit on the electrode. The conclusions are that the enhancement of the current to high value requires a large reactor [41].

2.5.2 Hydrocarbon Fuel in SOFCs

The high operating temperature of SOFCs has the advantage of being tolerant to a wide range of fuels [43]. The internal reforming is an important method of feeding hydrocarbons in SOFCs, which proceeds as in Equation 2.11.

$$C_nH_{n+2} + nH_2O \rightarrow nC + (n+1)H_2 \tag{2.11}$$

The following disadvantages prevent the commercialization of internal reforming: i) internal reforming requires a significant amount of water in fuel, which reduces the electromotive force of fuel cells through dilution, and ii) as steam reforming is an endothermic process, heat changes in the inlet area of stack leading to temperature gradient and reduce cell durability [44, 45]. The limitation of internal reforming is rectified by applying the widely used alternative technique of direct hydrocarbon feeding with a minimal amount of water and oxygen [46].

$$C_nH_{n+2} + (3n+1)O^{2-} \rightarrow nCO_2 + (n+1)H_2O + (6n+2)\ e^- \tag{2.12}$$

Another problem associated with direct hydrocarbon SOFCs is the significant potential drop caused by the deposition of carbon on the anode. When the temperature is exceptionally high, hydrocarbons form tar on the anode and lower the stability of the cell. These issues are increased in conventional Ni-based anodes [47]. The alternative anodes such as Cu-based cermet and doped (La, Sr)TiO_3 catalysts showed Carbon deposition (coking) tolerance for light hydrocarbons. Still, these electrodes have low conductivity, and low physical, chemical, and compact thermal ability with YSZ electrolytes at high temperatures [48]. A thin coating of Sn and BaO was applied on the Ni electrode (Ni-YSZ) to remove the coking effect. The developed electrode Cu-ceria-YSZ significantly reduces carbon deposition with excellent performance. Fuels such as butane, methane, ethylene, propane-1,2,3-triol, iso-octane, and ethoxyethane can be directly fed into SOFCs. The conducted study using butane as the fuel exhibits complete conversion at 800°C and none below 550°C [49]. Fuels such as n-decane, toluene, and fuel gas give the same power density under the same conditions. The main problem with SOFCs is carbon deposition on the electrode. The report says minimum carbon deposition is possible via injection of the fuel at a specific low flow rate and at a temperature of around 650°C [50]. Further research needs to be conducted before SOFC can be commercialized.

2.5.3 Hydrocarbon Fuel in Proton Exchange Membrane Fuel Cells (PEMFCs)

The PEM fuel cells have advantages such as high power density and solid cell compartments [51]. Hydrogen and hydrocarbons are the most common fuels of PEM fuel cells. Among them, hydrocarbons offer a better choice for PEM fuel cells. The hydrocarbon fuels can be directly fed into the cell or converted to hydrogen using an external reformer. The second method incurs the extra cost to fuel cells. The performance of PEM fuel cells using Pt, gold-lead, and Nafion as anode, cathode, and electrolyte were studied for various fuels [52]. The fuels such as butane, ethane, n-butane, methane, ethylene, propylene, and isobutylene have been tested at temperatures of >100°C. The electrode shows highest performance per unit surface area with propane and lowest with methane, ethylene, propylene, and isobutylene [53]. The Pt-based catalyst and Nafion are widely used electrodes and electrolytes for PEMFCs. These cells face a lot of kinetic constraints on anodic alcohol electro-oxidation reactions and relatively reduce the performance of alcohol fuel cells with fuel crossover. There are various ongoing efforts to develop portable direct methanol fuel cells. Most literature focuses on understanding the mechanism of electro-oxidation and improving cell performance [54, 55]. Other than this research, studies toward dispersion of catalyst on the electrode, humidity effect, optimum catalyst preparation method, and finding novel catalysts other than Pt and Ru are in demand.

2.5.4 Hydrocarbon Fuel in Molten Carbonate Fuel Cells (MCFCs)

Methane is a widely used fuel in direct MCFCs. The fuel undergoes internal steam reforming and the water gas shift reaction is as shown in Equations 2.13 and 2.14 [56].

$$CH_4 + H_2O \rightarrow CO + 3H_2 \tag{2.13}$$

$$CO + H_2O \rightarrow CO_2 + H_2 \tag{2.14}$$

The formed hydrogen undergoes electrochemical oxidation at the anode with the consumption of CO_3^{2-} ions. The CO_3^{2-} ions formation happens at the same time in the cathode by CO_2 reduction with oxygen (Equations 2.15 and 2.16).

Anode reaction

$$H_2 + CO_3^{2-} \rightarrow H_2O + CO_2 + 2e^- \tag{2.15}$$

Cathode reaction

$$\tfrac{1}{2}\,O_2 + CO_2 + 2e^- \rightarrow CO_3^{2-} \tag{2.16}$$

Mostly, the focused catalyst is based on Ni because of the high-temperature operation of MCFCs at around 650°C. The studies on Ni catalysts to reduce carbon deposition and enhance electrode poison resistance are carried out. Using supports like MgO, SiO_2, and TiO_2 has poor resistance to carbon poisoning [57], whereas perovskites exhibit good resistance for both carbon and alkali poisons. Recently, the concept of producing electricity from solid fuel coal has also been explored. This faces several technical challenges such as poor electrochemical activity, sulfur contamination, corrosion, and ash accumulation.

2.5.5 Hydrocarbon Fuel in Direct Alcohol Fuel Cells (AFCs)

The study of the alkaline electrolyte with the alcohol fuel cell started around 1960 [58]. The 4.5 M methanol alkaline fuel cell using the Pt-Ru/C electrode gives an operation voltage of 0.6–0.75 V at 0.53 A current loading. The ethanol electro-oxidation in ethanol/KOH mixture has been studied employing Pt/C as the anode. The cathode is made up of MnO_2 and Ni-supported carbon. The maximum current density is achieved at 3 M KOH electrolyte, and temperature enhances the cell performance [59, 60]. The maximum power density of 11 mW cm^{-2} is obtained at 60°C.

One of the major problems in ethanol fuel cells is C–C bond breaking at a temperature lower than 100°C and poor electron transfer kinetics [61]. Ethylene glycol (EG) is another fuel that rectifies this issue to some extent by undergoing partial oxidation with the production of electrons and oxalate ions [62]. However, it suffers from complete EG oxidation into CO_2 and a lack of knowledge about mass transportation.

2.6 Emerging Hydrocarbon Fuel Cell Technology

The usage of hydrocarbon fuels in portable devices is an emerging technology. The efforts are toward manufacturing comfortable portable devices with less device volume, weight, and high fuel density. Researchers are trying to develop portable fuel cells with ammonia, methanol, and ethanol direct fuels. Further developments of microbial fuel cells will soon be used for more wide-scale applications. In microbial fuel cells, the microbes decompose organics into small products through biochemical reactions and the electrons formed are utilized for energy production [63]. The direct carbon fuel cell is another emerging technology where the solid fuels such as carbon and high carbon containing hydrocarbons, coal, and biomass chars can be utilized as fuel. A broad fuel cell classification based on fuel type including future technologies is given by Badel et al. (Figure 2.7) [63].

2.6.1 Microbial Fuel Cells

In microbial fuel cells, the fuel undergoes electrochemical oxidation by microbes. In microbial fuel cells, the electrochemical reaction is catalyzed by micro-organisms such as *Escherichia coli, Pseudomonas fluorescens, Bacillus violaceus, Desulfuromonas acetoxidans, Methylovorus mays*, etc. The microbes catalyze the decomposition of organic matter resulting in the formation of electrons, CO_2, and H^+ ions. The electrons are transferred from the

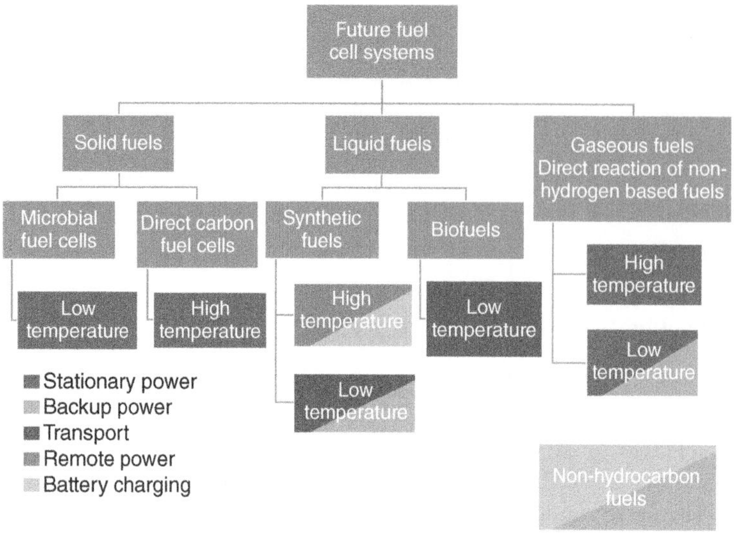

Figure 2.7 Broad classification of fuel cell based on fuel type and emerging technology. (Badwal, S. P. et al., 2014 / Frontiers Media S.A. / CC BY-4.0 [63].)

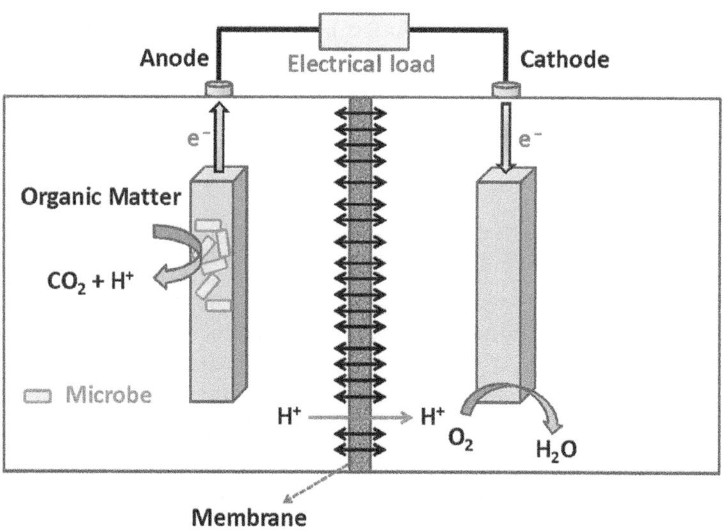

Figure 2.8 Schematic representation of microbial fuel cells.

micro-organisms to the electrode, where they further pass through the external load to the cathode. The H^+ ions are exchanged from the anodic to cathodic compartment by the ion exchange membrane. The O_2 undergoes reduction at the cathode with the exchanged H^+ ions (Figure 2.8).

2.6.2 Direct Carbon Fuel Cells

The direct carbon fuel cell comes under the class of MCFCs. The fuel carbon undergoes electrochemical oxidation at the anode with carbonate ions which leads to the formation of CO_2 and electrons. The CO_2 gas is collected and passed to the cathodic compartment with oxygen. In the cathode, CO_2 and oxygen are reduced into carbonate ions with the external electrons coming from the anode. The formed carbonate ions are transferred internally from cathode to anode (Figure 2.9) [64].

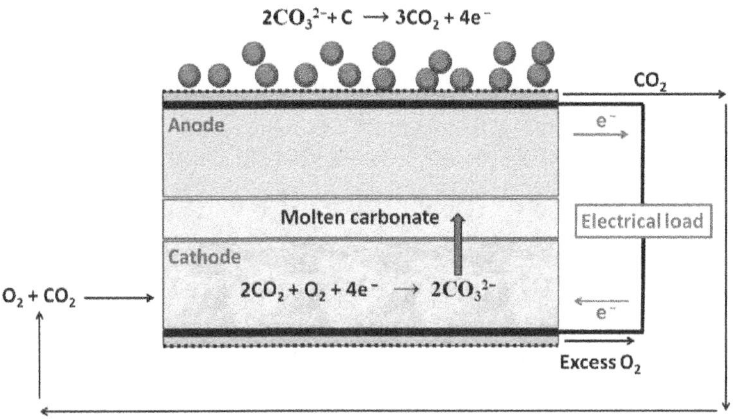

Figure 2.9 Schematic representation of carbon fuel cells.

2.6.3 Efficiency of Direct Hydrogen and Hydrocarbon Fuel Cells

The efficiency of fuel cells is defined as the ratio of work out/heat in. The amount of useful energy available in the system to work at constant pressure and temperature is called the Gibbs free energy (ΔG). Suppose the system is producing ΔH amount of total heat energy by fuel oxidation; out of this energy only ΔG amount of energy does the useful work and the rest of the energy is dissipated as waste. Then the efficiency of the fuel cell is equal to $\Delta G/\Delta H$.

The free energy of water formation at standard state is [65].

$$H_{2(g)} + \frac{1}{2} O_{2(g)} \rightarrow H_2O_{(l)} \tag{2.17}$$

ΔH_f = Enthalpy of water formation: −285.84 KJ mol^{-1}.

ΔG = Amount of energy doing useful work: 237.3 KJ mol^{-1}.

The efficiency of fuel cells (η) operating on pure hydrogen and oxygen at standard state is [65].

$$\eta = -(237.3/\ 285.84) \times 100 = 83\ \%.$$

Considering hydrocarbon fuel cells as an example, propane (C_3H_8) is based on the following reactions (Equations 2.18–2.20) [66]:

Anode

$$C_3H_8 + 6H_2O \rightarrow 3CO_2 + 20e^- + 20H^+ \quad E_a^0 = 0.137\ V \tag{2.18}$$

Cathode

$$5O_2 + 20e^- + 20H^+ \rightarrow 10H_2O \quad E_c^0 = 1.229\ V \tag{2.19}$$

Overall reaction

$$C_3H_8 + 5O_2 \rightarrow 3CO_2 + 4H_2O \quad E_{cell}^0 = 1.092\ V \tag{2.20}$$

The ΔG of the overall reaction is 2107 KJ mol^{-1} and the standard heat of reaction is −2219 KJ mol^{-1} [66]. The fuel cell efficiency is

$$\eta = -\left(2107/\ 2219\right) \times 100 = 95\ \%.$$

The direct hydrocarbon fuel cells have a theoretical efficiency of 95%, but obtained efficiency is 32%, wherein hydrogen fuel cells theoretical efficiency is 83%, but the practically obtained amount is 32% [15]. This result implies the actual efficiency of direct hydrocarbon fuel cells can be improved more than direct hydrogen fuel cells due to their higher theoretical efficiency.

2.6.4 High-Temperature Operation

The main problems associated with direct hydrocarbon fuel cells are low reaction rate and electrode poisons due to lower oxidation current density. The enhancement of operation temperature improves the CO tolerance and allows the usage of various fuels. The high operation temperature permits efficient generation and recovery of useful heat and reduces the liquid water accumulation at the cathode. In PEMFCs, proton conductivity depends on the hydration of electrolytes such as Nafion and other similar membranes. Hence, there is a need to develop a new fuel membrane class that can operate at high temperatures [67, 68].

2.6.5 Polarization Curve

The polarization curve helps to evaluate the performance of fuel cells. The polarization curve is obtained by a potentiostat/galvanostat that draws a fixed current from the fuel cell and measures the fuel cell output voltage. These current and potential values are plotted as the potential (V) vs. current density (mA m^{-2}) plot of a fuel cell, as shown in Figure 2.10. It is expected that the cell potential would be equal to the thermodynamic value of the cell with increasing reaction rate or current density. However, the polarization curve exhibits a drop in potential with an increase in current density. The potential drop is based on three potential losses such as activation, ohmic, and concentration polarization. The activation polarization is high in low current density. It is caused by surface diffusion, reaction, and desorption of reactive species.

The potential drop at moderate current density in the polarization curve is caused by ohmic polarization. This polarization arises through resistance to the flow of protons in electrolytes and electrons in the cell components and external wires. The total ohmic polarization can be explained by ohms law, potential drop due to ohmic polarization

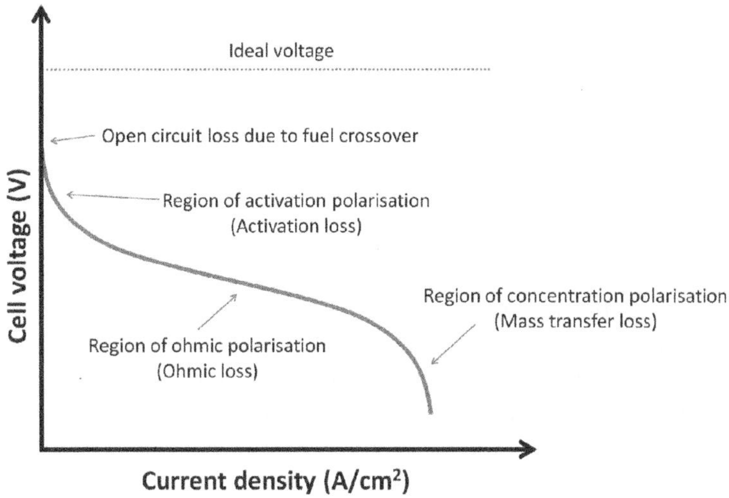

Figure 2.10 Polarization curve of the typical fuel cell.

$E_O = I\,(R_{ions} + R_{elect} + R_{ct})$, where R_{ions}, R_{elect}, and R_{ct} are ionic, electronic, and contact resistance, respectively. The R_{ions} and R_{elect} resistance can be reduced using electrolytes and electrodes with high conductivity. Further reducing the traveling distance of ions and electrons has an overall reduction effect on E_O.

The third potential drop is caused by concentration polarization. This drop was observed at the region of high current density in the polarization curve (Figure 2.10). The mass transfer limitations are responsible for this drop. For example, in PEMFC, the accumulation of water on the cathode restricts oxygen diffuses to the catalyst sites [69]. This effect becomes prominent when liquid water fills the pore structure of the cathode in PEMFC.

2.6.6 Exergy and Its Governing Equations

The maximum possible work obtained from a system when brought into thermal, mechanical, and chemical equilibrium with surroundings is called exergy. The concept of exergy analysis was brought into usage because energy analysis does not provide an actual picture of system performance. Further, it will not quantify those factors which cause thermodynamic losses and system deviation from ideality. Exergy analysis measures the degradation of available energy and identifies the cause, location, and magnitude of energy losses. To gain maximum work for input energy, the system's different processes should be reversible [70].

The total exergy of a system is the summation of the physical and chemical exergy of that same system. In fuel cells, the entering fuels (components) undergo a chemical reaction. As a result of a change in chemical composition, the chemical exergy of the system needs to be considered.

The specific exergy with chemical composition change is given by Equation 2.21.

$$e_i = (h_i - h_0) - T_0(s_i - s_0) + \Sigma_i x_i(\mu_i - \mu_0) \tag{2.21}$$

The system environment is usually defined as a reference state. h_0, s_0, T_0, and μ_0 are reference state enthalpy, entropy, temperature, and chemical potentials, respectively. The h, s, T, x, and μ symbols have the usual meaning, such as enthalpy, entropy, temperature, mole fraction, and chemical potential, respectively.

The specific exergy of a mixture or many component systems can be written as

$$e_i = (h_i - h_0) - T_0(s_i - s_0) + e_{ch} \tag{2.22}$$

$$e_{ch} = \Sigma_i \left(x_i e^0_{ch} + RT_0 x_i ln x_i\right) \tag{2.23}$$

where $e^0{}_{ch}$ is standard chemical exergy [71].

The fuel cell systems have both gases and liquid components, so their enthalpy and entropy changes have to be considered to know the exergy of the system.

The enthalpy (Δh) and entropy (Δs) changes of ideal gas are given in Equations 2.24–2.28.

$$h_i - h_0 = \Delta h = C_p(T_i - T_0) \tag{2.24}$$

$$s_i - s_0 = \Delta s = C_p ln(T_i/T_0) - R\, ln(p_i/p_0) \tag{2.25}$$

For a gas component with partial pressure $p_i = x_i p$ in a gas mixture of total pressure p, the entropy changes are defined as

$$s_i - s_0 = \Delta s = C_p ln(T_i/T_0) - R\, ln(x_i\, p/p_0) \tag{2.26}$$

The entropy and enthalpy changes for a liquid can be written as

$$h_i - h_0 = \Delta h = C_p(T_i - T_0) - \nu(p_i - p_0) \tag{2.27}$$

$$s_i - s_0 = \Delta s = C_p\, ln(T_i/T_0) \tag{2.28}$$

where ν is molar specific volume.

The exergy efficiency of a fuel cell is defined as the ratio of the useful work output ($W_{use.out}$) of system to reversible work output ($W_{rev.out}$).

$$\text{Exergy efficiency} = W_{use.out} / W_{rev.out}. \tag{2.29}$$

From the above equation, the exergy efficiency is written as

$$\text{Exergy efficiency} = (\Sigma_{inlet} ne - \Sigma_{outlet} ne - T_0 S_{gen}) / (\Sigma_{inlet} ne - \Sigma_{outlet} ne). \tag{2.30}$$

This equation is used in the literature to calculate the exergy efficiency of fuel cells.

2.6.7 Exergy Analysis of Fuel Cells

The exergy analysis of a 68 kW hydrogen fueled PEM fuel cell engine has been carried out by Mert et al. [72]. The thermodynamic analyses are performed for humidifiers, pressure regulator, cooling system, compressor, and fuel cell stack. The system shows an improvement in exergy efficiency of 8% with an enhancing temperature, pressure, and humidity of 10% [72]. Ratlamwala et al. have presented a new design for flow channels in hydrogen-oxygen fueled PEM fuel cells [73]. According to their observation, the temperature increment enhances cell efficiency from 33.8 to 47.7% and power output from 2.6 to 282.5 W. The lowering of cell efficiency from 45.5 to 28.4% is observed with improvement in current density and membrane thickness [73]. The comprehensive exergy analysis of direct methanol fuel cell was carried out by Bahrami et al. [74]. The result shows that fuel crossover and over-potential majorly contribute to exergy loss. Further, analysis of the methanol fuel cell by Joseph et al. shows that temperature and pressure increment leads efficiency up to 95%. The recent study reveals liquid methanol fuel gives better efficiency than its gas phase and that major exergy losses are from liquid fuel crossover [75].

The major energy loss of the solid oxide fuel cell (SOFC) is observed in the combustion chamber, when fueled with methane and ethanol [76]. Increase of cell pressure and lowering of temperature are found to enhance exergy efficiency. The exergy efficiency test of SOFC with fuels such as methanol, propane, and butane shows propane has highest exergy efficiency of 49% while methanol exhibits lower exergy with maximum irreversibility. The exergy analysis of molten carbonate fuel cell (MCFC) in integration with cooling, heating, and power generator setup, reveals major energy loss in the combustion chamber, and hence the process design for combustion chambers requires attention.

2.7 Conclusions

Among the various fuels used in fuel cells, hydrogen is the most common one and possesses several advantages and disadvantages. Currently, 95% of hydrogen is produced from fossil fuels (e.g., hydrocarbons) by steam reforming and is associated with a huge amount of CO_2 released into the atmosphere. The alternative method of water electrolysis has to be elevated to an industrial scale to avoid CO_2 dumping into the atmosphere. The present issues for commercialization of water electrolysis are due to limitations such as: i) lack of infrastructure for production, transportation, distribution, and storage of hydrogen; ii) hydrogen is highly inflammable and safety is problematic; iii) insufficient cooperation between political authorities and professional association in the field of hydrogen energy and economic operators; iv) high initial installation costs for production, distribution, and storage of hydrogen; and v) high production cost for fuel cells [77]. These issues can be solved by working with hydrocarbon fuel cells due to their advantages; high energy storage, global availability, and ease of transportation. In hydrogen fuel cell systems, the processes required to produce hydrogen takes up 30% of the capital cost. If hydrocarbons are directly fed to the anode of the fuel cell, costs will be lowered. The production, storage, and transportation of hydrocarbons are already well established, so the practical applications of direct hydrocarbon fuel cell can be implemented easily.

Hydrocarbon fuels also possess a few disadvantages. When the hydrocarbon fuel cells (PEMFCs) operate below 120°C, the electrode reaction rates are extremely slow. These effects are due to poor catalytic activity, CO poisoning effects on the precious metals catalyst (Pt), and high degree of hydrocarbon crossover through the Nafion membrane. The deposition of carbon on electrodes is another major issue in fuel cells operating at temperatures higher than 500°C (e.g., SOFC), so major modifications of catalyst and electrolyte membranes are required. The predicted theoretical efficiency of propane fuel cells is 95%, greater than for hydrogen fuel cells. This shows greater potential for the use of hydrocarbon fuel directly. The exergy analysis reveals the majority of exergy destruction occurs at the combustion chamber, but in some cases lower temperature increases exergetic efficiency. Even though lower temperature enhances exergy, the major loss in this condition happens due to fuel crossover, while usage of the fuel reformer in turn enhances the temperature. The overall conclusion is that usage of hydrocarbons directly maximizes the exergy efficiency. Hence, direct hydrocarbon fuel cells have a greater potential for commercialization than hydrogen fuel cells. The major developments on the anode catalyst and membranes have to be achieved before commercialization of such fuel cells.

References

1. Damo, U.M., Ferrari, M. L., Turan, A. et al. (2019). Solid oxide fuel cell hybrid system: A detailed review of an environmentally clean and efficient source of energy. *Energy* 168: 235–246.
2. Patil, V., Shastry, V., Himabindu, M. et al. (2016). Life-cycle analysis of energy and greenhouse gas emissions of automotive fuels in India. Part 2: Well-to-wheels analysis. *Energy* 96: 699–712.
3. St-Pierre, J. and Wilkinson, D.P. (2001). Fuel cells: A new, efficient and cleaner power source. *AIChE Journal* 47(7): 1482.
4. Kalamaras, C.M. and Efstathiou, A.M. (2013). Hydrogen production technologies: Current state and future developments. In: *Conference Papers in Science,* Limassol. Hindawi, Cyprus (19–21 November 2012).
5. LeValley, T.L., Richard, A.R., and Fan, M. (2014). The progress in water gas shift and steam reforming hydrogen production technologies: A review. *International Journal of Hydrogen Energy* 39(30): 16983–17000.
6. Giaconia, A. (2014). Thermochemical production of hydrogen. In: *Advances in Hydrogen Production, Storage and Distribution* (eds Julianelli, A. and Basile, A.), pp. 263–280. Sawston, Cambridge, UK: Woodhead Publishing.
7. Kumar Gupta, S., Kumari, S., Reddy, K. et al. (2013). Trends in biohydrogen production: Major challenges and state-of-the-art developments. *Environmental Technology* 34(13–14): 1653–1670.
8. Stropnik, R., Sekavčnik, M., Ferriz, A.M. et al. (2018). Reducing environmental impacts of the ups system based on PEM fuel cell with circular economy. *Energy* 165: 824–835.
9. Wang, J. (2015). Barriers of scaling-up fuel cells: Cost, durability and reliability. *Energy* 80: 509–521.
10. Rayment, C., and Sherwin, S. (2003). *Introduction to Fuel Cell Technology*. Notre Dame, IN: Department of Aerospace and Mechanical Engineering, University of Notre Dame. 46556: 11–12.
11. Perry, M.L. and Fuller, T.F. (2002). A historical perspective of fuel cell technology in the 20th century. *Journal of the Electrochemical Society* 149(7): S59.
12. Andújar, J.M. and Segura, F. (2009). Fuel cells: History and updating. A walk along two centuries. *Renewable and Sustainable Energy Reviews* 13(9): 2309–2322.
13. Barbir, F. (2012). *PEM Fuel Cells: Theory and Practice*. Academic Press.
14. Zhang, J. ed. (2008). *PEM Fuel Cell Electrocatalysts and Catalyst Layers: Fundamentals and Applications*. Springer Science & Business Media.

15 Ternan, M. (2006). The potential of direct hydrocarbon fuel cells for improving energy efficiency. In: *2006 IEEE EIC Climate Change Conference*, pp. 1–4. IEEE, Canada (10–12 May 2006).

16 Carrette, L., Friedrich, K.A., and Stimming, U. (2001). Fuel cells: Fundamentals and applications. *Fuel Cells* 1: 5–39.

17 Baldwin, R., Pham, M., Leonida, A. et al. (1990). Hydrogen–oxygen proton-exchange membrane fuel cells and electrolyzers. *Journal of Power Sources* 29(3–4): 399–412.

18 Wagner, N., Schulze, M., and Gülzow, E. (2004). Long-term investigations of silver cathodes for alkaline fuel cells. *Journal of Power Sources* 127(1–2): 264–272.

19 Bidault, F., Brett, D.J.L., Middleton, P.H. et al. (2009). Review of gas diffusion cathodes for alkaline fuel cells. *Journal of Power Sources* 187(1): 39–48.

20 Gülzow, E. and Schulze, M. (2004). Long-term operation of AFC electrodes with CO_2 containing gases. *Journal of Power Sources* 127 (1–2): 243–251.

21 Mekhilef, S., Saidur, R., and Safari, A. (2012). Comparative study of different fuel cell technologies. *Renewable and Sustainable Energy Reviews* 16(1): 981–989.

22 Nguyen, H.V.P., Song, S.A., Seo, D. et al. (2013). Hydrogen sulfide-resilient anodes for molten carbonate fuel cells. *Journal of Power Sources* 230: 282–289.

23 Arato, E., Audasso, E., Barelli, L. et al. (2016). Kinetic modelling of molten carbonate fuel cells: Effects of cathode water and electrode materials. *Journal of Power Sources* 330: 18–27.

24 Milewski, J., Futyma, K., and Szczęśniak, A. (2016). Molten carbonate fuel cell operation under high concentrations of SO_2 on the cathode side. *International Journal of Hydrogen Energy* 41(41): 18769–18777.

25 Chen, H., Cong, T.N., Yang, W. et al. (2009). Progress in electrical energy storage system: A critical review. *Progress in Natural Science* 19(3): 291–312.

26 Sammes, N., Bove, R., and Stahl, K. (2004). Phosphoric acid fuel cells: Fundamentals and applications. *Current Opinion in Solid State and Materials Science* 8(5): 372–378.

27 Murahashi, T. (2009). Fuel cells – phosphoric acid fuel cells: Electrolytes. In: *Encyclopedia of Electrochemical Power Source* (ed. Garche, J.), pp. 564–567. Newnes: Elsevier.

28 Yang, P., Zhang, H., and Hu, Z. (2016). Parametric study of a hybrid system integrating a phosphoric acid fuel cell with an absorption refrigerator for cooling purposes. *International Journal of Hydrogen Energy* 41(5): 3579–3590.

29 Singhal, S.C. (2000). Advances in solid oxide fuel cell technology. *Solid State Ionics* 135(1–4): 305–313.

30 Kusnezoff, M. (2009). Fuel cells – solid oxide fuel cells: Membranes. In: *Encyclopedia of Electrochemical Power Sources* (ed. Garche, J.), pp. 43–49. Newnes: Elsevier.

31 Stambouli, A.B. and Traversa, E. (2002). Solid oxide fuel cells (SOFCs): A review of an environmentally clean and efficient source of energy. *Renewable and Sustainable Energy Reviews* 6(5): 433–455.

32 Andersson, M., Yuan, J., and Sundén, B. (2013). SOFC modeling considering hydrogen and carbon monoxide as electrochemical reactants. *Journal of Power Sources* 232: 42–54.

33 Wang, Y., Chen, K.S., Mishler, J. et al. (2011). A review of polymer electrolyte membrane fuel cells: Technology, applications, and needs on fundamental research. *Applied Energy* 88(4): 981–1007.

34 Rahgoshay, S.M., Ranjbar, A.A., Ramiar, A. et al. (2017). Thermal investigation of a PEM fuel cell with cooling flow field. *Energy* 134: 61–73.

35 Vishnyakov, V.M. (2006). Proton exchange membrane fuel cells. *Vacuum* 80(10): 1053–1065.

36 Gnana Kumar, G., Farithkhan, A., and Manthiram, A. (2020). Direct urea fuel cells: Recent progress and critical challenges of urea oxidation electrocatalysis. *Advanced Energy and Sustainability Research* 1: 2000015.

37 Dagdougui, H., Sacile, R., Bersani, C. et al. (2018). *Hydrogen Infrastructure for Energy Applications: Production, Storage, Distribution and Safety*. London: Academic Press.

38 Cheddie, D. (2012). *Ammonia as a Hydrogen Source for Fuel Cells: A Review*, pp. 333–362. InTech.

39 Soltani, R., Rosen, M.A., and Dincer, I. (2014). Assessment of CO_2 captures options from various points in steam methane reforming for hydrogen production. *International Journal of Hydrogen Energy* 39(35): 20266–20275.

40 Mukherjee, A. and Basu, S. (2017). Direct hydrocarbon low temperature fuel cell. In: *Electrocatalysts for Low Temperature Fuel Cells: Fundamentals and Recent Trends* (eds Maiyalagan, T. and Saji, V.S.), pp. 113. Weinheim, Germany: Wiley-VCH.

41 Zhu, Y., Robinson, T., Al-Othman, A. et al. (2015). n-Hexadecane fuel for a phosphoric acid direct hydrocarbon fuel cell. *Journal of Fuels* 2015: 1–9.

42 Zhu, Y., Tremblay, A.Y., Facey, G.A. et al. (2015). Petroleum diesel and biodiesel fuels used in a direct hydrocarbon phosphoric acid fuel cell. *Journal of Fuels* 2015: 1–9.

43 Clarke, S.H., Dicks, A.L., Pointon, K. et al. (1997). Catalytic aspects of the steam reforming of hydrocarbons in internal reforming fuel cells. *Catalysis Today* 38(4): 411–423.

44 Mogensen, M. and Kammer, K. (2003). Conversion of hydrocarbons in solid oxide fuel cells. *Annual Review of Materials Research* 33(1): 321–331.

45 Meusinger, J., Riensche, E., and Stimming, U. (1998). Reforming of natural gas in solid oxide fuel cell systems. *Journal of Power Sources* 71(1–2): 315–320.

46 Hirschenhofer, J.H., Stauffer, D.B., Engleman, R.R. et al. (1996). *Fuel Cells: A Handbook*, 4th ed., pp. 7–4. Orinda, CA: Business Technology Books.

47 van Den Bossche, M. and McIntosh, S. (2013). Direct hydrocarbon solid oxide fuel cells. In: *Fuel Cells* (ed Kreuer, K.D.), pp. 31–76. New York: Springer.

48 Liu, M., Choi, Y., Yang, L. et al. (2012). Direct octane fuel cells: A promising power for transportation. *Nano Energy* 1(3): 448–455.

49 Kee, R. J., Zhu, H., and Goodwin, D.G. (2005). Solid-oxide fuel cells with hydrocarbon fuels. *Proceedings of the Combustion Institute* 30(2): 2379–2404.

50 Jamil, S.M., Othman, M.H.D., Rahman, M.A. et al. (2015). Recent fabrication techniques for micro-tubular solid oxide fuel cell support: A review. *Journal of the European Ceramic Society* 35(1): 1–22.

51 Daud, W.R.W., Rosli, R.E., Majlan, E.H. et al. (2017). PEM fuel cell system control: A review. *Renewable Energy* 113: 620–638.

52 Savadogo, O. and Varela, F.J. (2001). Low temperature direct propane polymer electrolyte membranes fuel cell (DPFC). *Journal of New Materials for Electrochemical Systems* 4: 93–97.

53 Perry, M.L. (1996). Exploratory fuel-cell research. Part I: Direct hydrocarbon polymer-electrolyte fuel cell. Part II: Mathematical modelling of fuel-cell cathodes. University of California, California Digital Library.

54 Barakat, N.A., Yassin, M.A., Al-Mubaddel, F.S. et al. (2018). New electrooxidation characteristic for Ni-based electrodes for wide application in methanol fuel cells. *Applied Catalysis A: General* 555: 148–154.

55 Abrego-Martínez, J.C., Moreno-Zuria, A., Cuevas-Muñiz, F.M. et al. (2017). Design, fabrication and performance of a mixed-reactant membrane less micro direct methanol fuel cell stack. *Journal of Power Sources* 371: 10–17.

56 Yuan, X.Z. and Wang, H. (2008). PEM fuel cell fundamentals. In: *PEM Fuel Cell Electrocatalysts and Catalyst Layers* (ed. Zhang, J.), pp. 1–87. London: Springer.

57 Pistonesi, C., Juan, A., Irigoyen, B. et al. (2007). Theoretical and experimental study of methane steam reforming reactions over nickel catalyst. *Applied Surface Science* 253(9): 4427–4437.

58 Badwal, S.P.S., Giddey, S., Kulkarni, A. et al. (2015). Direct ethanol fuel cells for transport and stationary applications: A comprehensive review. *Applied Energy* 145: 80–103.

59 Verma, A., Jha, A.K., and Basu, S. (2005). Evaluation of an alkaline fuel cell for multifuel system. *Journal of Fuel Cell Science and Technology* 2: 234–237.

60 Verma, A. and Basu, S. (2007). Direct alkaline fuel cell for multiple liquid fuels: Anode electrode studies. *Journal of Power Sources* 174(1): 180–185.

61 Oliveira, V.B., Pereira, J.P., and Pinto, A.M.F.R. (2017). Modeling of passive direct ethanol fuel cells. *Energy* 133: 652–665.

62 An, L. and Chen, R. (2016). Recent progress in alkaline direct ethylene glycol fuel cells for sustainable energy production. *Journal of Power Sources* 329: 484–501.

63 Badwal, S.P., Giddey, S.S., Munnings, C. et al. (2014). Emerging electrochemical energy conversion and storage technologies. *Frontiers in Chemistry* 2: 79.

64 Kouchachvili, L. and Ikura, M. (2011). Performance of direct carbon fuel cell. *International Journal of Hydrogen Energy* 36: 10263–10268.

65 Barnett, S.A. (2014). Solid oxide fuel cells: Direct hydrocarbon type. In: *Encyclopedia of Applied Electrochemistry* (eds Kreysa, G., Ota, Ki., and Savinell, R.F.). New York: Springer.

66 Varela, F.R. and Savadogo, O. (2006). The effect of anode catalysts on the behavior of low temperature direct propane polymer electrolyte fuel cells (DPFC). *Journal of New Materials for Electrochemical Systems* 9(2): 127.

67 Al-Othman, A., Tremblay, A.Y., Pell, W. et al. (2010). Zirconium phosphate as the proton conducting material in direct hydrocarbon polymer electrolyte membrane fuel cells operating above the boiling point of water. *Journal of Power Sources* 195(9): 2520–2525.

68 Zhang, H. and Shen, P.K. (2012). Recent development of polymer electrolyte membranes for fuel cells. *Chemical Reviews* 112(5): 2780–2832.

69 Sharaf, O.Z. and Orhan, M.F. (2014). An overview of fuel cell technology: Fundamentals and applications. *Renewable and Sustainable Energy Reviews* 32: 810–853.

70 Arshad, A., Ali, H.M., Habib, A. et al. (2019). Energy and exergy analysis of fuel cells: A review. *Thermal Science and Engineering Progress* 9: 308–321.

71 Kotas, T.J. (1995). *The Exergy Method of Thermal Plant Analysis*. Florida: Krieger.

72 Mert, S.O., Dincer, I., and Ozcelik, Z. (2012). Performance investigation of a transportation PEM fuel cell system. *International Journal of Hydrogen Energy* 37(1): 623–633.

73 Ratlamwala, T.A.H., El-Sinawi, A.H., Gadalla, M.A. et al. (2012). Performance analysis of a new designed PEM fuel cell. *International Journal of Energy Research* 36(11): 1121–1132.

74 Bahrami, H. and Faghri, A. (2011). Exergy analysis of a passive direct methanol fuel cell. *J. Power Sources* 196(3): 1191–1204.

75 Joseph, A.J., Abdulkareem, A.S., Jimoh, A. et al. (2015). Theoretical energy and exergy analyses of direct methanol fuel cell. *Advance Mater Science & Applications* 4: 63–75.

76 Wang, Z., Fan, W., and Zhang, G. (2017). Comparison of the exergy efficiency of four power generation systems from methane using fuel cells. *RSC Advance* 7(62): 39391–39402.

77 Felseghi, R.A., Carcadea, E., Raboaca, M.S. et al. (2019). Hydrogen fuel cell technology for the sustainable future of stationary applications. *Energies* 12(23): 4593.

3

Fundamental Understanding and Figure of Merits for Electrocatalytic and Photoelectrocatalytic H_2 Production

Swapna Pahra[1,#], Sweta Sharma[1,2,#], and Pooja Devi[1,2,]*

[1] *Materials Science and Sensor Application, Central Scientific Instruments Organisation, Chandigarh 160030, Punjab, India*
[2] *Academy of Scientific and Innovative Research, New Delhi, India*
[*] *Corresponding Author*
[#] *Equally contributed*

3.1 Introduction

In today's developing world, where resources are being increasingly depleted day by day, there is a crucial need for a renewable source of clean energy. As a result, sustainable-renewable energy resources such as solar, wave, and wind energy can be employed as alternative energy sources to address energy demand issues. Hydrogen as a fuel has the great potential to replace existing fuel sources due to its high energy density and clean combustion by-products. More than 95% of hydrogen is being produced through a different number of processes (Figure 3.1) [1]. Among various routes of hydrogen production, green hydrogen generation is the most sustainable approach.

Electrochemical (EC) or photoelectrochemical (PEC) water splitting is an efficient and green route to produce hydrogen, which includes simultaneous reactions: hydrogen evolution reaction (HER) and oxygen evolution reaction (OER) at cathode and anode, respectively. These processes are attained by an electro- or photoelectrocatalyst. Efficiency of the catalyst is the prime factor to obtain an efficient water splitting electrolyzer. In particular, the efficiency of the catalyst is decided by such factors as the following: i) abundant material; ii) high activity; iii) stability; iv) low overpotential; v) low cost; and vi) easily upscalable [2]. These factors can be decided on the basis of some of the fundamentals

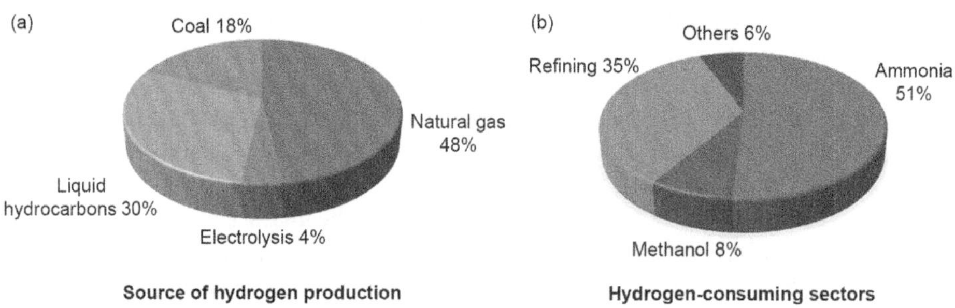

Figure 3.1 (a) Source of hydrogen production worldwide and (b) main sectors consuming hydrogen worldwide.

Green Energy Harvesting: Materials for Hydrogen Generation and Carbon Dioxide Reduction, First Edition. Edited by Pooja Devi.

discussed in this chapter, which will provide an understanding of the basics of HER and OER and factors affecting the performance of the catalyst.

Currently, state-of-the-art noble electrocatalysts for water splitting with low overpotential, such as Pt group metals and Ru/Ir-based oxides, are employed for HER and OER, respectively, but their high cost and scarcity limits their utilization [3]. Alternatively, considerable effort has been put into the design and synthesis of earth-abundant electrocatalysts for the HER (such as transition-metal phosphides, chalcogenides, nitrides, and carbides) as well as for the OER (such as transition-metal phosphates, oxides, hydroxides, perovskites, sulfides, selenides). Two-dimensional (2D) materials such as graphene, metal oxides, metal-organic frameworks, transition metal dichalcogenides (TMDs), transition metal nitrides/carbides, MXenes, and their heterostructure have gained interest for efficient H_2 production due to their high specific surface area, excellent electrical and thermal conductivity, remarkable mechanical and optical qualities, and good chemical stability. The following properties of 2D are largely responsible for high performance as electro-catalysts. First, the 2D electro-catalyst has a wide surface area due to the large number of exposed surface atoms, which exposes many catalytic active sites to the medium. Second, promising 2D scaffolds make defect and structure engineering simple, allowing for additional active sites to be created. Third, a 2D platform can be used to encapsulate other compounds with desired properties and build heterogeneous composite catalysts. Another factor that effects the catalyst's performance is the choice of substrate. Two-dimensional nanomaterial (TMDs, oxides, MOFs, etc.) coated onto substrates like carbon cloth, metal foams, carbon fiber paper, silicon, etc., have also been explored, as these supports provide large active surface area and stable standing structure [4]. Recently, transition metal carbides and nitrides have been widely studied for energy-related application due to their excellent electronic structure and metal-like characteristics. In addition, several modification approaches have been achieved; for example, doping with heteroatoms, defect engineering, nano-size modulation, construction of heterostructure, and establishment of a bimetallic system. Even some electro-catalysts such as phosphides act as bifunctional catalysts, i.e., they can be used as anode and cathode simultaneously. Transition metals, when compared to other 2D electrocatalysts, have superior properties such as low cost and quick availability, as well as high efficiency toward HER and OER, which is comparable to precious metals [5].

In 1998, Khaselev and Turner reported the first PEC solar-to-hydrogen conversion efficiency of 12.4% demonstrating the huge potential for a PEC technology that combines solar energy harvesting and water electrolysis into a single device where photon energy is transformed to electrochemical energy, which can split water into hydrogen and oxygen immediately (chemical energy).

Two-dimensional materials, especially graphene, have also found applications in catalysis for solar hydrogen generation. Other 2D materials, such as transition metal dichalcogenides (2D-TMDs), which have graphene-like structures, have also been widely explored. Each layer of the TMD material, MX_2 (M: transition metal atom, X: chalcogenide atom (S, Se, Te, etc.)), is made up of an M atom sandwiched between two X-atom layers and held together by weak Van der Waals interactions. Thus, they can easily be exfoliated to a few single layers, which offer superior characteristics over bulk materials in this size range. The conductivity, strength, and stability of 2D-TMDs are all improved when their thickness is reduced. Thin-layer 2D-TMDs have found a wide range of applications in solar cells,

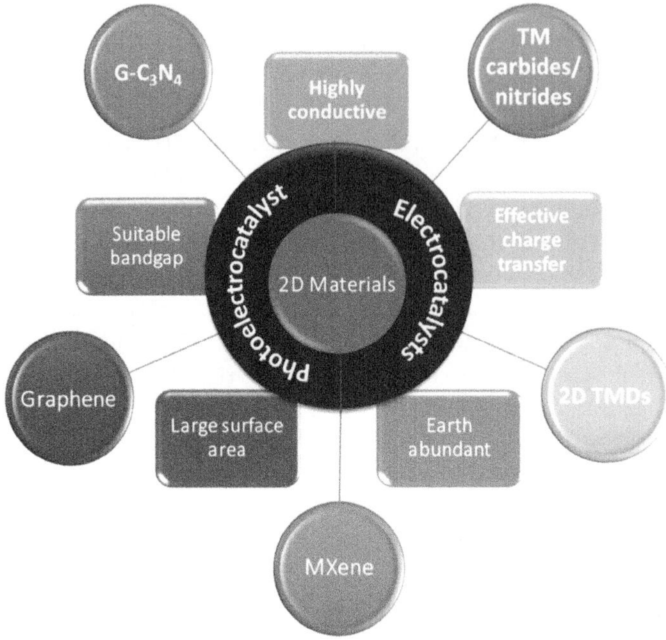

Figure 3.2 Advantages of 2D materials over other materials as electrocatalysts and photoelectrocatalysts.

organic light-emitting diodes, transistors, photodetectors, and gas sensors, thanks to their unique features. Typical 2D materials, such as graphene analogs, transition metal compounds, and other layered-structure compounds, have demonstrated excellent PEC activity. Properties of 2D materials and their advantages over other existing materials for electrocatalysis/photoelectrocatalysis are summarized in Figure 3.2.

3.2 Part I: Fundamentals of Electrocatalytic Hydrogen Production

3.2.1 Mechanism and Reaction Kinetics of OER/HER

Electrolytic water splitting comprises of two half reactions: HER and OER at cathodic and anodic electrodes, respectively (Figure 3.3). The overall water splitting reaction is as in Equation 3.1.

$$2H_2O \rightarrow 2H_2 + O_2 \quad (3.1)$$

The cathodic potential region is roughly divided into two regions: low and high potential region (below and above 80 mV). In the lower region, there is strong bonding between the active site and hydrogen atom. At high overpotential, formation of a hydrogen molecule occurs and new vacant sites are formed for further proton absorption. The nature of proton donor changes with change in pH of the electrolyte, i.e., hydronium ion in case of acidic media and water molecule in case of alkaline media, are shown in Equations 3.2 and 3.3 [2].

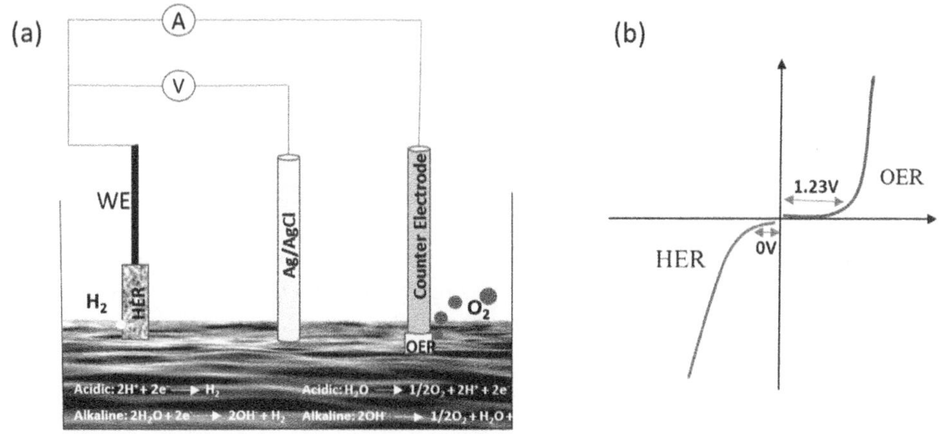

Figure 3.3 (a) HER and OER water-splitting mechanism in acidic and alkaline media. (b) I-V curve for overall water splitting.

The HER mechanism is a two-electron process consisting of three steps:

Volmer or discharge step reaction

$$2H_3O^+ + 2e^- \rightarrow H_2\,(\text{acidic media}) \tag{3.2}$$

$$2H_2O + 2e^- \rightarrow H_2 + 2OH^-\,(\text{alkaline media}) \tag{3.3}$$

Heyvorsky or atom + ion step reaction

$$H_3O^+ + e^- \rightarrow H_{ads} + H_2O\,(\text{acidic media}) \tag{3.4}$$

$$H_2O + e^- \rightarrow H_{ads} + OH^-\,(\text{alkaline media}) \tag{3.5}$$

Tafel or atom + atom step reaction

$$H_{ads} + H_{ads} \rightarrow H_2 \tag{3.6}$$

These three steps are rate determining steps (RDS) and the cathodic current increases exponentially with each of these steps.

On the other hand, OER is a complicated four-electron process as it consists of formation of several intermediates such as O*, HO*, and HOO* [2]. Its reactants also vary with pH of the electrolyte as shown in Equations 3.7 and 3.8.

In acidic media

$$2H_2O \rightarrow 4H^+ + O_2 + 4e^- \tag{3.7}$$

In alkaline media

$$4OH^- \rightarrow 2H_2O + O_2 + 4e^- \tag{3.8}$$

3.2.2 Parameters Determining Efficiency of OER/HER Electrocatalyst

3.2.2.1 Overpotential

In electrochemistry, overpotential (η) is the difference between the thermodynamic potential value and the experimental potential value that we apply to achieve a certain current density. It is the first and most important criteria both in HER and OER to judge the performance of a particular electrode in electrochemistry. The lower the value of overpotential (η), the better the performance of the electrode. In HER, generally the performance of different reporting electrodes is measured at 10 mA cm^{-2} or in some cases at 100 mA cm^{-2}. Overpotential (η) is generally reported in the form of V vs. RHE, where RHE stands for reversible hydrogen electrode which can be calculated by applying the Nernst equation (Equation 3.9).

$$E_{RHE} = E_{ref} + 2.303RT/F.pH + E^{o}_{ref} \tag{3.9}$$

where E_{ref} and $E^{o}{}_{ref}$ correspond to the applied potential and standard electrode potential of the reference electrode, respectively. The reference electrode can be Ag/AgCl or Hg/HgO, or Hg/Hg_2Cl_2.pH, which is the electrolyte/medium in which the reaction is taking place. In some cases, cell geometry and position of electrodes can also affect the overpotential values. To remove such an error, the ohmic potential drop (iR drop), which arises between the working electrode and reference electrode, should be subtracted from E_{RHE} to obtain the corrected overpotential, as shown in Equation 3.10 [6].

$$E_{corrected} = E_{RHE} - iR \tag{3.10}$$

3.2.2.2 Tafel Slope

The Tafel slope is another and most important criteria to compare the performance of electrode with the ideal Pt electrode. The Tafel slope (Equation 3.11) is basically derived from the Butler–Volmer equation.

$$\eta = b \log j + a \tag{3.11}$$

where η = overpotential, b = Tafel slope, j = current density, and a = Tafel constant.

The Tafel slope can be calculated for a particular electro-catalytic reaction by plotting a graph between overpotential (η) and logarithmic values of current density (mA cm^{-2}) at low scan rate. The linear part in the graph gives the Tafel slope (b), the value of which tells much about the reaction mechanism and about the rate determining step. There are three steps in the HER mechanism: Volmer, Heyrovsky, and Tafel. The Tafel slope values of 120 mV dec^{-1}, 40 mV dec^{-1}, and 30 mV dec^{-1}, correspond to the Volmer, Herovsky, and Tafel steps as the rate determining steps respectively. The smaller the value of b, the faster the kinetics of the reaction. Sometimes, ohmic resistances contribute to the potential in the general method of Tafel analysis. To avoid this issue, Vrubel et al. applied the impedance to analyze the Tafel slope for molybdenum sulfide [7]. The impedance method measures the charge-transfer resistance (R_{ct}) of the HER at various overpotentials. A plot between the logarithm of the inverse charge-transfer current $\log(1/R_{ct})$ and the corresponding overpotential, results in a linear relationship and directly gives the Tafel slope [2].

3.2.2.3 Exchange Current Density

In theoretical terms, exchange current density is the anodic and cathodic components of current density at equilibrium. Exchange current density (j_o) is related to the Tafel equation and can be easily calculated from this equation. When the overpotential, i.e., η is equal to zero, the value of current density at that point is considered as the exchange current density. The higher the value of j_o, the faster the electron transfer [8].

3.2.2.4 Faradic Efficiency

The Faradic efficiency is basically the efficiency of utilization of electrons in an electrochemical reaction. It can be calculated as the ratio of the number of moles of hydrogen produced to the theoretical number of moles of hydrogen to be produced. Experimental values can be found by florescence sensor or volumetric methods [8]. Also, quantification of gas produced can be easily done with gas chromatography by applying a constant current over a range of time. To measure the theoretical number of moles, the formula can be applied as follows:

$$n_{H_2} = I.t/2F$$

where I is the current, t is the time of reaction, and F is the Faraday constant. Hence the formula of Faraday efficiency will become

$$FE\% = \frac{\text{No. of moles } H_2 \text{ produced}}{I.t / 2F}$$

3.2.2.5 Stability

To make an efficient electrode that can be commercialized, the most challenging point is the stability of the electrode. Stability can be calculated in three ways: (1) performing Chronoamperometry, i.e., applying a constant voltage to achieve a constant current over a certain time period; (2) performing Chronopotentiaometery, i.e., applying a constant current (not less than 10 mA) for a certain amount of time to achieve a constant potential; and (3) running CV (cyclic voltammetry) or LSV (linear sweep voltammetry) cycles (not less than 1000 cycles). LSV is taken before and after these experiments and overlapping of the LSV graph suggests stability of the material/electrode.

3.2.3 Factors Determining OER/HER Activity

3.2.3.1 Turn Over Frequency

Turn over frequency (TOF) is the efficiency of the electrocatalysts to generate the desired product per catalytic site as a function of time. There are different ways of calculating the TOF in catalysis:

TOF in terms of turn over number (TON)

$$TOF = TON / \text{reaction time}$$

where TON = moles of desired product formed/number of active site or surface area of the electrode.

TOF in terms of currents density using the following equation:

$$TOF = j.A/2F.n$$

where j is current density at a given potential, n is the number of moles of catalyst reacting, A is the working electrode area, and F is the Faraday constant.

TOF in terms of current density and number of active sites with the equation:

$$TOF = j/nFN$$

where n is the stoichiometric number of electron consumed in the reaction and N is the number of active sites [9].

$$N = R_f \left(N_A d / M_f\right)^{2/3}$$

where M_f and N_A are formula weight of the catalyst and Avogadro's number respectively. R_f is a roughness factor of the electrode surface, which is the ratio of actual surface area to the geometric surface area of the electrode [10].

3.2.3.2 Electrochemical Active Surface Area

The electrochemical active surface area (ECSA), as the term itself suggests, is the actual active surface area of the electrode which is taking part in the reaction. It can be calculated with double layer capacitance (C_{dl}) and specific capacitance (C_s) according to Equation 3.12 [11].

$$ECSA = C_{dl}/Cs \tag{3.12}$$

C_{dl} is obtained by taking cyclic voltammograms in the non-faradic region at different scan rates and then C_{dl} is the positive slope of graph plotted between scan rate and the current densities taken on a particular potential value. C_s can also be calculated with Equation 3.13.

$$Cs = \frac{\text{Area under the curve}}{\text{scan rate} \times \Delta V} \tag{3.13}$$

3.2.3.3 Charge Transfer Resistance

The charge transfer resistance (R_{ct}) is the resistance at infinite or higher resistance, except for C_{dl} and electrolyte resistance. It can be easily taken by EIS at higher frequency. The diameter of the semicircle formed is the charge transfer resistance. The lower the value of R_{ct}, the better the material or electrode. The higher the voltage applied, the lower the R_{ct}. Hence, to have an efficient electrode/material, it should show lower R_{ct}, even when a small potential is applied.

3.3 Transitional Metals for Electrocatalytic HER

In addition to precious metal-based HER electrocatalysts, earth-abundant materials have been widely used in alkaline media. Mostly, transition metals-based catalysts reported for electrocatalysis are in the form of 2D-hybrids of TMDs. In this section we will discuss the one-dimensional (1D) transition metal-based nanomaterial (nanowires, nanotubes, composite nanoarrays). In a 1D structure, charge transfer is along the length of the structure.

As nickel (Ni) is a good conducting material, its hybrid has been synthesized as NiP nanowires which have a large surface area (15.9 mF m^{-2}) which results in more active sites for the adsorption of the water molecules and the HER activity of the catalyst, as it shows an HER performance of 71 mV at a 10 mA cm^{-2} current density, and the stability of 40 h at a current density of 360 mA cm^{-2} in alkaline media. Another hybrid of Ni with molybdenum results in a very good HER performance. The active sites on NiMo nanowire array and interfacial activity due to enhanced electron transfer through molybdenum, showed an HER property of 17 mV at a current density of 10 mA cm^{-2}.

Other than nanowires, hollow 1D nanoarrays exhibit better interconnectivity and shorter ion diffusion length and more electrochemically active sites because of the larger surface area. Hollow 1D nanoarray can be either mesoporous or porous nanowires exhibiting excellent HER properties. Among transition metal phosphides, the Co phosphide nanoarray, CoP_3 exhibited excellent performance that was synthesized on Ti mesh (np-CoP_3/TM) via an acid etching method. Because of the large active area, np-CoP_3/TM showed an enhanced HER performance of 76 mV at a current density of 10 mA cm^{-2}. On the basis of the mesoporous structure, oxygen vacancies were also introduced in $NiCo_2O_4$ to modify the catalysts. $NiCo_2O_4$ delivers a current density of 360 mA cm^{-2} at 317 mV due to oxygen vacancies and its porous structure [4]. Another electrocatalyst based on oxygen and boron vacancies, V_OB-Co_3O_4, has been reported. Self-supported V_OB-Co_3O_4/NF nanowire arrays, which were directly grown on nickel foam, act as a trifunctional catalyst for producing oxygen and hydrogen via electrolysis of water in alkaline media or producing hydrogen via hydrolysis of alkaline $NaBH_4$ solution. The boron and oxygen defects that are produced in Co_3O_4 nanowires boost the electrical conductivity and large number of electroactive sites. In 1.0 M KOH, the self-supported V_OB-Co_3O_4/NF electrode showed a current density of 50 mA cm^{-2} at overpotentials of 184 mV for HER and 315 mV for OER [12]. Another Co TMDs-based nanowire has been reported which is a ternary electrocatalyst. $CoS_{2x}Se_{2(1-x)}$ nanowires synthesized on carbon fiber with hydrothermal and chemical vapor deposition (CVD) methods possess excellent catalytic activity for electrochemical HER, achieving current densities of 10 mA cm^{-2} at an overpotential of 129.5 mV in acidic media [13]. Similarly, other Co-based nanowires were synthesized by treating Co_3O_4 nanowires with N_2RF plasma at room temperature to obtain CoN nanowires, which act as very good anodic electrocatalysts showing an overpotential of 290 mV at current density 10 mA cm^{-2}. Another category of transition metals-based electrocatalysts is transition metal carbides and nitrides. A bimetallic nitrogen-based nanowires NiMoN electrocatalyst showed an overpotential of 38 mV at current density of 10 mA cm^{-2} and 290 mV at 50 mA cm^{-2} for HER and OER in alkaline media, respectively [14].

3.4 2D Materials for Electrocatalysis

We now discuss substrate type and role for materials.

3.4.1 MXenes/TMDs for HER

Transition metal-based hybrid electrocatalysts have proven to be superior and successful catalysts for water spitting. One of the most studied and excellent hybrids are 2D-transition metal dichalcogenides or simply transition metal 2D hybrids. As an example, porous MoC_X nano-octahedrons showed good performance toward HER both in acidic and alkaline solution. An overpotential of 151 mV is required for the MoC_X nano-octahedrons to reach the current density of 10 mA cm^{-2} and the Tafel slope of 59 mV dec^{-1}. Also, some tungsten carbides and nitrides have also been explored. Like W_2N_3, they showed an overpotential of 98.2 mV at current density of 10 mA cm^{-2} and in comparison with carbide of tungsten, W_2C showed an even lower overpotential of 75 mV at 10 mA cm^{-2} current density [14].

Another category of 2D-TMDs hybrid is with Mxene. As MXene provides a good reactive surface area, binding with TMDs results in better HER/OER performance of the catalyst. Ti_3C_2 is a very common MXene and its nanosheets, which are fully functionalized with oxygen, have been reported for HER activity which shows 190 mV of overpotential at 10 mA cm^{-2} as a result of increased active sites due to oxygen functionalization [15]. Doping of $Ti_3C_2T_X$ MXene with Nb moves the Fermi energy level to the conduction band, which results in enhancement of conductivity and charge transfer. Further, surface modification with Ni/Co alloy further enhances HER activity by moderating the surface M-H affinity. $Ni_{0.9}Co_{0.1}$@NTM nanohybrid showed an extraordinary HER activity in alkaline solution, as it needs 43.4 mV overpotential to reach 10 mA cm^{-2} of current density [16]. In another study, Ni-foam is used as a substrate which helps to increase the conductivity of the catalyst. A hybrid of Ni_2P/MXene has been coated onto the Ni-foam which results in a huge boost in the HER activity showing an overpotential of 135 mV at 10 mA cm^{-2} and Tafel slope around 86.6 mV dec^{-2} (Figure 3.4) as compared to pristine Ni_2P, due to distinctive

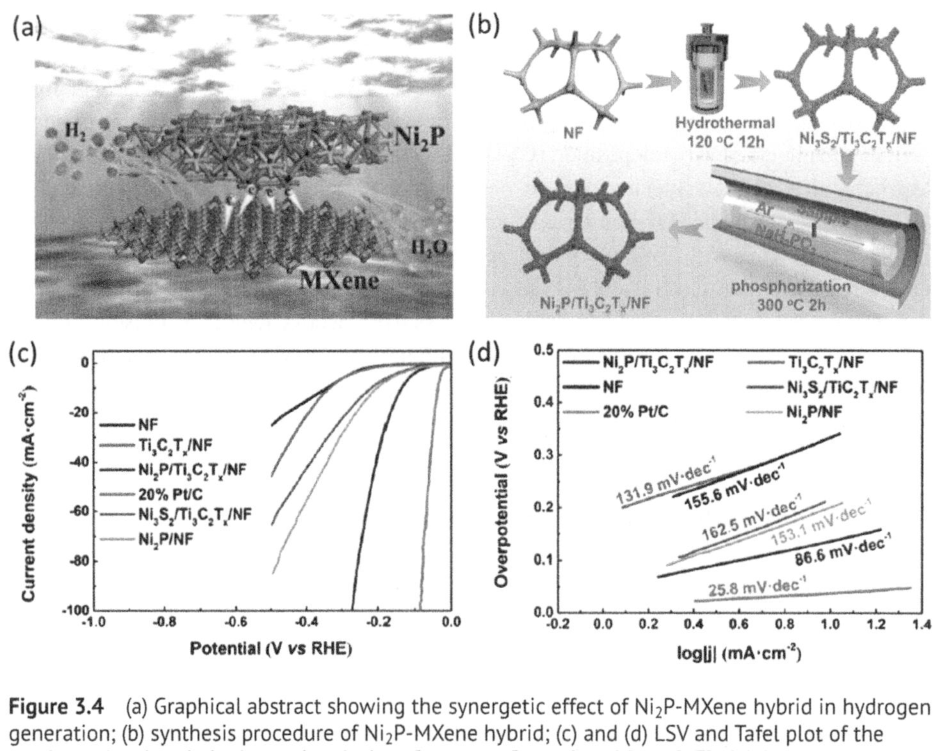

Figure 3.4 (a) Graphical abstract showing the synergetic effect of Ni_2P-MXene hybrid in hydrogen generation; (b) synthesis procedure of Ni_2P-MXene hybrid; (c) and (d) LSV and Tafel plot of the catalysts showing their electrochemical performance. Reproduced from [17] / With permission of American Chemical Society.

structure and synergic coupling of Ni_2P and $Ti_3C_2T_x$ MXene [17]. Another similar hybrid of Ni and MXene is MXene/Ni_3S_2 nanosheets over three-dimensional (3D) Ni-foam, which showed even better results than Ni_2P/MXene (72 mV at 10 mA cm^{-2}) [18]. As Pt is well known for HER activity, it has been incorporated with Ni and MXene to achieve better HER activity. For these Pt_xNi, ultrathin nanowires has been grown *in-situ* on Ni-foam. Such PtxNi@Ti_3C_2 electrocatalysts exhibit excellent HER performance in both acidic and alkaline solutions, and achieve record-breaking performance in terms of lowest overpotential of 18.55 mV which is attributed to the decrease in Gibbs free energy and additional active sites provided by the nanowires [19].

3.4.2 2D Materials (with Transition Metals) for OER

Efficient electrocatalysts at OER for water splitting are mostly transition metal-based oxides. Molybdenum nitrides and carbides are very good OER independent catalysts. Their hetero-interface, Mo_2N–Mo_2C, has a total density of states (TDOS) at the N–Mo–C interface, which are higher than those of their single phases; therefore, the electron mobility at the hetero-interface of Mo_2N–Mo_2C is intensified, which is beneficial for the OER reaction. In alkaline media, the electrochemical test revealed that 2D transition metal carbide/nitride hetero-structure nanosheets (h-TMCN) have a good OER reactivity requiring only 276 mV of over-potential at an anodic current density of 50 mA cm^{-2}. Also, nitrides of bimetallic combination of transition metals have proven to be good OER catalysts. In the NP Au/$CoMoN_x$ the OER activity is enhanced due to increased electron transfer and mass transport, which showed an ultra-high current density of 1156 mA cm^{-2} at overpotential of 370 mV [14].

3.4.3 2D Materials Hybrid (with Transition Metals) as Bifunctional Catalyst

There are many research groups who are focusing on the electrocatalysts that can both act as bifunctional, i.e., cathodic and anodic catalysts for water splitting. Many such 2-D TMDs are reported to date but we will discuss only a few of them; mainly classified into two groups, TMDs carbides/nitrides hybrids and TMDs/MXene hybrids.

TMD-based 1D nanotubes hybrid, WS_2/CNTs synthesized by spray-drying method on Ni-foam, acts as an excellent bifunctional catalyst for water splitting. The enhanced catalytic performance of WS_2/CNTs composite is mainly due to the strong polarized coupling between CNTs and WS_2 nanosheets, which significantly promotes the charge redistribution on the interface of CNTs and WS_2. The overpotential at 10 mA cm^{-2} for HER was measured as 146 mV and for the OER process was 316 mV (at 50 mA cm^{-2}) for OER, which outperforms almost all reported 2H WS_2-based catalysts [20].

Shalom et al. designed and synthesized Ni_2N@NF by growing Ni_2N layer on Ni-foam using supramolecular chemistry, which exhibits high electrocatalytic performance and stability in HER, OER, and ORR. Similarly, Zhang et al. synthesized nickel nitride with different stoichiometry by treating the Ni-foam with microwave plasma to obtain nitrogen vacancies rich in nickel nitride. These vacancies lead to better adsorption of H_2O molecules which promotes better electrochemical activity. The Ni_3N_{1-x}/NF electrode showed a superior HER activity requiring only 55 mV overpotential to reach a current density of 10 mA cm^{-2} [14].

A recent similar study has been done by Li et al., where they synthesized Ni_3N from $Ni(OH)_2$ nanosheets on the Ni-foam by using an N_2-H_2 glow discharge plasma. This resulted in a further lower overpotential of 44 mV at the current density of 10 mA cm^{-2}. The W_2N is usually known as an OER and ORR catalyst with WC for HER. The reported W_2N/WC interfaces showed a synergetic effect which promotes the easy charge transport and separation resulting in an excellent electrocatalytic activity in ORR, OER, and HER. The overpotential measured at current density of 10 mA cm^{-2} is 320 mV and 148.5 mV for OER and HER, respectively [14].

Carbon nanotubes have proven to be a good conductive material. Many researchers are now focusing on using them as a substrate. A trimetallic-based electrocatalyst $Ni_3ZnC_{0.7}$/NCNT has been synthesized which exhibited efficient HER and OER activities. The potential achieved was 1.66 V at a current density of 10 mA cm^{-2} for overall water splitting. In another similar study, TMDs are incorporated with carbon nanotubes. Mo_2C was induced into carbon nanotubes which promote the complete sulfidation of Ni by increasing the amount of pyrrolic-N. This fully sulfurated N-CNTs/NiS_2@Mo_2C shows remarkable electrocatalytic performance in both HER and OER, such that for overall water splitting it requires low overpotential of 1.52 V to achieve a current density of 10 mA cm^{-2}. On the other hand, FeOOH-based electrocatalyst show superior HER and OER catalytic performance, having an overpotential of 67 and 244 mV at 10 mA cm^{-2} for HER and OER, respectively. The Fe–O bonds from FeOOH result in easily adsorbance of the OH^- ions and H_2O molecules, while the Ni_3N tends to adsorb the H^+ ions from the H_2O molecule due to its negative charge. This special ability of the catalysts facilitating the water or OH-dissociation improves its electrochemical activity [14].

Another category is TMDs-MXene hybrids. MXene has become popular for its various applications. Its hybrids have proven to be very efficient electrocatalysts for overall water splitting. A transition metal phosphide hybrid with MXene, Co_2P@$Ti_3C_2T_x$, has proven to be an efficient catalyst for water splitting. The low overpotential of 42 mV at 10 mA cm^{-2} is attributed to the better conductivity and adsorption of H* and H_2O on Co_2P@$Ti_3C_2T_x$. Additionally, the OER performance of the catalyst has also been strengthened in a similar way by the synergic effect of Co_2P and MXene with a overpotential of 267 mV at 10 mA cm^{-2} and of 1.46 V at 10 mA cm^{-2} for overall water splitting [5]. Ni-based 0D–2D nanohybrids bimetal phosphorus trisulfide decorated on MXene nanosheets consisting of bimetallic phosphorus trisulfide ($Ni_{1-x}Fe_xPS_3$) nanoarrays were decorated on the surface of MXene nanosheets (denoted as NFPS@MXene) through a facile self-assemble process and *in situ* solid-state reaction step at low temperature. Different Ni-Mo stoichiometric ratios were studied for overall water splitting. Best activity was achieved with the Ni:Fe ratio of 7:3, obtained as $Ni_{0.7}Fe_{0.3}PS_3$@MXene nanohybrid showing lowest overpotential of 196 mV at 10 mA cm^{-2} for the HER, which is better than other Ni-Co@MXene-based catalysts. Ti_3C_2@mNiCoP NS showed an overpotential of 237 mV for HER [21]. Meanwhile, the $Ni_{0.9}Fe_{0.1}PS_3$@MXene showed an overpotential of 282 mV at 10 mA cm^{-2} for OER in 1 M KOH solution. For overall water splitting, the coupling of $Ni_{0.7}Fe_{0.3}PS_3$@MXene || $Ni_{0.9}Fe_{0.1}PS_3$@MXene required only 1.65 V to obtain a current density of 10 mA cm^{-2} [22]. Another hybrid of 1T/2H $MoSe_2$ and MXene perform remarkably as an electrocatalyst for water splitting. On one hand, the 1T/2H $MoSe_2$ possesses abundant active sites which results in high electrocatalytic activity and, on the other hand, the MXene nanosheets serve as high-conductive 2D substrates capable of providing high charge transfer ability. Also, MXene nanosheets prevent 1T/2H $MoSe_2$ from aggregating. For HER performance, the 1T/2H $MoSe_2$/MXene reveals an HER overpotential of 95 mV and for OER performance of 340 mV [23].

Another subcategory which, is being explored is double metal MXene. Its composite with Ni-Fe is FeNi@MXene, where MXene is Mo_2TiC_2Tx. The $FeNi@Mo_2TiC_2Tx@NF$ is obtained by introducing Ni ions from the substrate itself, in this case Ni-foam. This catalyst showed excellent activity due to the synergetic effect of Mo_2TiC_2Tx and FeNi alloys with overpotentials of 165 and 190 mV for the HER and OER at a current density of 10 mA cm^{-2}, respectively. MXene provides a high active surface area which promotes HER and FeNi nanoalloys to promote the OER. It was proved from theoretical simulation that electrons are transferred from the Mo surface to the interface between FeNi and MXene, which facilitate the formation of intermediate NiOOH. The catalyst showed an overpotential of 1.74 V at a current density of 50 mA cm^{-2} for overall water splitting in alkaline solution (Table 3.1) [24].

Table 3.1 Comparison of various catalysts for electrochemical water splitting on the basis of overpotential.

Catalyst	Category	Overpotential (mV at 10 mA cm^{-2})	References
HER			
NixP@NF	1D nanowires	71	[4]
NiMo	1D nanowires	17	[4]
np-CoP_3/TM	Hollow 1D nanoarray	76	[4]
$CoS_{2x}Se_{2(1-x)}$	1D nanowire	129.5	[13]
MoC_x	2D nanooctahedrons	151	[14]
W_2N_3	2D TM nitrides	98.2	[14]
W_2C	2D TM carbides	75	[14]
Ti_3C_2 (O-functionalized)	MXene	190	[15]
$Ni_{0.9}Co_{0.1}$@NTM	Bimetallic MXene hybrid	43.4	[16]
Ni_2P/MXene	TM-MXene hybrid	135	[17]
MXene/Ni_3S_2	MXene hybrid nanosheets	72	[18]
MXene/Ni_3S_2	1D-2D MXene hybrid	18.55	[15]
OER			
CoN	1D nanowire	290 at 50 mA cm^{-2}	[14]
Mo_2N–Mo_2C	2D TM hybrid	276 at 50 mA cm^{-2}	[14]
Au/$CoMoN_x$	TM bimetallic hybrid	370 at 1156 mA cm^{-2}	[14]
Bi-functional			
V_OB-Co_3O_4/NF	1D nanowires	HER-184 OER-315 At 50 mA cm^{-2}	[4]
NiMoN	1D nanowire	HER-38 at 10 mA cm^{-2} OER-290 at 50 mA cm^{-2}	[14]
WS_2/CNTs	1D-2D TMD hybrid	HER-146 at 10 mA cm^{-2} OER-316 at 50 mA cm^{-2}	[20]
Ni_3N_{1-x}/NF		HER-55	

(Continued)

Table 3.1 (Continued)

Catalyst	Category	Overpotential (mV at 10 mA cm^{-2})	References
W_2N/WC	TM* carbide-nitride hybrid	HER-148.5 OER-320	[14]
$Ni_3ZnC_{0.7}/NCNT$	Bimetallic-CNT hybrid	1.66 V for overall water splitting	[14]
N-CNTs/NiS_2@Mo_2C	Transition bimetallic hybrid	1.52 V for overall water splitting	[14]
Ni_3N-FeOOH	Transition metal nitride hybrid	HER-67 OER-244	[14]
Co_2P@$Ti_3C_2T_x$	Transition metal-MXene hybrid	HER-42 OER-267	[5]
$Ni_{1-x}Fe_xPS_3$@MXene	Transition metal-MXene hybrid	HER-196 (7:3) OER-282 (9:1)	[22]
1T/2H $MoSe_2$/MXene	TMDs-MXene hybrid	HER-95 OER-340	[23]
FeNi@$Mo_2TiC_2T_x$@NF	Double metal MXene hybrid	HER-165 OER-190	[24]

*TM = Transition Metals

3.5 Part II: Fundamentals of Photoelectrocatalytic Hydrogen Production

3.5.1 Mechanism and Reaction Kinetics in PEC HER/OER

In a semiconductor, when the light falls into the valence band the electron absorbs the energy and jumps to the conduction band leaving behind the hole. These two types of charge carriers are responsible for creating an electric current in a semiconductor material. If charge transport across the semiconductor–electrolyte interface is excessively sluggish, carriers will accumulate at the surface, resulting in an electron–hole recombination. Surface states may potentially be able to trap the carriers. The charge-transfer mechanisms in most photoanodes, however, are exceedingly slow due to weak surface kinetics at the photoelectrode–electrolyte surfaces. As a result, efficient catalysts are needed to improve surface kinetics.

For PEC material, it is critical to figure out what is causing the slow surface kinetics and choose the right catalysts. Electrochemical impedance spectroscopy (EIS), a widely used method, may examine the kinetics of an electrochemical process, allowing better understanding of the charge-transfer process at the semiconductor–electrolyte interface or the semiconductor–catalyst–electrolyte interface. In p-type semiconductors, the majority of charge carriers are holes. Holes move with positive charge in the crystal lattice, producing an electric current in a p-type semiconductor that has a Fermi level near the valence band. In an n-type semiconductor, the majority of charge carriers are electrons that move with a negative charge in the crystal lattice (solution) producing an electric current. In this semiconductor, the Fermi level is present near a conduction band. The PEC study is all about the semiconductor electrolyte interface. The Fermi level of both types of semiconductors is an electronic version of chemical potential. The electrolyte charge dominates the chemical potential of the system and so the Fermi level matches the chemical potential of the solution. Thus, the band bending separates the electron and holes at the electrolyte interface. Therefore, the electron and hole cannot recombine, and charge carrier transport takes place. Basically, in a semiconductor, the distribution of the valence electron wave function creates an energy band that separates the valence band (where the electron is filled) and the conduction band (where the electron is unoccupied).

In a photoelectrolysis process, when photons of visible light are irradiated to the semiconductor, the electrons in the valence band receive the photon energy to form electron–hole pairs. These electron–hole pairs will be separated into electrons and holes, which in turn take part in the electrocatalysis of water to generate hydrogen and oxygen gas at the cathode and anode, respectively (Figure 3.5). The photoelectrochemical–hydrogen evolution reaction (PEC-HER) mechanism is where the photo-induced electrons and holes participate in HER to produce O_2 and H_2 on the surface of the catalysts by the following process. In the PEC system, the electrons and holes are collected via connecting with a conducting electrode. Therefore, the recombination of the electron–hole is significantly reduced and the efficiency of the photocatalytic process is improved. In addition, the external potential is generally employed to manipulate the flow of hole or electron, thus only the reduction or oxidation process occurs on the working electrode to generate hydrogen or oxygen, respectively. On the counter electrode, the rest of the process of oxidation or reduction is operated which corresponds to oxygen or hydrogen generation. Thermodynamically, the minimum energy required for the redox process of water is 1.23 eV. At the valence band, the holes contribute to the oxidization of water, giving rise to O_2; meanwhile, at the conduction band, the electrons reduce water to produce H_2. To achieve a high performance of the water-splitting process, the semiconductor requires bandgap energy that is close to the redox potential of water [25].

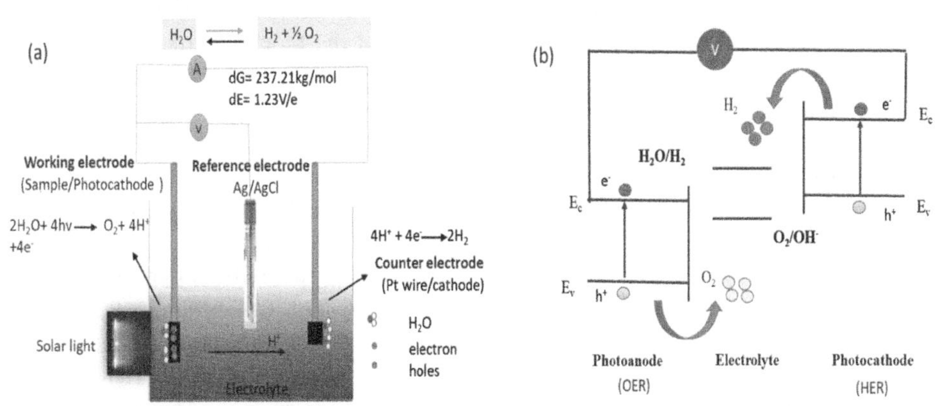

Figure 3.5 (a) Diagram of basic principles of photoelectrochemical water splitting; (b) illustrative band diagram of photoelectrochemical water splitting using a photoanode.

3.5.2 Parameters for Efficient PEC Electrode Design and Benchmarking

3.5.2.1 Flat Band Potential

Minimum voltage is required to bring the energy bands to a flat band condition in semiconductor electrodes or p-n junction. After applying the potential between substrate and electrolyte or p-n junction there is no band bending or charge depletion that can take place. One-dimensional relationship between charge density and potential difference are shown in Equation 3.14.

$$\frac{d^2\varphi}{dx^2} = -\frac{\rho}{\varepsilon\varepsilon_o} \tag{3.14}$$

where ρ is the charge density, φ is the potential difference, ε is the dielectric constant of semiconductor, and ε_o is the permittivity of free space. By using Equation 3.14 we can determine Equation 3.15.

$$\frac{1}{C^2} = \frac{2}{\varepsilon\varepsilon_o A^2 e N_D}\left(V - V_{fb} - \frac{k_B T}{e}\right) \tag{3.15}$$

The plot between $1/C^2$ vs. V is a straight line and the tangent is drawn to cut the *x*-axis. The potential where the tangent cuts the axis is called the flat band potential. The n-type material has a positive slope, while the p-type material possesses a negative slope. It helps to elucidate the band structure of semiconductors as photocathode (p-type conductivity) in HER and for the photoanode (n-type conductivity). In the case of OER in the experiments, Equation 3.15 represents the Mott–Schottky equation, where ε and ε_0 are the dielectric constant and free space permittivity, respectively. The applied potential is denoted by the letter V, and the k_BT/e is so small at room temperature, it may be ignored. We may determine V_{fb} from the intercept of C^{-2} with the axis that measures potential using Equation 3.15. Another essential metric, the carrier concentration (n), can be calculated using Equation 3.15 and the slope of the linear region. However, we should point out that the Mott–Schottky approach has been exploited in the literature, and the premise given above may not be correct. The global value of the measured capacitance includes information from the electrolyte. The capacitance may complicate its variation as a function of the applied potential. Direct application of the Mott–Schottky method may produce erroneous information on V_{fb} and carrier concentrations unless these parameters can be effectively excluded (n) [26].

3.5.2.2 Ohmic Contact

The ohmic contact refers to the optimal charge carrier transfer between the semiconductor and the substrate and is vital to form a low-loss electronic contact. An ohmic contact follows Ohms law in which the current is directly proportional to the voltage and gives a linear graph of current vs. voltage. Ohmic contact is a non-rectifiable contact and cannot become an electric current and inject minority charge carriers into the bulk of the semiconductor material, but can have a restrained resistance. Mostly, for the contact to be ohmic, the substrate must improve the majority of carriers at the interface to a level greater than the bulk of the semiconductor.

In the near ohmic contact, the graph is not exactly a straight line but depicts a good flow of current in both directions; for example, the n-n$^+$, p-p$^+$ junction. In both cases, forward bias and reverse bias, the I-V characteristic is almost identical, therefore it is near the ohmic contact. At the n-n$^+$ junction, the electrons are the major charge carriers and in the p-p$^+$ case, the holes are the major charge carriers after applying the bias in both directions. In the forward and reverse bias, a similar current with a good flow of carriers is obtained. In general, the substrate must enrich the major carriers at the interface to a level greater than in the semiconductor bulk for the contact to be ohmic. A conductive substance having a work function (ϕ_M) greater than the semiconductor work function (ϕ_S) is often required for a p-type semiconductor. A contender for a large work function ohmic contact is gold ($\phi_M = 5.3$ eV). In this structure, holes form an accumulation layer at the surface, and the semiconductor acts as a metal at the junction. Similar arguments may be made for n-type materials, which require a conductive material with a lower work function than the semiconductor. Aluminum ($\phi_M = 4.3$ eV) is a candidate for a tiny work function. It is worth noting that a material's work function is influenced by a variety of parameters, including chemical composition and crystallographic orientation. Thus, stated work function values can be used to guide contact material selection, but a thorough evaluation of the contact–semiconductor interface's electrical characteristics is recommended.

Theoretically, there are two ways to make a good ohmic contact: (1) deposition of metal with a sufficiently low work function on an n-type crystal (or a high work function on a p-type material, and (2) application of a metal that can act as an electron donor in the semiconductor when used as a doping material. In the first situation, an accumulation layer forms at the interface of the semiconductor. This approach is not relevant to wide-gap semiconductors because a metal with a low enough work function to provide a tiny barrier height does not exist in most cases. As a result, the second procedure must be used. The sample is heated to the point where the metal diffuses into it, generating a heavily doped area immediately below the semiconductor surface (n' in n-type and p$^+$ in p-type). The Fermi level then passes very close to the appropriate band edge through the semiconductor–metal interface (degenerated surface), or the electrons tunnel through the very thin n$^+$ area. In practice, however, difficulties with ohmic contact technology arise regularly, and many of them are overcome empirically. Forming two contacts and putting a voltage across them, then measuring a current-voltage curve that should not only demonstrate a linear dependence but also a slope that corresponds to the material's resistivity, is a simple way to assess the quality of ohmic contacts [27].

3.5.2.3 Schottky Barrier

The Schottky barrier refers to the junction between a metal and a semiconductor. It could be a rectifying contact or ohmic contact. It is important because it provides a contact electrode to a semiconductor device. For p-type semiconductors, conductive material with a work function (φ_M) larger than the semiconductor work function (φ_s) is required, and for n-type semiconductors, conductive material with a work function of less than the semiconductor work function is required [27].

3.5.2.4 Solar-to-Hydrogen Conversation Efficiency (STH)

STH efficiency is the most overreaching mainstream benchmarking efficiency. In general, there are two ways to empirically calculate the overall STH efficiency in PEC water splitting.

First, it is calculated by considering the rate of hydrogen production. In a second method, the efficiency is estimated by taking the short circuit photocurrent density into account. The device is lighted with AIR MASS 1.5 global spectra to accomplish this. The device is frequently used in two-electrode configurations (WE and CE), with OER and HER occurring in each electrode only when exposed to sunlight and a zero bias voltage (no external voltage applied). The WE and CE are operated under short-circuited circumstances for direct STH efficiency measurements, which are not possible when the WE and CE are compartmentalized and immersed in solutions of various pH. As a result, for a proper STH efficiency test, both the WE and CE must be immersed in the same pH solution (though compartmentalization is still allowable). Further, there should be no sacrificial donors or acceptors in the electrolyte because the redox (reduction–oxidation) reactions would no longer reflect genuine water splitting. STH efficiency is defined as the chemical energy of the hydrogen produced divided by the solar energy input from sunlight incident on the process for direct solar-to-hydrogen processes, as shown in Equation 3.16.

$$\frac{\mathrm{Jsc}\left(\frac{\mathrm{mA}}{\mathrm{cm}^2}\right) * 1.23\mathrm{V} * \mathrm{nF}}{P_{\mathrm{total}}\left(\frac{\mathrm{mW}}{\mathrm{cm}^2}\right)} \tag{3.16}$$

3.5.2.5 Applied Bias Photon to Current Efficiency (ABPE)

A new efficiency value is ABPE when a bias is introduced between the working electrode and the counter electrode. It is distinct from STH because it does not represent a true solar-to-hydrogen conversion mechanism. An "applied bias photon-to-current efficiency" should be defined as a result of this (ABPE). The application of a bias increases the current drawn from the device in general; however, it should be noted that adding a bias that exceeds the thermodynamic splitting potential can cause the device to overheat (1.23 V). The ABPE measurement is diagnostic in material measurement since it is not a real solar-to-hydrogen measurement (Equation 3.17).

$$\mathrm{ABPE} = \frac{\mathrm{Jph}\left(\frac{\mathrm{mA}}{\mathrm{cm}^2}\right) * \left(1.23\mathrm{V} - \mathrm{Vb\,V}\right) * \mathrm{nF}}{P_{\mathrm{total}}\left(\frac{\mathrm{mW}}{\mathrm{cm}^2}\right)} \tag{3.17}$$

3.5.2.6 Incident Photon-to-Current Efficiency (IPCE)

IPCE describes the greatest efficiency with which incoming radiations can split water to form hydrogen. To current efficiency, it is a wavelength-dependent incident photon. IPCE considers the efficiencies of three key PEC processes: i) photon absorbance; ii) the fraction of electron–hole (e-/h^+) pairs created per incoming photon flux charge transport to the solid–liquid interface; and iii) interfacial charge transfer efficiency.

IPCE is commonly produced from a potentiostat measurement in a PEC system. While measuring the current and the PEC electrode to monochromatic light at various wavelengths, a bias can be imposed between the working electrode vs. a counter electrode (2-electrode experiment) or a reference electrode (3-electrode experiment). The photocurrent that emerges owing to redox reactions occurring at the surface of the working electrode is the difference between the steady-state current under monochromatic illumination and the steady-state background current (Equation 3.18).

$$\mathrm{IPCE} = \frac{\mathrm{Jph}\left(\frac{\mathrm{mA}}{\mathrm{cm}^2}\right) * 1239.8\,(\mathrm{V} * \mathrm{nm})}{P_{\mathrm{mono}}\left(\frac{\mathrm{mW}}{\mathrm{cm}^2}\right) * \lambda(\mathrm{nm})} \tag{3.18}$$

3.5.2.7 Absorbed Photon-to-Current Efficiency (APCE)

The losses from imposing photons that are reflected or transmitted are implicitly included in PEC device efficiency as evaluated by IPCE/EQE or STH. To gain a better understanding of a material's fundamental performance, remove these losses and measure efficiency only on photons absorbed. The photocurrent collected per incident photon absorbed is known as the absorbed photon-to-current efficiency (APCE). Internal quantum efficiency is the same as APCE (IQE). When investigating thin films, this is a very helpful metric to measure, since it aids in determining the best balance between maximum photon absorption route length and minimum effective e-/h^+ transport distance within the material (Equation 3.19).

$$\mathrm{APCE} = \frac{\mathrm{Jph}\left(\frac{\mathrm{mA}}{\mathrm{cm}^2}\right) * 1239.8\,(\mathrm{V} * \mathrm{nm})}{P_{\mathrm{mono}}\left(\frac{\mathrm{mW}}{\mathrm{cm}^2}\right) * \lambda(\mathrm{nm}) * (1 - 10 - 4)} \tag{3.19}$$

3.5.2.8 Faradic Efficiency of H_2/O_2 Quantification

Faraday efficiency (also known as coulombic efficiency, current efficiency, or faradaic yield) refers to how efficiently charge (electrons) is transmitted in a system to facilitate an electrochemical reaction. Two electrons are required for the reduction of two protons, resulting in one H_2 molecule. As a result, the following formula can be used to calculate the Faradaic efficiency.

$$\mathrm{FE}(\%) = \frac{\text{No moles of hydrogen produced}}{\mathrm{Q}/2\mathrm{F}}$$

The theoretical value of hydrogen production is Q/2F. Q is the value of charge and F is the Faraday constant:

$$\mathrm{FE}(\%) = \frac{\text{No moles of hydrogen produced}}{\mathrm{I} * \mathrm{t}/2\mathrm{F}}$$

$$\mathrm{FE}(\%) = \frac{2 * \text{produced hydrogen}\left(\frac{\mathrm{mol}}{\mathrm{cm}^2}\right) * 96485\left(\mathrm{s}.\frac{\mathrm{A}}{\mathrm{mol}}\right)}{\text{photo current density}\left(\frac{\mathrm{A}}{\mathrm{cm}^2}\right) * \text{time}}$$

Faradic efficiency of oxidation evolution:

$$\mathrm{FE}(\%) = \frac{\text{No moles of Oxidation produced}}{\mathrm{Q}/4\mathrm{F}}$$

where Q is the charge in the form of Q = photocurrent (I) * Time (t), and F is the Faraday charge.

3.6 2D Material as PEC Catalyst

Two-dimensional TMD materials with the general formulae MX_2 (M: transition metal atom, X: chalcogenide atom) contains two X-atom layers sandwiched with an M atom in between, and a weak Van der Waals bond interaction between the layers [28]. Much recent research has discovered that the indirect bandgap can be turned into a direct bandgap on reduction of the size (or number of layers) of the 2D TMDs. It results in a significant increase in optical characteristics, which supports their use in photocatalysts [29]. Further, the oxidation and reduction potentials of water are well-suited to the conduction band (CB) and valence band (VB) of 2D-TMDs, suggesting that 2D-TMDs are capable of catalyzing the water-splitting reaction [30]. Photoelectroncatalysts/co-catalysts increase the chemical stability of semiconductor photoelectrodes and, in most cases, eliminate photo corrosion. In the future, earth-abundant, highly efficient, stable electrocatalysts will be required to ease the PEC-HER and PEC-OER, as well as boost the reaction kinetics of water splitting, which will enhance overall energy-conversion efficiency. Pt-based compounds, for example, have the maximum HER activity, while Ir (Ru)-group metals function best for the OER. Both metals suffer from high costs and limited natural availability, which prevents them from being used in large-scale commercial applications [31].

3.6.1 2D Material in PEC HER

The reduction in the atomic layer of 2D-TMDs usually results in enhanced strength, stability, and conductivity. The unique functionalities of 2D TMDs allow them to transform the indirect bandgap in bandgap-layered TMDs to a direct bandgap (an active light absorption component) in single-layered TMDs. It was revealed that by decreasing of atomic layer number, the semiconducting 2H-MoS_2 transforms to metallic 1T-MoS_2 [32]. This feature is helpful for good carrier mobility and accelerating the movement of photogenerated charge carriers. The optoelectronic properties of TMDs inject more charge carriers for catalytic reaction at the electrocatalyst electrolyte interface. Chen et al. reported that the PEC-HER MoS_2 nanostructured possessed a higher photocurrent density in comparison with bulk MoS_2. Based on the theoretical and experimental studies, it was described that the monolayer MoS_2 has a direct bandgap of 1.9 eV, while the corresponding bulk MoS_2 has an indirect bandgap of 1.3 eV. So, the nanostructured TMDs has been used to improve the concentration of catalytically active edges, but the bulk form of TMDS has not been very efficient in HER. Density functional theory (DFT) calculations have also indicated that the edges of MoS_2 nanosheets are active for HER. Thin-layer 2D-TMDs have a wide range of applications, including solar cells, organic light-emitting diodes photodetectors, transistors, and gas sensors, thanks to their characteristics [33, 34]. The bandgaps of 2D-TMDs (CB and VB) are well-matched to the redox potential of water, implying that they can catalyze the water-splitting reaction. The photocathode sulfur-doped molybdenum phosphide (S: MoP) photocathode obtained a photocurrent density of –33 mA cm^2 at 0 V vs. RHE [35]. For the PEC-HER, the coupled 2D-TMDs form a p–n heterojunction. Due to the abundance of electrons in chalcogenide atoms, the 2D-TMDs exhibit n-type behavior. WS_2 does, however, exhibit p-type behavior. As a result, a p–n heterojunction can be formed using only 2D materials. In the PEC-HER, Xiao et al. successfully produced an effective photocathode MoS_2/WSe_2 bilayer heterojunction. These findings reveal that 2D-TMDs may generate p–n heterojunctions, indicating that they can be used as efficient photocatalysts in the PEC-HER [36].

Two-dimensional material graphene shows extraordinary electronic, optical, and physical properties. It has been used widely in various applications due to the excellent charge kinetics. Reduced graphene oxide (an imperfect graphene structure) can be combined with a catalyst to reduce charge recombination and improve the material's conductivity. In practice, reduced graphene oxide has been composited with MoS_2 and WS_2 to improve electrocatalytic and photocatalytic performance in hydrogen production. The Pt/NG/Si photocathode showed a photocurrent density (J_{ph}) of –10 mA cm^{-2} at potential of 0.25 V vs. RHE and the solar to hydrogen efficiency (STH) of 3.05% [37]. The graphene quantum sheets (GQS)-based N/GQS/Si photocathode showed a photocurrent density (Jph) of –35 mA cm^{-2} at 0 V vs. RHE [38]. The main electronic role of graphene at the SC–graphene interface is to serve as a transporter, electron acceptor, and mediator in the graphene-based composites owing to its conductive 2D structure. Charge separation at the interface of 2D materials (e.g., MoS_2 and WS_2 nanosheets) with other SCs is a key mechanism in PEC water splitting. One of the most important discoveries is the photo stabilization of non-stable photocatalysts such as Cu_2O, CdS, and $GaInP_2$. The photocurrent density of p-Cu_2O coated with optimized MoS_2 was seven times higher than that of bare Cu_2O [39]. Further, after 9 h of continuous photo-irradiation, MoS_2@Cu_2O showed good photo stability, losing only 7% of its original photocurrent. Excellent photocurrent stability was reported after the deposition of vertically oriented MoS_2 nanosheets on the surface of an Al_2O_3/n^+ p-Si photocathode. After 120 h of illumination, an optimized $MoS_2/Al_2O_3/n^+$ p-Si photocathode showed high photocurrent (36 mA cm^{-2}) with very little loss [40] (Figure 3.6).

Graphitic carbon nitride, i.e., g-C_3N_4, is a semiconductor that can be used as a catalyst for PEC. Particularly, g-C_3N_4 reveals several benefits like physical, chemical stability, a wide range of visible light absorption for water splitting, appropriate bandgap, low cost, and stability. Pramoda et al. have successfully shown the 1T MoS_2/g-C_3N_4 and nitrogenated reduced graphene oxide as a co-catalyst for PEC-HER [41]. Likewise, TMDs/MXene composite nanostructures upholds fast ion transfer between the sheets and interlayer space, which makes these composites promising electrode materials. Two-dimensional MoS_2 provides great catalytically active edges in PEC with layered dynamic morphology, hydrophobicity, and tuneable bandwidth. The sizeable limitations of MoS_2 include its low intrinsic conductivity, oxidation of the edges, inadequate voltage window, structural disintegration, poor cycling stability, and difficult synthesis with controllable size, crystal phase, and defects [42, 43]. Due to superior metallic conductivity and hydrophobicity combined with cyclability, 2D MXene behaves as an efficient electrode material. Consequently, MXene-based MoS_2 hybrid composites have been broadly explored to overcome the above limitations [44].

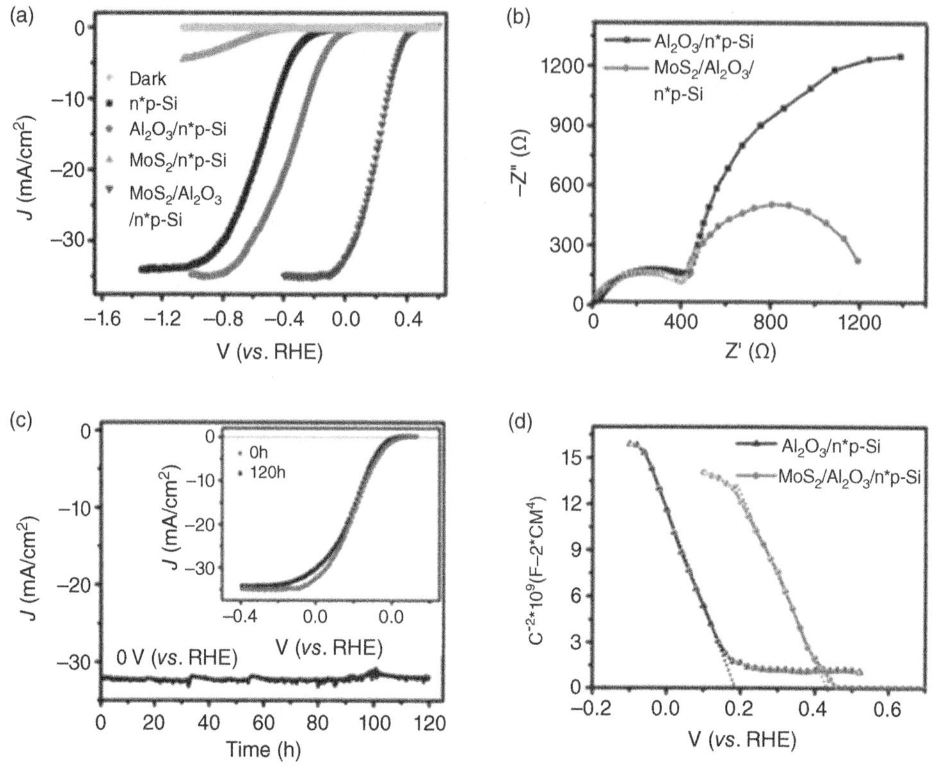

Figure 3.6 (a) Photocurrent density vs. potential curve of $MoS_2/Al_2O_3/n^+p$-Si photocathode; (b) EIS spectra; (c) stability study of photocurrent density of photocathode for 2 h and (d) Mott–Schottky plot. Reproduced from [40] / With permission of American Chemical Society.

3.6.2 2D Materials in PEC OER

Doping of carbon can cause electron modulation, changing the charge distribution of adjacent carbon atoms, affecting their interaction with oxygen intermediates and, ultimately, their electrocatalytic activities, due to the excellent electrical conductivities, large specific surface areas, diverse morphologies, and different electronegativities of transition metals compared to carbon atoms. In the oxygen evolution reaction, the photoanode should be an n-type semiconductor material because the band bending will generate an electric field, which resulta in driving holes to the surface of the materials. These two main factors need to be considered before choosing photoanode to improve water splitting efficiency and reduced overpotential stability under water oxidation conditions and the physicochemical properties like conductivity. The level of VB and CB and the width of the bandgap are also vital because the position of VB decides that the water oxidation can proceed and the range of solar light absorption is influenced by the bandgap. Thus, the synthesis of oxygen from water is enabled by a suitable bandgap (~2.0–3.2 eV) and an appropriate position of VB that is more positive than the potential of O_2 [45]. Hematite, $BiVO_4$, ZnO, TiO_2, and other photoanodes are commonly used for water oxidation. Due to inherent limitations in performance and defects in properties, such as low solar light absorption, poor charge–carrier transportation, and severe photo corrosion under illumination, a photoanode made up of a single semiconductor will not be able to match the commercial application's photocurrent or conversion efficiency requirements [46]. Therefore, effective strategies such as surface modifications need to be applied to improve these properties of photoanodes in PEC water splitting.

The slow kinetics of OER, as well as the other important half-reaction of water splitting, limit the efficiency of water splitting electrolysers. So, for the OER, we are looking for the most efficient and long-lasting electrocatalysts. The water-splitting system must be efficient and widely available. For example, using a sputtering strategy, a functional NiO_x cocatalyst was prepared that not only effectively protected the n-type InP photoanode from photo corrosion, which is a serious problem for semiconductors with narrow bandgaps, but also maintained excellent PEC performance during long-term PEC-OER tests (>48 h) with 100% Faradaic efficiency under 1 sun irradiation in 1.0 M KOH solution. The photocurrent density of the p^+n^-InP/NiO_x photoanode was 0.86 V at the onset potential [47]. The photoanode $CQDs/C_3N_4/TiO_2$ showed photocurrent density of 2.2 mA cm^{-2} at 1.23 V vs. RHE [48].

It is difficult to construct a TMD-based OER catalyst for PEC water splitting for most TMDs nanomaterials in basic solutions. During the OER process, metal atoms on the surface of TMDs were first oxidized into metal oxides/hydroxides, which could then protect the TMDs from further corrosion, and oxidized metal species forming *in situ* on the surface of the TMDs served as primary active phases for the OER, according to a few reports. Two types of nanostructured TMDs, $CoSe_2$ and $NiSe_2$ nanocrystals, were produced by laser photolysis of metal precursors for evaluation of their catalytic performance toward the OER, as an example. The Ni oxide–hydroxide layer of $NiSe_2$ nanocrystals was shown to be more active than the Co oxide–hydroxide layer, resulting in increased catalytic activity. The $NiSe_2$ nanocrystals were then deposited as OER electrocatalysts onto a Si-NW array photoanode for water splitting. For solar-driven OERs, a consistent photocurrent density of 5.8 mA cm^{-2} was reached at 1.23 V with a photocurrent start of 0.7 V. Because of their excellent electrical conductivities, large specific surface areas, diverse morphologies, and different electro-negativities of transition metals compared to carbon atoms, doping of carbon can cause electron modulation, changing the charge distribution of adjacent carbon atoms, affecting their interaction with oxygen intermediates and, ultimately, their electrocatalytic activities for the OER [49].

In addition, a few metal-free carbon nanomaterials have been proven to improve OER. To improve the catalytic performance of the Si photoanode for the PEC-OER, a carbon nanotube (CNT)–graphene composite was developed as an effective hole-extraction layer and a protective layer to pass using a simple coating process. The graphene capping layers worked as hole-exchange layers with the electrolyte and shielded the CNTs and the Si photoanode in this hybrid system, and the semiconducting CNTs acted as efficient hole acceptors and transporters to the graphene. As a result, the Si/CNTs–graphene photoanode demonstrated photoactivity with an 8 mA cm^{-2} current density at 1.83 V and exceptional endurance under 1 sun illumination. While PEC activity is still inferior to that of transition-metal compound electrocatalysts, modified carbon materials have been shown as viable alternatives to transition-metal compounds for effective PEC-OER [50].

Many metal oxides have been employed as photocatalysts for solar water splitting, such as TiO_2, Fe_2O_3, ZnO, WO_3, and SnO_2. Non-layered oxide nanosheets (TiO_2 and Fe_2O_3) have been extensively investigated and compared to layered 2D nanosheets like graphene and

TMD, despite growing interest in these 2D materials. The strong interaction between metal cations and oxygen anions makes the synthesis of 2D non-layered nanomaterials difficult. TiO_2-based 2D nanosheets photocatalysts have been systematically researched in these 2D non-layered oxide nanosheets photocatalysts due to their nontoxicity, low cost, good stability, and readily available supplies in nature. In PEC water-splitting processes, 2D metal oxide nanosheets have been widely employed. When exposed to visible light, Yao et al. produced a porous hybrid structure comprising Fe_2O_3 nanothorn/TiO_2 nanosheet (Fe_2O_3NT/TiO_2NS) photoanodes that had significant PEC activity [51]. Then the photocurrent density was obtained six times higher than that of naked TiO_2 photoanodes (2.50 mA cm^{-2}). Using the layer-by-layer (LBL) assembly method, Wang et al. produced coated ultrathin TiO_2 films on FTO (conductive surface) substrates. LBL-deposited TiO_2 films were found to improve the PEC water-splitting performance of hematite films by limiting the back transfer of electrons from the FTO to the hematite films [52].

3.6.3 2D Materials as a Functional PEC Catalyst

The two most prevalent phases are a triangular prismatic phase (2H) and an octahedral phase (1T). The thermodynamically stable 2H-MoS_2 has six-siemens atoms prismatically encircling each of the Mo atoms, while the metastable 1T MoS_2 has six-siemens atoms in the form of a deformed octahedron around each of the Mo atoms. Because 2H-MoS_2 is semiconducting, but 1T-MoS_2 is a metallic and more active hydrogen evolution catalyst, each of these phases has distinct electrical characteristics. The most essential crystal phases of WS_2 and MoS_2 nanosheets are 2H and 1T. The 2H phase is the semiconducting SC phase that can absorb light to generate electron–hole pairs, while the 1T phase has metallic characteristics which do not contribute to the bandgap excitation. In this phase, the charge-carrier mobility is substantially higher. Due to the synergetic effect of both phases, the cohabitation of the 1T phase and 2H-MoS_2 was good for PEC performance. To overcome the charge recombination process, 2H-MoS_2 acts as a light absorber and photosensitizer, while the 1T phase acts as an electron acceptor and transporter in MoS_2 heterostructures. The charge transfer resistance R_{ct} of the 1T@2H-MoS_2 structure was substantially smaller than 2H-MoS_2, and its photocurrent was double compared to the 2H-MoS_2, supporting the favorable role of the metallic phase [53].

Many TMDs form a graphite-like layered structure, resulting in significant anisotropy in their electrical, chemical, mechanical, and thermal characteristics. TMDs are mostly stacked; however, some of them can also be found in non-layered forms. Each layer in layered formations is generally 67% thick, with a hexagonally packed layer of metal atoms sandwiched between two layers of chalcogen atoms. The interlayer M–X connections are mostly covalent, whereas the sandwich layers are only weakly linked by van der Waals forces, allowing the crystal to easily fracture along the layer surface. As in the case of graphene, studies have demonstrated that the single layers are stabilized by the creation of a ripple structure. The metal atoms contribute four electrons to fill the bonding states of TMDs, resulting in metal (M) and chalcogen (X) atoms with oxidation states of +4 and −2, respectively. The chalcogen atoms' lone-pair electrons terminate the layers' surfaces, and the absence of dangling bonds makes the layers resistant to interactions with ambient species. Depending on the size of the metal and chalcogen ions, the M–M bond length varies between 3.15 and 4.03. These values are 15–25% longer than the bond lengths seen in elemental transition metal solids, demonstrating that the d-orbitals in TMD compounds have little energy and spatial overlap.

For solar absorption, 2D WS_2 and MoS_2 nanosheets have appropriate energy bandgaps (Eg). Depending on thickness, they can be tweaked in the 1.2–2 eV range. Due to the quantum confinement effect, different bandgaps can be generated in 2D TMD nanosheets by varying the thickness and lateral dimension [54]. These materials with a low bandgap can be used as a sensitizer to boost the visible absorption of other SCs. The visible-light absorption efficiency was improved by heterojunction of TMDs with the basal SC material [55].

In photocatalytic applications, platinum (Pt) is the most efficient and widely utilized co-catalyst. Despite this, attempts have been made to seek other materials to replace Pt due to its high cost and scarcity. Two-dimensional materials research (e.g., graphene and TMDs and their composites) has opened up new possibilities for finding a viable alternative for Pt. DGH* (the Gibbs free energy for atomic hydrogen adsorption) on the edge sites of nanoscale MoS_2 has been proven to be comparable to that of Pt [2]. The surface binding energy of atomic hydrogen on the MoS_2 surface is near to zero, which is identical to that of Pt, according to DFT simulations. As a result, much effort has gone into improving TMD catalytic activity, notably for H_2 generation. Because of the quantum confinement effect, the CB position of bulk MoS_2 is insufficient for H^+ reduction, while that of the nanosized 2D MoS_2 structure is sufficient. It has been demonstrated that WS_2 and MoS_2 nanosheets can be used as a more efficient co-catalyst for photocatalytic H_2 synthesis than Pt [56].

As previously stated, boosting the catalytic activity of TMD 2D nanosheets requires increasing the unsaturated active edge sites. Charge transport condition, active surface area, and light absorption capability, on the other hand, are all important factors. As a result, boosting the unsaturated active edge sites, as well as improving charge transfer, surface area, and light absorption capacities should be prioritized. Although amorphous TMDs with varied flaws give more unsaturated active edge sites with a larger surface area, the presence of defects reduces overall conductivity significantly. Because a large potential (0.12 V) is required for electron hopping between adjacent layers, conductivity along the basal plane of MoS_2 is 2000 times higher than in the out-of-plane direction, and the catalytic activity of MoS_2 toward H_2 evolution decreases by a factor of 4.5 (examined via exchange current density) as a layer is added. To increase the surface area and active edge sites with better conductivity, ultrathin 2D nanosheets of TMDs are desired. The ability to absorb light is dramatically reduced when the bulk TMDs are thinned. As a result, as compared to bulk TMDs, the photocatalytic activity of 2D nanosheets cannot be as effective as the dark catalytic activity [57].

TMD is a rapidly growing material for PEC water splitting because it has suitable bandgap values (1–2 eV) for effective visible-light absorption and active catalytic sites for hydrogen evolution. TMDs come in a variety of compositions, allowing for variable bandgap tuning via modification. However, the actual implementation of TMDs is still a long way off due to a slew of issues that must be addressed. The main problem is TMDs limited efficiency and durability in PEC-HER water splitting applications. Because only the edge sites of TMDs in the 2H phase are catalytically active, developing 2D morphologies with more active edge sites is very desirable. More work should be put into making chemically active 2H-TMD basal planes and improving the stability of 1T-TMD nanosheets. Due to the less electrical conductivity of the 2H phase and the limited stability of the 1T phase, the

Table 3.2 Comparison of various catalysts for photo electrochemical water splitting on the basis of Photocurrent density at onset potential.

Catalyst	Photocurrent Density at Onset Potential	Reference
Photocathodes		
$MoS_2/Al_2O_3/Si$	-32 mA cm^{-2} at 0 V vs. RHE	[40]
S:MoP/Si sulfur-doped molybdenum phosphide/Silicon	-33.1 mA cm^{-2} at 0 V vs. RHE	[35]
The Pt/NG/Si	-0 mA cm^{-2} at 0.25 V vs. RHE	[37]
N/GQS/Si	-35 mA cm^{-2} at 0 V vs. RHE	[38]
Photoanodes		
Si/CNTs/graphene	8 mA cm^{-2} at 1.83 V	[50]
Fe_2O_3NT/TiO_2NS	2.50 mA cm^{-2} at 0 V vs. Ag AgCl	[51]
$CQDs/C_3N_4/TiO_2$	2.2 mA cm^{-2} at 1.23 V vs. RHE	[48]
Si–$NiSe_2$ NW	5.8 mA cm^{-2} at 0.7 V vs. RHE	[49]

PEC activity of bare TMDs is insufficient for practical applications. To overcome this, it is necessary to comprehend the correlations between PEC activities and the important optical, electrical, and surface-related properties of TMDs. Further, the effects of structural and surface changes on PEC phenomena should be explored more completely. TMDs with adjustable size, surface imperfections, and a crystalline phase, on the other hand, are extremely difficult to synthesise and modify. For practical and large-scale PEC applications, more simple synthesis methods for TMDs are still required (Table 3.2) [58].

3.7 Conclusions

Hydrogen, a green, environmentally-friendly fuel, is a possible alternative for replacing old fossil fuels like charcoal and oils. Electrocatalysis and photoelectrocatalysis is a viable approach for creating hydrogen in the near future, since it uses a limitless natural energy resource. Thus, a lot of work has gone into developing low-cost and commercial electrocatalysts. In this chapter, we have discussed hydrogen production through water splitting, divided into two parts: Electrochemical and Photoelectrochemical Water Splitting. We also covered the most recent significant advancements in the design and use of transition metal-based and 2D TMDs materials as electrocatalysts and photoelectrocatalysts in the HER/OER. For commercialization of a catalyst, a fabrication method with excellent reproducibility and scalability at a lower cost is critical. However, most of the catalyst material development and system build-up for electrochemical and photoelectrochemical water splitting is still limited to the lab environment.

Thus for future development of 2D-TMDs in the PEC/EC-HER should concentrate on: i) improving visible light absorption to maximize light to electricity efficiency; ii) improving the catalytic activity of 2D materials via functionalization, structure engineering, and compositional engineering; iii) improving the stability of materials under working conditions; and iv) improving the scalability and reproducibility of synthetic methods targeting commercializability.

References

1 Armaroli, N. and Balzani, V. (2011). The hydrogen issue. *ChemSusChem* 4: 21–36.

2 Theerthagiri, J., Lee, S.J., Murthy, A. et al. (2020). Fundamental aspects and recent advances in transition metal nitrides as electrocatalysts for hydrogen evolution reaction: A review. *Curr. Opin. Solid State Mater. Sci.* 24: 100805.

3 Mohammed-Ibrahim, J. and Xiao Ming, S. (2019). Recent progress on earth abundant electrocatalysts for hydrogen evolution reaction (HER) in alkaline medium to achieve efficient water splitting: A review. *J. Energy Chem.* 34: 111–160.

4 Xiang, Q., Wang, J., Miao, Q. et al. (2021). Recent progress in self-supported nanoarrays with diverse substrates for water splitting and beyond. *Mater. Today Nano* 15.

5 Lv, Z., Ma, W., Dang, J. et al. (2021). Induction of Co_2P growth on a MXene (Ti_3C_{2tx})-modified self-supporting electrode for efficient overall water splitting. *J. Phys. Chem. Lett.* 12: 4841–4848.

6 Niu, S., Li, S., Du, Y. et al. (2020). How to reliably report the overpotential of an electrocatalyst. *ACS Energy Lett.* 5: 1083–1087.

7 Vrubel, H., Moehl, T., Grätzel, M. et al. (2013). Revealing and accelerating slow electron transport in amorphous molybdenum sulphide particles for hydrogen evolution reaction. *Chem. Commun.* 49: 8985–8987.

8 Chen, Z., Qing, H., Zhou, K. et al. (2020). Metal-organic framework-derived nanocomposites for electrocatalytic hydrogen evolution reaction. *Prog. Mater. Sci.* 108: 100618.

9 Costentin, C., Passard, G., and Savéant, J.M. (2015). Benchmarking of homogeneous electrocatalysts: Overpotential, turnover frequency, limiting turnover number. *J. Am. Chem. Soc.* 137: 5461–5467.

10 Douglass, E.F., Driscoll, P.F., Liu, D. et al. (2008). Effect of electrode roughness on the capacitive behavior of self-assembled monolayers. *Anal. Chem.* 80: 7670–7677.

11 Li, Y., Dastafkan, K., Sun, Q. et al. (2021). Ni-based 3D hierarchical heterostructures achieved by selective electrodeposition as a bifunctional electrocatalyst for overall water splitting. *Electrochim. Acta* 379.

12 Yuan, H., Wang, S., Ma, Z. et al. (2021). Oxygen vacancies engineered self-supported B doped Co_3O_4 nanowires as an efficient multifunctional catalyst for electrochemical water splitting and hydrolysis of sodium borohydride. *Chem. Eng. J.* 404: 126474.

13 Liu, K., Wang, F., Xu, K. et al. (2016). $CoS_2xSe_{2(1-x)}$ nanowire array: An efficient ternary electrocatalyst for the hydrogen evolution reaction. *Nanoscale* 8: 4699–4704.

14 Chen, P., Ye, J., Wang, H. et al. (2021). Recent progress of transition metal carbides/nitrides for electrocatalytic water splitting. *J. Alloys Compd.* 883.

15 Jiang, Y., Sun, T., Xie, X. et al. (2019). Oxygen-functionalized ultrathin Ti_3C_2 TX MXene for enhanced electrocatalytic hydrogen evolution. *ChemSusChem* 12: 1368–1373.

16 Du, C.F., Sun, X., Yu, H. et al. (2019). Synergy of Nb doping and surface alloy enhanced on water–alkali electrocatalytic hydrogen generation performance in Ti-based MXene. *Adv. Sci.* 6: 1–7.

17 Lv, Z., Wang, M., Liu, D. et al. (2021). Synergetic effect of Ni_2P and MXene enhances catalytic activity in the hydrogen evolution reaction. *Inorg. Chem.* 60: 1604–1611.

18 Tie, L., Li, N., Yu, C. et al. (2019). Self-supported nonprecious MXene/Ni_3S_2 electrocatalysts for efficient hydrogen generation in alkaline media. *ACS Appl. Energy Mater.* 2: 6931–6938.

19 Jiang, Y., Wu, X., Yan, Y. et al. (2019). Coupling PtNi ultrathin nanowires with MXenes for boosting electrocatalytic hydrogen evolution in both acidic and alkaline solutions. *Small* 15: 1–9.

20 Chen, G., Zhang, C., Xue, S. et al. (2021). A polarization boosted strategy for the modification of transition metal dichalcogenides as electrocatalysts for water-splitting. *Small* 17: 1–10.

21 Yue, Q., Sun, J., Chen, S. et al. (2020). Hierarchical mesoporous MXene-NiCoP electrocatalyst for water-splitting. *ACS Appl. Mater. Interfaces* 12: 18570–18577.

22 Du, C.F., Dinh, K.N., Liang, Q. et al. (2018). Self-assemble and *in situ* formation of $Ni_1{-}xFe_xPS_3$ nanomosaic-decorated MXene hybrids for overall water splitting. *Adv. Energy Mater.* 8: 1–9.

23 Li, N., Zhang, Y., Jia, M. et al. (2019). 1T/2H $MoSe_2$-on-MXene heterostructure as bifunctional electrocatalyst for efficient overall water splitting. *Electrochim. Acta* 326: 1–9.

24 Wang, J., He, P., Shen, Y. et al. (2021). FeNi nanoparticles on $Mo_2TiC_2T_x$ MXene@nickel foam as robust electrocatalysts for overall water splitting. *Nano Res.* 14: 3474–3481.

25 Ma, P. and Wang, D. (2018). The principle of photoelectrochemical water splitting. In: *Nanomaterials for Energy Conversion and Storage, pp.* 1–61. Europe: World Scientific. doi: 10.1142/9781786343635_0001.

26 Gelderman, K., Lee, L., and Donne, S.W. (2007). Flat-band potential of a semiconductor: Using the Mott–Schottky equation. *J. Chem. Educ.* 84: 685.

27 Ghita, R.V., Logofatu, C., Negrila, C. et al. (2005). Studies of ohmic contact and Schottky barriers on Au-Ge/GaAs and Au-Ti/GaAs. *J. Optoelectron. Adv. Mater.* 7: 3033–3037.

28 Xu, M., Liang, T., Shi, M. et al. (2013). Graphene-like two-dimensional materials. *Chem. Rev.* 113: 3766–3798.

29 Huang, X., Zeng, Z., and Zhang, H. (2013). Metal dichalcogenide nanosheets: Preparation, properties and applications. *Chem. Soc. Rev.* 42: 1934.

30 Kang, J., Tongay, S., Zhou, J. et al. (2013). Band offsets and heterostructures of two-dimensional semiconductors. *Appl. Phys. Lett.* 102: 012111.

31 Hou, Y., Zhuang, X., and Feng, X. (2017). Recent advances in earth-abundant heterogeneous electrocatalysts for photoelectrochemical water splitting. *Small Methods* 1: 1700090.

32 Chhowalla, M., Shin, H.S., Eda, G. et al. (2013). The chemistry of two-dimensional layered transition metal dichalcogenide nanosheets. *Nat. Chem.* 5: 263–275.

33 Li, Y., Wang, H., Xie, L. et al. (2011). MoS_2 nanoparticles grown on graphene: An advanced catalyst for the hydrogen evolution reaction. *J. Am. Chem. Soc.* 133: 7296–7299.

34 Li, T. and Galli, G. (2007). Electronic properties of MoS_2 nanoparticles. *J. Phys. Chem. C* 111: 16192–16196.

35 Chang Kwon, K., Choi, S., Lee, J. et al. (2017). Drastically enhanced hydrogen evolution activity by 2D to 3D structural transition in anion-engineered molybdenum disulfide thin films for efficient Si-based water splitting photocathodes. *J. Mater. Chem. A* 5: 15534–15542.

36 Xiao, J., Zhang, Y., Chen, H. et al. (2018). Enhanced performance of a monolayer MoS2/WSe2 heterojunction as a photoelectrochemical cathode. *Nano-Micro Lett.* 10: 60.

37 Sim, U., Moon, J., An, J. et al. (2015). N-doped graphene quantum sheets on silicon nanowire photocathodes for hydrogen production. *Energy Environ. Sci.* 8: 1329–1338.

38 Sim, U., Yang, T.-Y., Moon, J. et al. (2013). N-doped monolayer graphene catalyst on silicon photocathode for hydrogen production. *Energy Environ. Sci.* 6: 3658.

39 Zhao, Y.-F., Yang, Z.-Y., Xhang, Y.-X. et al. (2014). Cu O decorated with cocatalyst MoS_2 for solar hydrogen production with enhanced efficiency under visible light. *J. Phys. Chem. C* 118: 14238–14245.

40 Fan, R., Mao, J., Yin, Z. et al. (2017). Efficient and stable silicon photocathodes coated with vertically standing nano-MoS_2 films for solar hydrogen production. *ACS Appl. Mater. Interfaces* 9: 6123–6129.

41 Pramoda, K., Gupta, U., Chhetri, M. et al. (2017). Nanocomposites of C_3N_4 with layers of MoS_2 and nitrogenated RGO, obtained by covalent cross-linking: Synthesis, characterization, and HER activity. *ACS Appl. Mater. Interfaces* 9: 10664–10672.

42 He, Z. and Que, W. (2016). Molybdenum disulfide nanomaterials: Structures, properties, synthesis and recent progress on hydrogen evolution reaction. *Appl. Mater. Today* 3: 23–56.

43 Wang, G., Zhang, J., Yang, S. et al. (2018). Vertically aligned MoS_2 nanosheets patterned on electrochemically exfoliated graphene for high-performance lithium and sodium storage. *Adv. Energy Mater.* 8: 1702254.

44 Hu, Z., Kuai, X., Chen, J. et al. (2020). Strongly coupled MoS_2 nanocrystal/Ti_3C_2 nanosheet hybrids enable high-capacity lithium-ion storage. *ChemSusChem* 13: 1485–1490.

45 Li, J. and Wu, N. (2015). Semiconductor-based photocatalysts and photoelectrochemical cells for solar fuel generation: A review. *Catal. Sci. Technol.* 5: 1360–1384.

46 Gan, J., Lu, X., and Tong, Y. (2014). Towards highly efficient photoanodes: Boosting sunlight-driven semiconductor nanomaterials for water oxidation. *Nanoscale* 6: 7142.

47 Sun, K., Kuang, Y., Verlage, E. et al. (2015). Sputtered NiO_X films for stabilization of P^+n–InP photoanodes for solar-driven water oxidation. *Adv. Energy Mater.* 5: 1402276.

48 Wei, Q., Yan, X., Kang, Z. et al. (2017). Carbon quantum dots decorated C_3N_4/TiO_2 heterostructure nanorod arrays for enhanced photoelectrochemical performance. *J. Electrochem. Soc.* 164: H515–H520.

49 Kwak, I.H., Im, H.S., Jang, D.M. et al. (2016). $CoSe_2$ and $NiSe_2$ nanocrystals as superior bifunctional catalysts for electrochemical and photoelectrochemical water splitting. *ACS Appl. Mater. Interfaces* 8: 5327–5334.

50 Yoon, K., Lee, J.-H., Kang, J. et al. (2016). Metal-free carbon-based nanomaterial coatings protect silicon photoanodes in solar water-splitting. *Nano Lett.* 16: 7370–7375.

51 Yao, H., Liu, L., Fu, W. et al. (2017). Fe_2O_3 nanothorns sensitized two-dimensional TiO_2 nanosheets for highly efficient solar energy conversion. *FlatChem* 3: 1–7.

52 Wang, D., Zhang, X.-T., Sun, P.-P. et al. (2014). Enhanced photoelectrochemical water splitting on hematite thin film with layer-by-layer deposited ultrathin TiO_2 underlayer. *Int. J. Hydrogen Energy* 39: 16212–16219.

53 Qi, Y., Xu, Q., Wang. Y. et al. (2016). CO_2-induced phase engineering: Protocol for enhanced photoelectrocatalytic performance of 2D MoS_2 nanosheets. *ACS Nano* 10: 2903–2909.

54 Zhao, W., Liu, Y., Wei. Z. et al. (2016). Fabrication of a novel P–n heterojunction photocatalyst n-$BiVO_4$@p-MoS_2 with core–shell structure and its excellent visible-light photocatalytic reduction and oxidation activities. *Appl. Catal. B Environ.* 185: 242–252.

55 Wang, Q.H., Kalantar-Zadeh, K., Kis, A. et al. (2012). Electronics and optoelectronics of two-dimensional transition metal dichalcogenides. *Nat. Nanotechnol.* 7: 699–712.

56 Hinnemann, B., Moses, P.G., Bonde, J. et al. (2005). Biomimetic hydrogen evolution: MoS_2 nanoparticles as catalyst for hydrogen evolution. *J. Am. Chem. Soc.* 127: 5308–5309.

57 Yu, Y., Huang, S.-Y., Li, Y. et al. (2014). Layer-dependent electrocatalysis of MoS_2 for hydrogen evolution. *Nano Lett.* 14: 553–558.

58 Faraji, M., Yousefi, M., Yousefzadeh, S. et al. (2019). Two-dimensional materials in semiconductor photoelectrocatalytic systems for water splitting. *Energy Environ. Sci.* 12: 59–95.

4

Single Atom Catalysts for Hydrogen Production from Chemical Hydrogen Storage Materials

Rajani Kumar Borah[1,2], *Adarsh P. Fatrekar*[1,2], *Panchami R.*[1], *and Amit A. Vernekar*[1,2,*]

[1] *Inorganic and Physical Chemistry Laboratory (IPCL), Council of Scientific and Industrial Research (CSIR) – Central Leather Research Institute (CLRI), Chennai 600020, India*
[2] *Academy of Scientific and Innovative Research (AcSIR), Ghaziabad 201002, India*
[*] *Corresponding Author*

4.1 Introduction

Rapid population growth, technological development, and urbanization have led to the elevation of energy demand worldwide [1]. Most of the world's energy demand is met by nonrenewable fossil fuel resources, mainly coal, oil, and natural gas. Increasing energy demand leads to the fast depletion of these limited fossil fuels [2]. Also, the combustion of fossil fuels emits tremendous greenhouse gases, especially carbon dioxide (CO_2), which is a major contributor to increasing global temperatures [3]. Therefore, in the perspective of sustainable and environmental concerns, it is now widely accepted that replacing fossil fuels with cleaner and renewable alternatives can be an effective solution to these problems. Hydrogen has drawn enormous attention as a clean and promising alternative energy carrier to meet the upcoming global power demand due to its higher energy density. Hydrogen has three times more energy density (142 MJ kg^{-1}) than gasoline (46 MJ kg^{-1}), and water is the only end-product upon its combustion [4]. Hence, widespread usage of hydrogen may play a significant role in designing an eco-friendly and sustainable energy model in the near future.

Hydrogen can be used in high-efficiency onboard power generators such as fuel cells for automobiles and disseminated electricity generation systems [5]. Hydrogen or hydrogen-rich fuels may be delivered directly into proton exchange membrane-based fuel cells with an oxidant, which produces electricity via a low-temperature electrochemical process. Proton exchange membrane fuel cells were commercialized in recent years, and their demand may exponentially grow in the near future. However, secure and efficient hydrogen storage and release is still a challenge for the hydrogen economy [6]. Although hydrogen has enormous energy density, it faces storage challenges due to its lower volumetric energy density than gasoline, diesel, and other fuels [7]. Hydrogen can be stored in a physical or chemical form. In physical form, it is stored as molecular hydrogen, whereas in chemical form, it is stored as a hydridic/protonic source in combination with elements like boron (B), carbon (C), and nitrogen (N) [8]. The volumetric energy density of hydrogen can be improved during physical storage by using several techniques like compression, liquefaction, and physisorption of hydrogen in solid materials [6]. But these techniques

Green Energy Harvesting: Materials for Hydrogen Generation and Carbon Dioxide Reduction, First Edition. Edited by Pooja Devi.

have their own set of limitations, such as electricity consumption and safety issues, the requirement of high energy input, specific container, specialized management, low adsorption capacity, and sluggish dynamics blocking their practical applications. In chemical hydrogen storage, stable compounds of hydrogen are used that, upon thermal or catalytic decomposition, produce gaseous hydrogen [9]. Chemical hydrogen storage is categorized into two types, namely solid chemical hydrogen storage (hydrides of metal and their alloys, borohydrides, complex transition metal hydrides, imides, amides, alanates, etc.) and liquid hydrogen storage media (water, alcohols, hydrazine, etc.) [8].

In the last few decades, solid-state hydrogen storage compounds have exhibited a greater efficiency for scaled-up applications due to their relatively high stability, hydrogen density, and safe storage [10]. However, several of these hydrogen storage materials possessing high storage capacities have limitations such as elevated temperature requirement for release of hydrogen, deterioration with successive cycles, severe heat dissipation, and slow kinetics, creating a barrier to their broad range of applications. Among these solid-state chemical hydrogen storage compounds, ammonia borane (NH_3BH_3, AB) shows great potential for future applications because of its high gravimetric (19.6 wt%) and volumetric densities (146 g L^{-1}) of hydrogen, non-toxicity, high solubility in polar solvents, and a moderate decomposition temperature [11]. Liquid hydrogen storage materials, on the other hand, have gained substantial research attention because of the simplicity with which hydrogen can be released under mild circumstances with the use of a suitable catalyst. Another promising hydrogen storage material is formic acid (HCOOH, FA), a liquid organic hydrogen carrier (LOHC) with a hydrogen content of 4.4 wt%. The volumetric hydrogen density of FA is 53 g L^{-1} which is a very high value compared to the 2020 target of 40 g L^{-1} fixed by the U.S. Department of Energy (DOE) for light-duty fuel cell vehicles having onboard hydrogen storage [8]. The advantages of FA compared to other LOHCs are low temperature of catalytic decomposition, low toxicity, and inflammability.

In this chapter, our main emphasis is on the production of hydrogen from chemical hydrogen storage compounds, mainly AB and FA, because they are considered as promising hydrogen storage materials for onboard hydrogen production. In general, hydrogen generation from AB and FA can be achieved using homogeneous and heterogeneous catalytic systems [12, 13]. The homogeneous catalysts are difficult to separate, recycle, and deactivate rapidly, which impedes their practical applications. The heterogeneous catalytic systems like nanoparticles (NPs) and support stabilized NPs can be used to overcome these limitations. The NPs with reduced particle size considerably enhance the hydrogen production rate by providing easily accessible scattered active sites for the reaction [14]. On further reducing the size of metals to the atomic level, the activity of the catalyst can be maximized. These catalysts are known as single-atom catalysts (SACs), and they consist of atomically distributed isolated active metal atoms on an appropriate support surface. By exposing each active metallic site for the reaction to occur, the SACs enhance the utilization efficiency of metal atoms. Another advantage of SACs over nanoparticle catalysts is that the homogeneity of single atom active sites in SACs leads to high selectivity toward a specific product [15]. On the other hand, all the particle sizes in conventional heterogeneous catalysts are not the same. They contain a mixture of different size particles ranging from NPs to subnanometer clusters. Only a small portion of these particles having appropriate size distribution can act as active sites of the catalyst. The particles with other sizes are either not involved in the reaction or may catalyze some other side reactions [16]. Because of such heterogeneity in the catalytic site, some nanoparticle catalysts show less selectivity and low metal utilization efficiency. Here, we briefly discuss SACs for hydrogen production from chemical hydrogen storage, namely AB and FA.

4.2 Single-Atom Catalysts (SACs)

SACs have revolutionized the area of heterogeneous catalysis and have attracted enormous research interest in recent years [17]. In 2003, Flytzani-Stephanopoulos and coworkers reported that ionic gold (Au) or platinum (Pt) species supported over the surface of ceria catalyze the water-gas shift reaction (WGSR) [18]. They observed that the metallic NPs present in the catalyst has no role in the catalyst's activity, as the elimination of metallic Pt or Au particles via cyanide leaching did not change the activity of the catalyst. Subsequently, Bashyam et al. proposed that cobalt (Co) atoms linked with polypyrrole via N atoms in functionalized carbon exhibited high performance as a non-expansive cathode catalyst in the electrochemical oxygen reduction reaction (ORR) [19]. The exclusive presence of singly dispersed atomic species in the catalyst was confirmed later by Zhang and coworkers in 2011. For the first time, they coined the term "single atom catalysis" for their work on isolated Pt single atoms supported on iron oxide (FeO_x) for oxidation of carbon monoxide (CO) at room temperature [20]. SAC has established itself as a new frontier in catalysis research, in the last decade opening up a new route in the discipline.

As most chemical reactions utilize catalysts, developing more efficient ones is vital [21]. Generally, homogeneous catalysts exhibit excellent activity and exclusive selectivity compared to heterogeneous catalysts [22]. However, they are rapidly deactivated and difficult to separate and recycle from the reaction mixture, which limits their wide practical applications. On the other hand, conventional heterogeneous catalysts are often stable and can be separated easily. However, they exhibit lower atom efficiency during the catalytic process. SACs are promising for bridging the gap between homogeneous and heterogeneous catalysts [23]. Uniform distribution of catalytically active single atomic sites possesses superior selectivity and catalytic activity almost equal to their homogeneous analogs and also inherits properties of heterogeneous catalysts like high stability and reusability. The highest atom economy of heterogeneous catalysis can be achieved by employing SACs as it fully utilizes all metal atoms. This property helps in the construction of cost-effective catalysts from precious metals.

Dispersion of metal on the support provides a very high surface area that ultimately enables the SAC to show high activity [24]. The metal and support interactions play a critical role in maintaining single atoms stablility and activity in SACs. The metal atoms in SACs interact with the support surface by forming covalent, coordinate, or ionic bonds with heteroatoms such as C, N, and oxygen (O) [25]. These metal support interactions provide thermodynamic stability to SACs by lowering the chemical potential compared to the metal NPs and may also offer kinetic stability due to the high aggregation barrier that suppresses agglomeration [26]. Zeng and coworkers studied the metal support interactions using a phase-transition support, vanadium dioxide (VO_2), and how it affects the activity when the phase change occurs in VO_2 [27].

4.2.1 Synthesis and Characterization of SACs

The surface energy of an isolated single atom is quite high, and it can spontaneously organize into larger structures such as clusters or NPs. Therefore, the synthesis of a free single atom is a challenging task. In single atom catalysis, the most common approaches adopted for reducing the surface energy are stabilizing single atoms on support, incorporating them in a framework or the lattice of other metals. The amount of metal loading is decreased to ensure the formation of single atomic sites at a sufficient distance that prevents their contact and aggregation. The composition of support determines the catalytic activity of SAC through its electronic interactions.

The synthetic methods of SACs can be mainly classified into two approaches, namely "bottom-up" and "top-down," which are based on the difference in the starting precursor material used. In the bottom-up approach, starting precursors like mononuclear metal complexes are chosen and dispersed as individual atoms on suitable support having a high surface area. The SACs obtained through this approach contain low metal loading, limiting their practical applications in certain reactions. To increase the metal loading content and stability of SACs, the surface of the support can be modified by constructing nitrogen (N)-, phosphorus (P)-, and sulfur (S)-based anchoring sites on the support surface, by spatial confinement strategy using porous materials, or by defect design strategy. However, in the top-down approach, the bulk metal or metal NPs are used as starting materials for the construction of SACs. By this route, metal atoms are separated from the bulk by providing heat energy to cleave the metal–metal bond. These newly formed metal single atoms bind strongly with the anchoring sites of the supports, which provides stability to SACs against migration and aggregation. In a recent review, Ji et al. discussed several synthesis and design strategies of SACs, including impregnation and coprecipitation, spatial confinement, coordination site construction, defect design, transforming metal NPs or bulk metal into SACs, chemical etching, electrochemical, photochemical, atomic layer deposition (ALD), microwave-assisted, freezing assisted, and ball-milling methods [28].

The characterization of SACs requires sophisticated imaging tools that can operate at atomic precision. Atomic resolution aberration-corrected (ac) scanning transmission electron microscopy (STEM) is a technique which can be directly employed to visualize the dispersed single atom species and atomic structure of SACs [29]. High-angle annular dark-field (HAADF) is a type of STEM imaging technique that can view an individual single metal atom of an SAC as a bright spot. The technique is based on detecting scattered electrons at high angles when an electron beam is scattered by the nucleus of a sample atom [30]. Characterization of SACs is also done using X-ray absorption near-edge spectroscopy (XANES) and extended X-ray absorption fine structure (EXAFS). XANES can offer information on the sample atom's coordination geometry and formal oxidation state, whereas EXAFS provides knowledge regarding the type of neighboring atoms, bond distances, and coordination number of the single atom [31]. X-ray photoelectron spectroscopy (XPS) can be employed to know the oxidation state, elements present, and coordination environment of SACs. The specific catalytic mechanisms of SACs are further studied by using density functional theory (DFT). The detailed synthetic pathways and characterization techniques utilized to synthesize and characterize the SACs used for hydrogen generation from AB and FA are discussed later in this chapter.

4.3 SACs for Dehydrogenation of AB

AB is a B–N–hydride compound having a molecular weight of 30.7 g mol^{-1} and 19.6 wt% hydrogen content [11]. It is a nontoxic colorless solid at ambient temperature, having a density of 0.74 g cm^{-3}. AB is highly soluble in polar solvents, like alcohols, water, and possesses high air stability. These properties make AB a promising H_2 storage candidate. Thermolysis and solvolysis are the two methods by which hydrogen can be released from AB [8]. The major obstacles of thermolysis are high dehydrogenation temperature and a long induction period [9]. Only two equivalents of hydrogen can be liberated from AB at 150–200°C, but an extremely high temperature is required to release the third equivalent of hydrogen. Solvolysis of AB can be performed in polar protic solvents, like water and methanol, by utilizing an appropriate catalyst. When solvolysis of AB is performed in water, it is called hydrolysis, and if it is performed in methanol, it is called methanolysis. In both cases, three equivalents of hydrogen is generated according to Equations 4.1 and 4.2.

$$\text{Hydrolysis reaction}: NH_3BH_3 + 2H_2O(l) \rightarrow NH_4 \cdot BO_2 + 3H_2(g) \quad (4.1)$$

$$\text{Methanolysis reaction}: NH_3BH_3 + 4MeOH(l) \rightarrow NH_4 \cdot B(OMe)_4 + 3H_2(g) \quad (4.2)$$

Among them, hydrolysis is the most extensively investigated method [8]. Several homogeneous and heterogeneous catalysts are extensively studied for the hydrolysis of AB [12, 32, 33]. In a recent review, Yu and coworkers discussed the metal nanocatalysts supported on porous materials for hydrogen production from AB and some other liquid chemical hydrogen storage compounds [14]. Here, we will discuss the SACs used for the dehydrogenation of AB. Although the field of SACs emerged quickly, only a few studies reported dehydrogenation of AB by SACs.

4.3.1 Noble Metal Containing SACs for Dehydrogenation of AB

SACs containing Rh and Pt metals have been studied for dehydrogenation of AB to date. In 2017, Zeng and coworkers reported a quantitative study investigating the interactions between metal and support where the highest occupied energy state of the SAC defined their catalytic activity [27]. The phase-transition materials, such as VO_2, have been found to be appropriate supports for the dispersion of single atoms in order to investigate the importance of metal–support interactions in the catalytic process. The bandgap of the support can be tuned without altering the spatial arrangement of the active single atomic sites. They prepared uniformly dispersed Rh SACs on the surface of VO_2 nanorods (Rh_1/VO_2) by injecting a solution of Na_3RhCl_6 into VO_2 nanorods solution while stirring at room temperature.

The HAADF-STEM image (Figure 4.1a) of the Rh_1/VO_2 catalyst showed uniformly dispersed individual Rh atoms on the surface of the VO_2 nanorods as bright spots. According to inductively coupled plasma atomic emission spectroscopy (ICP-AES) data, the mass loading of Rh was found to be 0.5 wt%. The similar XANES spectra of Rh_1/VO_2 with Rh_2O_3 reveal that the Rh is in an oxidized state in Rh_1/VO_2 (Figure 4.1b). From the EXAFS spectrum, a prominent peak of Rh–O interaction having a coordination number of 6.4 was observed without any Rh–Rh interactions (Figure 4.1c). The positive shift in the binding

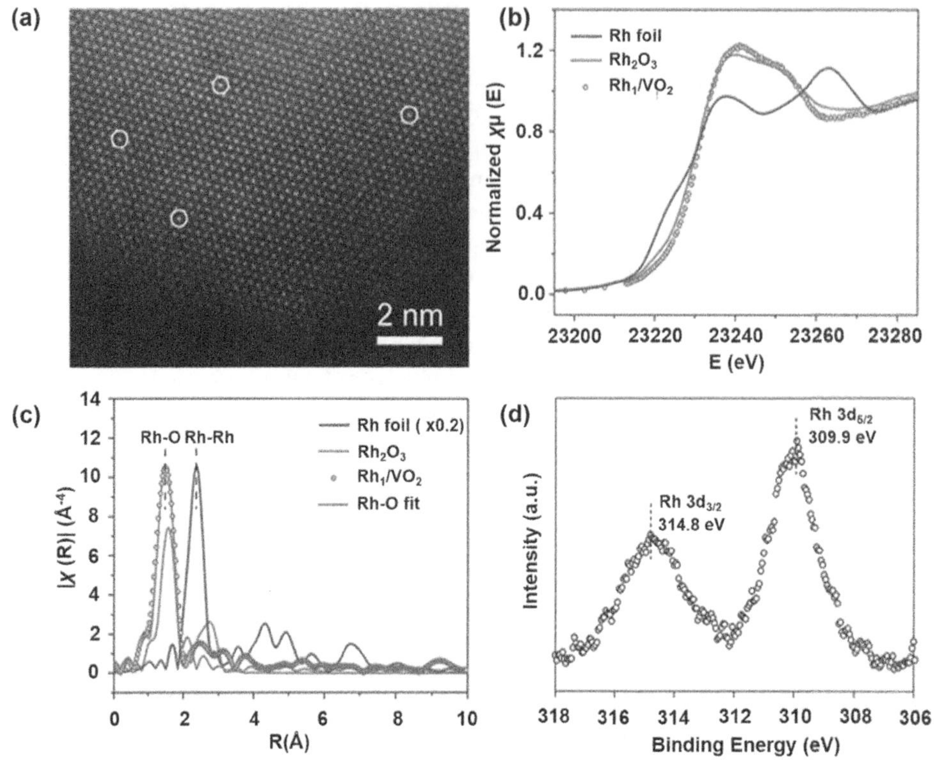

Figure 4.1 (a) HAADF-STEM image of Rh_1/VO_2; (b) Rh K-edge XANES spectra for Rh foil, Rh_2O_3, and Rh_1/VO_2; (c) Rh K-edge EXAFS spectra for Rh_1/VO_2 with fit, Rh foil, and Rh_2O_3; and (d) XPS spectrum of 3d Rh Rh_1/VO_2. *Source*: Wang et al. (2017), Reproduced with permission of John Wiley & Sons.

energy from 307.5–308.0 eV to 309.9 eV for Rh $3d_{5/2}$ in the case of Rh_1/VO_2 in the XPS measurement also indicates the strong metal–support binding interactions between Rh single atoms and VO_2 supports, which stabilizes the single Rh atoms from migration and aggregation (Figure 4.1d).

To study the role of metal–support interactions in the Rh_1/VO_2 catalyst, the authors performed hydrolysis of AB. The Rh_1/VO_2 catalyst produces hydrogen with a turnover frequency (TOF) of 0.8 and 1.2 s^{-1} at 333.2 and 353.2 K, respectively. The hydrogen generation rate increases upon elevation of temperature (from 323.2 to 358.2 K), but it is not linearly increasing like other conventional catalysts. The data fit well in the linear plot in section I (345.7–358.2 K) and section II (323.2–335.7 K). The activation energy (E_a) for section I (E_{aI}) and section II (E_{aII}) were 13.6 ± 0.9 and 52.3 ± 2.4 kJ mol^{-1}, respectively. The sharp change in the E_{aI} and E_{aII} is due to the changing of VO_2 support's phase from monoclinic (M) to rutile (R) phase during the transition temperature region (337.2–344.2 K) [34, 35]. The reaction rate is directly proportional to Rh loading, but the unchanged TOF number demonstrated that only single Rh atoms are acting as active sites in the reaction.

The authors proposed two possible pathways for the hydrolysis of AB over Rh_1/VO_2 catalyst. The first path occurs via proton activation, whereas the second path occurs by activating both water and AB. The stable reaction rate and TOF of the reaction with varied concentrations of AB rule out the second path. The decrease in TOF number with the increase in pH indicated that the proton activation is the rate-limiting step, and the reaction follows the first path. Mechanistic investigation revealed that the phase transition of the substrate from M to R modulates its band structure. The electron from the highest occupied state of the single Rh atom in the Rh_1/VO_2 (R) atom can easily activate the proton compared to Rh_1/VO_2 (M). As a result, Rh_1/VO_2 (R) has a higher catalytic activity. On substituting Rh with other metals, like Fe, Ni, Au, Ag, Cu, and Co, the resulting E_a for hydrolysis of AB is roughly identical, which indicates the crucial activity of M_1/VO_2. The E_a ranges from 38.7–55.7 kJ mol^{-1} (Table 4.1), which relates E_a with the single atom's highest occupied state.

Zeolites are an excellent support to stabilize single metal atoms due to their uniform micropores and exceptional thermal stability. Yu and coworkers prepared a catalyst by encapsulating Rh single atoms with aluminosilicate ZSM-5 zeolites and MFI-type silicalite-1 (S-1) [36]. $[Rh(NH_2CH_2CH_2NH_2)_3]Cl_3$ complex was used as a starting material in the one-pot hydrothermal synthesis process, followed by ligand-protected direct hydrogen reduction at 500°C. Direct hydrogen reduction is vital for the encapsulation of Rh atoms inside zeolite frameworks, while in the case of conventional calcination-reduction, sub-nanometric Rh clusters are formed within zeolites. No peaks in powder XRD measurements for Rh species in the synthesized catalyst, Rh@S-1-H, indicate that the metal atoms are encapsulated inside S-1 zeolites without large Rh particles on the surface. XPS analysis also showed that there are no signals for Rh in Rh@S-1-H. The amount of Rh loading in the catalyst is 0.28 wt% as measured by ICP-AES. XANES spectra of Rh K-edge indicate that Rh atoms encapsulated by zeolites are in higher oxidation states. O atoms stabilize Rh atoms via the formation of Rh–O bond with coordination number 4.6 ± 0.2. There is no Rh–Rh bond formation as indicated by EXAFS spectra. *In situ* CO-diffuse reflectance infrared Fourier transform spectroscopy (DRIFTS) measurements further confirmed that the Rh atoms in Rh@S-1-H are atomically distributed. They also synthesized Rh single atoms encapsulated

Table 4.1 Comparison of E_a for various noble or non-noble metal catalysts.

Catalyst	E_{all} (kJ mol^{-1})	E_{al} (kJ mol^{-1})	ΔE_a (kJ mol^{-1})
Rh_1/VO_2	52.3 ± 2.4	13.6 ± 0.9	38.7 ± 2.6
Fe_1/VO_2	176.1 ± 4.5	122.2 ± 2.6	53.9 ± 5.2
Co_1/VO_2	179.5 ± 5.2	128.1 ± 2.7	51.4 ± 5.9
Ni_1/VO_2	170.1 ± 4.6	122.3 ± 2.3	47.8 ± 5.1
Cu_1/VO_2	154.2 ± 4.7	107.2 ± 3.2	47.0 ± 5.7
Ag_1/VO_2	164.5 ± 4.1	110.6 ± 4.3	53.9 ± 5.9
Au_1/VO_2	163.1 ± 4.8	107.4 ± 2.2	55.7 ± 5.3

Source: [27] (table S4). Reproduced with permission of Wiley.

in aluminosilicate ZSM-5 zeolite by varying Si/Al ratios, like Rh@ZSM-5-H with Si/Al = 105 and Rh@ZSM-5-H with Si/Al = 245 containing 0.32 and 0.30 wt% of Rh, respectively.

The catalysts show good catalytic performance for AB hydrolysis at 298 K. The TOF values for Rh@S-1-H and Rh@ZSM-5-H with Si/Al = 105 are 432 and 699 mol_{H2} mol_{Rh}^{-1} min^{-1}, respectively. In Rh@ZSM-5-H, a higher hydrolysis rate was observed because the Rh atoms and Brønsted acidic sites of support synergistically activate the AB and water molecules, respectively [37]. The apparent E_a of 75.5 kJ mol^{-1} was observed for Rh@S-1-H. With respect to catalyst concentration, the reaction follows first-order kinetics, and the rate is unaffected by AB concentration. The catalyst exhibited excellent recycling stability as it remained active after five consecutive cycles.

Pt SACs also exhibit good catalytic efficiency for AB dehydrogenation. A single atom alloy (SAA) with atomically dispersed Pt over the Ni surface was reported by Li et al., which shows excellent hydrogen evolution activity from aqueous AB [38]. Initially, 5% Ni-containing catalysts were synthesized on various supports (activated carbon (AC), carbon nanotubes (CNT), and covalent triazine framework (CTF)). The introduction of Pt to Ni-based catalysts replaced some atoms of Ni, as revealed by the XRD measurements. Lower loading of Pt is essential for the maximum utilization of Pt. The HAADF-STEM images of 1/1000 Pt + Ni-CTF showed Pt atoms are atomically dispersed (red circles) on the Ni (Figures 4.2a,b). Pt species are in a higher oxidation state in the catalyst, as indicated by the XANES spectra (Figure 4.2c). On higher loading of Pt; e.g., 1/20 Pt + Ni-support or 1/100 Pt + Ni-support, there was a Pt–Pt bond formation in the catalyst with coordination numbers 4.3 and 3.5, respectively, as shown by EXAFS spectra (Figure 4.2d). The coordination number of Pt in 1/500 Pt + Ni-support is 1.3, which indicates the atomic distribution of Pt over the surface of Ni particles on lower loading. For 1/1000 Pt-based catalyst, Pt content is too low to get a spectrum. So, based on the coordination number of 1/20, 1/100, 1/500 Pt + Ni-support catalyst, it was believed that 1/1000 Pt + Ni-support catalyst would retain the atomically dispersed morphology. The electronic interaction between these two metals was studied with the help of XPS measurements. Downshift in the binding energy of Pt indicates the transfer of electrons from Ni to Pt (Figure 4.2e). Therefore, Pt and Ni have slight negative and positive charges, respectively, which is also observed in the Bader charge simulation.

A single Pt atom supported on an Ni surface effectively catalyzes the hydrolysis of AB. The catalysts with low Pt loading (160 ppm), 1/1000 Pt + Ni/CFT, and 1/1000 Pt + Ni/CNT show high TOF of 10,900 and 12,000 mol_{H2} mol_{Pt}^{-1} min^{-1}, respectively. The 1/1000Pt + Ni/CNT catalyst exhibit TOF of almost 27 and 428 times higher than that of 0.16% Pt/CNT and 5% Ni/CNT catalyst, respectively. The TOF of 1/1000Pt + Ni/CNT is about 21-fold higher than the Pt-based catalyst (Pt/CNT-HT) with the highest activity reported to date [39]. Further studies revealed that the hydrolysis reaction shows zero-order and first-order kinetics with respect to AB concentration and catalyst amount, respectively, confirming that the AB activation is not a rate-determining step. The kinetic isotope effect studies determined the activation of H in the H_2O molecule as the rate-determining step. The E_a of the catalyst 1/1000 Pt + Ni/CTF (17 kJ mol^{-1}) is lower than that of 5% Ni/CTF (27.4 kJ mol^{-1}). The catalysts' activity is boosted by the synergistic effect of Pt and Ni. According to the DFT studies, Ni ($Ni^{\delta+}$) and Pt ($Pt^{\delta-}$) in Pt/Ni alloy interacted with the OH^- and H^+ of H_2O molecule, respectively, and reduced the kinetic energy barrier of the reaction.

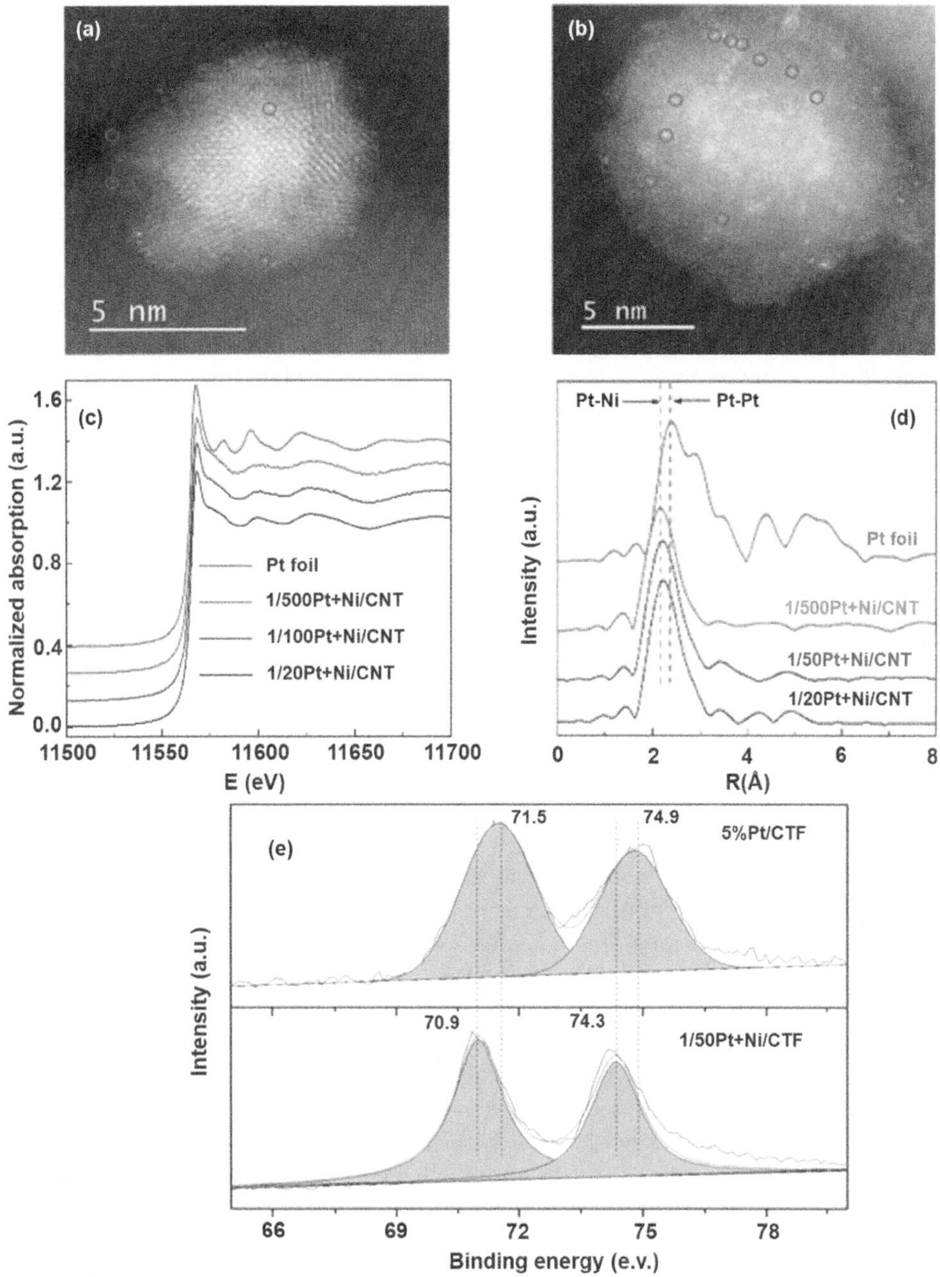

Figure 4.2 HAADF-STEM images of (a) 1/500Pt + Ni/CNT; (b) 1/1000Pt + Ni/CTF; (c) normalized XANES spectra; (d) EXAFS spectra; and (e) Pt 4 f XPS spectra. *Source*: Li et al. (2017), Reproduced with permission of American Chemical Society.

Strong electronic interactions between metal and support in SACs are crucial to keep the catalyst stable and active against sintering and leaching. Li et al. reported that designing the Pt_1 single atoms on Co_3O_4 with an unoccupied 5d state through electronic metal–support interactions enhances the catalytic activity almost 68 times higher as compared with other supports (CeO_2, ZrO_2, and graphene) used for AB dehydrogenation at room temperature [40]. They tailor the interaction between metal and support by varying the supports, which is an efficient way to design electronic metal–support interactions [41]. Single Pt atoms supported on Co_3O_4, ZrO_2, CeO_2, and graphene, with different 5d electronic states, were prepared by the ALD method with a Pt loading of 0.5, 1.1, 0.2, and 0.4 wt%, respectively. Atomically dispersed Pt was observed on all these supports through HAADF-STEM imaging (Figures 4.3a,b). Figure 4.3(b) showed atomically dispersed Pt atoms Co_3O_4 support; no NPs/clusters were observed. The white line peak with strong intensity in XANES spectra for Pt_1/Co_3O_4 compared to other catalysts implies that Pt is in a higher oxidation state (+4), and Pt_1 in Pt_1/Co_3O_4 has a large population of unoccupied 5d states of Pt (Figure 4.3c) [42]. But in other supports, the oxidation state of Pt_1 is lower, and they have less population in unoccupied Pt 5d states, as shown by the XANES spectra. The formation of a Pt–O bond with a coordination number of 5.7 and Pt-Co coordination results in strong binding between the metal atom and support, as revealed by EXAFS spectra. The absence of a Pt-Pt coordination peak implies that all Pt species are atomically dispersed in all the catalysts. Infrared (IR) spectroscopy of the chemisorption of CO tool is applied to determine the metal 5d electronic states population. The higher frequency of the linear CO peaks for Pt_1/Co_3O_4 in DRIFTS of CO chemisorption unveils that the Pt_1 atom has less populated 5d electronic state electrons in Pt_1/Co_3O_4 compared to other supported Pt_1 catalysts. XPS measurements showed that most Pt_1 atoms are in +4 state in Pt_1/Co_3O_4, which agrees with the XANES and IR results. In the case of other supports, Pt_1 atoms are mostly in the +2 state.

They performed hydrolysis of AB with all the prepared catalysts to investigate the effect of electronic metal–support interactions on SACs. Pt_1/Co_3O_4 catalyst exhibits an excellent rate of 1220 $mol_{H2}\ mol_{Pt}^{-1}\ min^{-1}$ (with E_a of 37.4 kJ mol^{-1}) compared to other Pt_1 supported catalysts at room temperature (Figure 4.3d). Recyclability studies of the catalysts showed that Pt_1/Co_3O_4 catalyst is active for 15 cycles; after that, the Pt_1 atom reduced to +2, while the other catalysts became highly deactivated only after three cycles. The strong metal–support interactions in Pt_1/Co_3O_4 contribute to the higher activity of the catalyst and excellent stability. DFT calculations revealed that, in Pt_1/Co_3O_4, the unoccupied Pt_1 5d states are comparatively at higher energy which is responsible for average AB adsorption and much lower adsorption of gaseous H_2 product, thus enhancing the H_2 desorption rate. On the other hand, in the CeO_2 and ZrO_2, the gap between unoccupied Pt_1 5d states and the Fermi level is too high to adsorb AB effectively. Thus, the electronic metal–support interactions significantly tune the vacant Pt_1 5d state, that changes the adsorption of AB and H_2 molecules, thus enhancing the catalytic activity.

Although Zeng and coworkers performed comparative dehydrogenation of AB using some first row transition metal-based SACs supported on VO_2 [27], there are no specific reports on such SACs anchored on other supports.

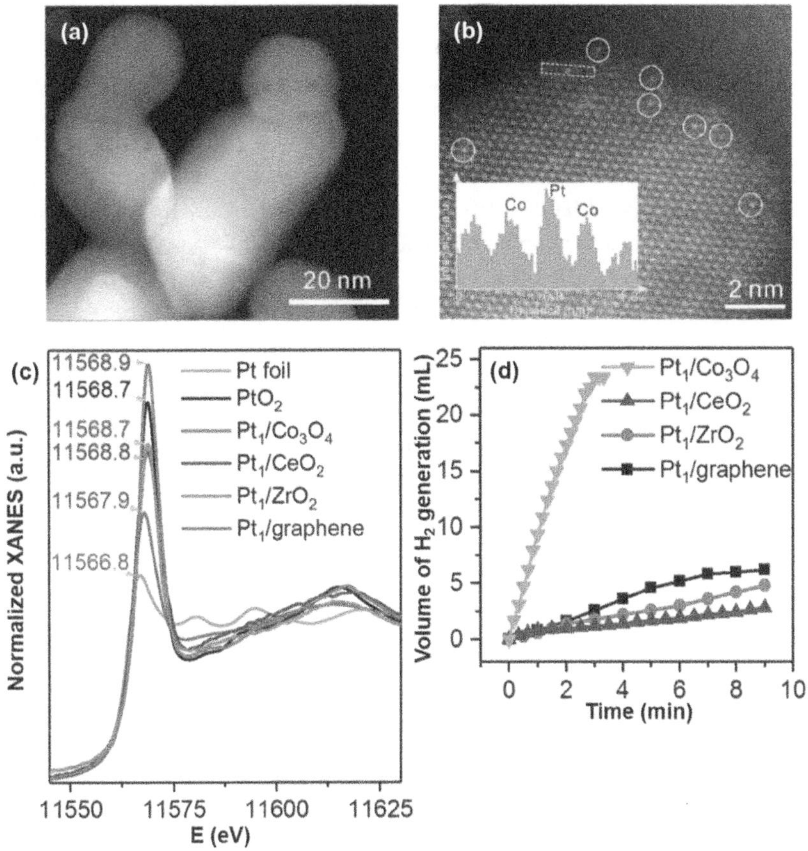

Figure 4.3 Ac HAADF-STEM images of Pt_1/Co_3O_4, at (a) low magnification; (b) high magnification; (c) normalized XANES spectra; and (d) catalytic performance of Pt_1 SACs. *Source*: Li et al. (2019), Reproduced with permission of American Chemical Society.

4.4 SACs for Decomposition of FA

FA, the simplest carboxylic acid, mainly originating from the biomass process [43], has shown great potential as LOHC since it has volumetric and gravimetric hydrogen densities of 53 g L^{-1} and 4.4 wt% and also because of its suitability for easy handling, transportation, and refueling [44]. In 1978, Williams et al. recognized FA as an efficient hydrogen storage system for the first time [45]. They generated FA electrochemically by reducing CO_2 and utilized it as a hydrogen source by decomposition over palladium (Pd) supported on fine carbon particles. The FA decomposition can follow two pathways: (1) the dehydrogenation pathway forming hydrogen and CO_2, and (2) the dehydration pathway forming water and CO, as shown by Equations 4.3 and 4.4.

$$\text{Dehydrogenation reaction}: HCOOH \leftrightarrow H_2 + CO_2 \tag{4.3}$$

$$\text{Dehydration reaction}: HCOOH \leftrightarrow CO + H_2O \tag{4.4}$$

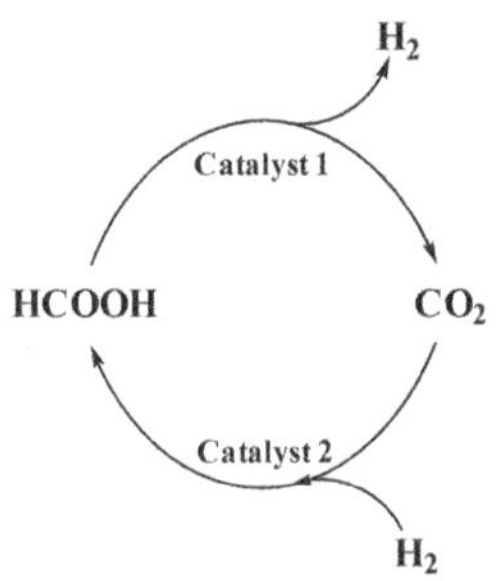

Scheme 4.1 Hydrogen storage and production from FA.

The decomposition of FA should proceed through the dehydrogenation route to use it as fuel, and the dehydration route should be strictly eliminated by choosing an appropriate catalyst and reaction conditions. Even small traces of CO may deteriorate the properties of catalysts used in fuel cells during the FA decomposition reaction. CO_2 generated during the dehydrogenation reaction can be converted back to FA by H_2 reduction (Scheme 4.1) [46].

Several homogeneous and heterogeneous catalysts are used for the decomposition of FA [13, 47]. The difficulties in the separation and recycling process limit the practical application of homogeneous catalysts. Along with supported metal nanocatalysts, SACs are also used for the decomposition of FA. SACs containing Pd, Pt, Au, Ni, and Co as metal atoms are reported for FA decomposition. SACs supported on carbon and graphene have also been studied in recent years [48, 49]. Doping of these supports with O, P, S, N, and B changes the catalytic activity of the SACs by binding these heteroatoms with the active sites. For example, porous N-doped carbon supports strongly bind single atoms to form some basic N-sites, stabilizing the SACs and enhancing their catalytic performance. Carbon materials are gaining special attraction as supports for FA decomposition reaction because they provide ease of control of the size and electronic state of single atoms supported on it. Recently, Bulushev et al. published a review on SACs utilized for the generation of hydrogen from FA [50].

4.4.1 Noble Metal Containing SACs for Dehydrogenation of FA

4.4.1.1 Pd SACs

Pd SACs anchored on N-doped carbon are extensively investigated for both liquid- and gas-phase decomposition of FA. Catalysts containing Pd are found to be the most efficient for hydrogen production from FA. Podyacheva et al. reported a Pd catalyst synthesized by depositing 0.2 – 2 wt% of Pd on the CNTs and N-doped CNTs (N-CNTs) by wet impregnation strategy using Pd acetate in acetone [51]. From transmission electron microscopy (TEM) analysis, it was found that in N-free CNTs, upon lowering the Pd loading from 2 to 0.2%, the average particle size of Pd NPs decreased from 2.3 to 1.2 nm, respectively. However, in N-CNTs on low loading of Pd (0.2 wt%), Pd dispersed as single atoms. The ac HAADF-STEM image of the 0.2 wt% Pd/N-CNTs catalyst showed atomically dispersed Pd as bright dots (Figure 4.4a). XPS spectra of 0.2 wt% Pd/CNTs and 0.2 wt% Pd/N-CNTs catalysts revealed that Pd is in a +2 oxidation state in the N-doped catalysts, and +2 and metallic state in N-free catalyst. The peaks at 335.9 and 337.9 eV in the spectrum (Figure 4.4b) are assigned to metallic Pd components and Pd^{2+} species, respectively. Isolated Pd^{2+} are coordinated with two pyridinic N atoms, providing stability to the catalyst. CO chemisorption experiments at room temperature have been performed with these Pd samples after their

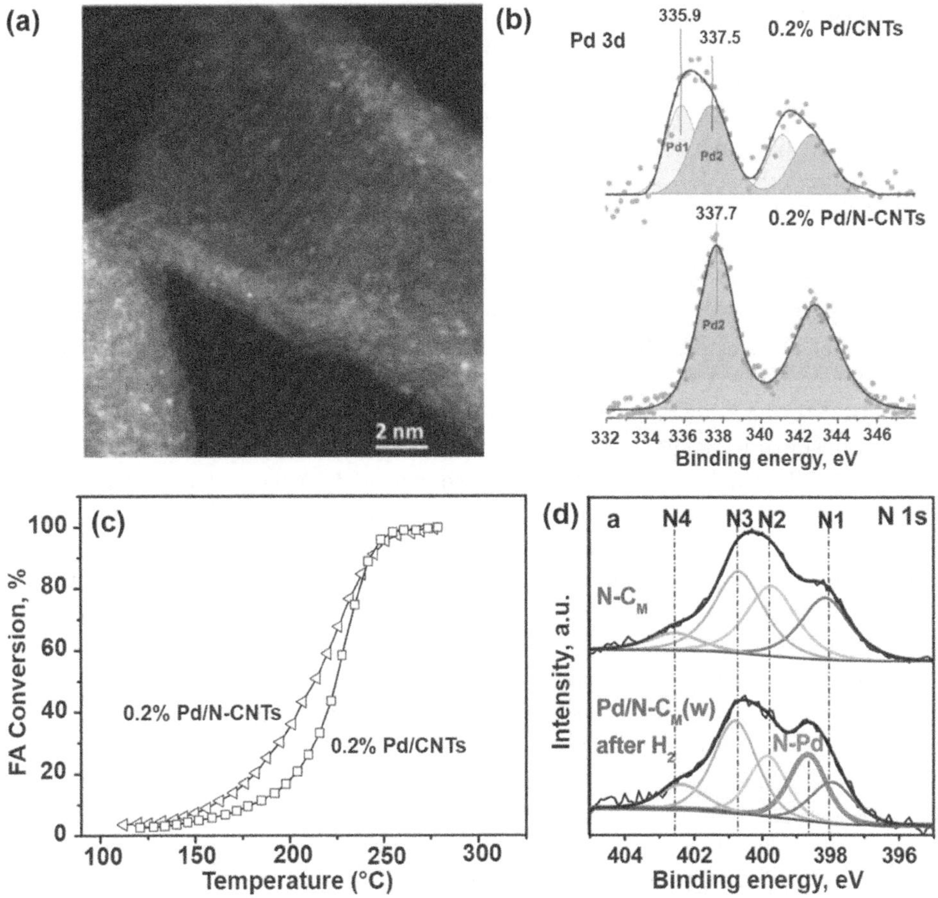

Figure 4.4 (a) ac HAADF-STEM images of Pd/N-CNTs with 0.2 wt% Pd; (b) Pd 3d XPS spectra; and (c) conversion-temperature curve of FA decomposition; and (d) N 1s XPS spectra of the 1% Pd/N-C catalyst and N-carbon support, after reduction in H_2. *Source*: Podyacheva et al. (2018), Reproduced with permission of John Wiley & Sons.

reduction in hydrogen at 473 K. The CO/Pd ratios were equal to 33% and 16% for the N-free and N-doped Pd catalysts, respectively. Hence, the chemisorption of CO on the single-atom sites is weaker than on the surface metallic sites of NPs. The 0.2%Pd/N-CNTs catalyst performed excellent efficiency for the gas phase decomposition of FA with a TOF of 0.081 s^{-1} at 125°C, which is 1.5–2-fold higher than the N-free catalysts, as shown in Figure 4.4c. According to mechanistic studies, Pd^{2+} coordinated with two pyridinic N-atoms at the ends of graphite planes of N-doped carbon nano-materials (N-CNMs) functioned as active sites. Such active sites break the C–H bonds of FA molecules forming the adsorbed H_2 atom and carboxyl fragment, further facilitating their interaction to yield hydrogen molecule and CO_2. Doping of N to CNTs improves the catalytic activity as well as the catalyst's selectivity toward hydrogen.

The quantity of Pd in the catalyst has a considerable impact on the formation and activity of SACs. Bulushev et al. compared 1 wt% Pd catalysts on N-doped porous carbon supports (Pd/N-C) and N-free carbon (Pd/C) supports [52]. In both cases, NPs of 2.5 nm mean sizes were formed. But a significant number of single atoms were observed in N-doped samples in contrast to the N-free samples. The ratio of the single atoms to NPs in the 1 wt% Pd/N-C estimated from the ac HAADF/STEM image was equal to 200 [53]. Pyridinic moieties on the N-C support coordinates with the Pd to stabilize the Pd single atoms. In the N 1s XPS spectra of the N-C support and 1 wt% Pd/N-C catalyst (Figure 4.4d), it is seen that deposition of Pd significantly changes the binding energy of the line related to pyridinic N species from 398.1 to 398.8 eV. However, only slight changes in the intensity are observed for graphitic (400.7 eV) and pyrrolic (399.8 eV) N species. Arrigo et al. also reported identical changes in near edge X-ray absorption fine structure (NEXAFS) spectra of N upon deposition of Pd [54]. The N-doped catalysts of Pd showed more activity for the FA decomposition than the unsupported Pd powder and the N-free catalysts, which contain mainly NPs. The selectivity of the hydrogen production was also increased for the N-doped catalyst compared to the N-free catalyst (96 vs. 94%). For the 1 wt% Pd N-doped catalyst, the CO/Pd ratio was lower by a factor of 2 than the N-free catalyst (35 and 18%). Moreover, the 1 wt% Pd/N-C catalyst showed excellent stability in the reaction conditions at 448 K for 35 h [52].

4.4.1.2 Pt SACs

Pt is regarded as the highly active noble metal for FA decomposition. Therefore, Pt SACs are also under extensive focus for their performance in the dehydrogenation of FA. Podyacheva et al. prepared Pt catalysts with CNTs and N-CNTs as supports. They studied the role of N doping on the activity of Pt catalysts toward FA decomposition reactions [55]. By using incipient wetness impregnation (im) and homogeneous precipitation (pr) methods, they deposited Pt (0.2–2 wt%) on the N-CNTs and CNTs. The catalysts were assigned as 0.2–2%Pt/CNTs-im, 0.2–2%Pt/N-CNTs-im, 0.2–2%Pt/CNTs-pr, and 0.2–2%Pt/N-CNTs-pr. The ac HAADF-STEM studies of prepared catalysts revealed that these Pt deposition techniques on supports produced dispersed catalysts consisting of NPs (range of 1.1–1.9 nm) and single Pt atoms. On the low deposition of Pt (0.2 wt%), the quantity of single Pt atoms distributed on the support's surface is considerably higher in the N-doped catalysts as compared to N-free catalysts. The ac HAADF-STEM image of 0.2%Pt/N-CNTs showed the well-dispersed single Pt atoms in Figure 4.5a. The deconvoluted XPS spectrum of Pt 4f in 0.2%Pt/CNTs and 0.2%Pt/N-CNTs showed that Pt is present in three states, Pt^0(71.7 eV), Pt^{2+}(72.9 eV), and Pt^{4+}(74.2 eV) (Figure 4.5b). The Pt^{4+} species resulted due to the atmospheric oxidation of Pt NPs during the transfer of the catalyst to the spectrometer. The relative contribution of the Pt^{2+} species (at 72.9 Ev) in N-CNTs increases compared to CNTs, as shown in Figure 4.5b. This implied that the majority of the Pt atoms are in Pt^{2+} state in 0.2%Pt/N-CNTs. According to DFT calculations, the pyridinic sites of N-CNMs interact with Pt, creating electron-deficient Pt^{2+} [56]. CO chemisorption measurements showed that CO/Pt value (15%) is lower for 0.2%Pt/N-CNTs than N-free catalysts (CO/Pt = 67%). This is because of the existence of a considerable amount of ionic (Pt^{2+}) single metal atoms in 0.2%Pt/N-CNTs, which do not adsorb CO. In the N-free catalysts, Pt is present as ~1.3 nm particles, which adsorbed CO.

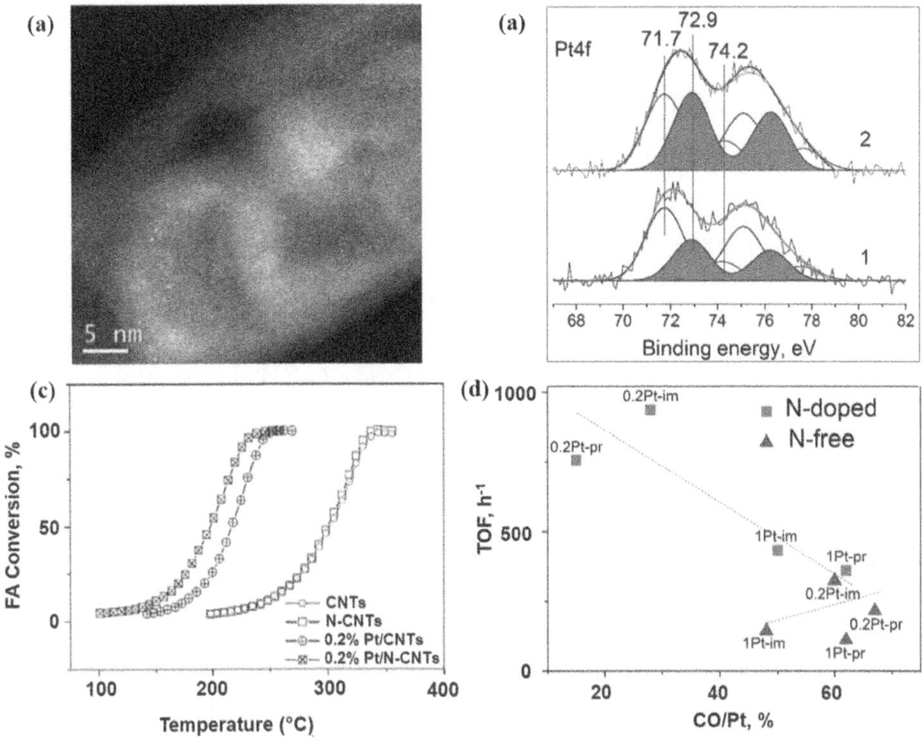

Figure 4.5 (a) ac HAADF-STEM images of Pt/N-CNTs with 0.2 wt% Pt; (b) Pt 4f deconvoluted XPS spectra; (c) conversion-temperature curve of FA decomposition; and (d) change in TOF of FA decomposition at 125 °C with CO/Pt ratio. *Source*: Podyacheva et al. (2019), Reproduced with permission of Multidisciplinary Digital Publishing Institute.

Figure 4.5c depicts the catalyst's activity in the FA decomposition reaction. One can see the negative shift by more than 100°C in the temperature curve of FA conversion upon loading 0.2 wt% Pt on both the supports. It was even more shifted for N-CNTs indicating the different properties of Pt species loaded on CNTs and N-CNTs. Irrespective of the deposition method, N-CNTs showed better activity with TOF values 3–4-fold higher than N-free CNTs. N-CNTs catalysts are more selective (98%) toward hydrogen formation than N-free catalysts (92.6%). For the N-free catalysts, TOF increases slightly with a rise in CO/Pt ratio, which may be because of the uniform deposition of metallic Pt in the catalysts. For the N-doped catalysts, TOF was found to be decreasing with an increase in the CO/Pt value. As shown in Figure 4.5d, a significant change was observed in the amount of CO chemisorbed by the catalysts with a decrease in the Pt content from 1 to 0.2%. This is because, in low Pt content, most Pt is present as single Pt ions, as revealed by the ac HAADF-STEM image and XPS spectra (Figure 4.5a,b).

The authors also studied the activity of Pt catalyst by using herring-bone N-doped carbon nanofibers (N-CNFs) as supports to study the effect of different N-doped carbon structural supports on catalytic activity. They concluded that the interaction of Pt with pyridinic N-sites of the N-CNFs or N-CNTs is mainly responsible for its catalytic activity irrespective of different carbon-based supports used.

Bulushev et al. found that by increasing the amount of Pt loading up to 1 wt% on both N-CNFs and CNFs, the catalysts containing single-atom species and ~1 nm-sized NPs are formed [56]. The N-doped catalysts showed significantly higher TOFs and selectivities for hydrogen formation than the N-free catalyst and unsupported Pt powder. The E_a value for the 1 wt% Pt/N-CNFs (43 kJ mol^{-1}) was less than the 1 wt% Pt/CNFs (51 kJ mol^{-1}) [57]. The different E_a values may indicate a different reaction mechanism on these catalysts. The pyridine-type N atom pair situated at the ends of the graphene sheet provides a strong stabilization to individual metal atoms, as revealed from the DFT studies. The formation of such active sites resulted in improved H_2 selectivity and a 10-fold improvement in reaction rate compared to N-free catalysts.

4.4.1.3 Au SACs

Au SACs are less explored for decomposition of FA as compared to Pd and Pt. Flytzani-Stephanopoulos and coworkers studied atomically dispersed Au on ceria (Au/CeO_2) as an efficient catalyst for FA decomposition at near-room temperatures. The catalyst shows a TOF of nearly 0.63 s^{-1} at 130°C. It shows selectivity only for hydrogen and CO_2. Mechanistic studies revealed that Au-O_x species catalyze the FA decomposition, and the selection of support is crucial for the stabilization and distribution of the atomically dispersed Au species [58].

In another report, Bulushev et al. prepared Au catalysts supported on N-doped (0.7 wt% Au) and N-free (1.9 wt% Au) porous carbon by deposition of $HAuCl_4$ without using chloride as a stabilizer [59]. Both the catalysts contained Au single atoms and NPs, as obtained by the ac HAADF-STEM imaging. High distribution of the Au NPs was observed for N-doped carbon support (Au/N-C), having an ~2.2 nm average size, as compared to catalyst with N-free support (Au/C) having size of ~10 nm. According to the N 1s XPS analysis of Au/N-C and Au/C, there is no strong interaction between Au and the support's N functionality. From the analysis of Au 4f spectra it is found that Au species are present mainly in the Au^0 state with a small amount of Au^{1+}. The presence of metallic Au indicates that the interaction of Au^{3+} ions with supported carbon reduced Au^{3+} ions to Au^0 [60]. The Au/N-C catalyst showed higher activity than Au/C. The selectivity of Au/C catalyst to H_2 is higher (99.5%) compared to Au/N-C catalyst (96.6%) [61]. The N-doped Au catalysts also showed more activity than Al_2O_3, and SiO_2 supported catalysts of 2.2 and 1.6 nm average particle sizes, respectively.

The DFT calculations revealed that Au single atoms bind strongly to the edge or vacancies of the graphene through its carbon atoms. The coordination of Au with N atoms of support is weaker compared with Pd or Pt SACs. The FA interacts with Au single atoms supported on graphene through its O atoms by forming a formate intermediate.

4.4.2 Non-noble Metal Containing SACs for Dehydrogenation of FA

SACs of non-noble metals are also able to attract significant research attention because of the high cost of noble metal catalysts. Nickel (Ni) and Co SACs are studied for the decomposition of FA. However, their lower activity is a matter of concern for their widespread applications. Here we will discuss supported non-noble metal catalysts with single-atom active sites for the FA decomposition reaction.

4.4.2.1 Ni SACs

Ni forms SACs over the surface of N-doped carbon. Nishchakova et al. synthesized Ni-based catalysts on N-doped (Ni/N-C) and N-free (Ni/C) carbon supports synthesized at various temperatures (973 K, 1073 K, and 1173 K) by depositing 1 wt% Ni using Ni acetate [62]. After the reduction in gaseous FA at 623 K, the samples were studied by conventional HAADF/STEM, EXAFS, and XPS. Through the HAADF-STEM imaging study, it was difficult to visualize any atomic-scale Ni in Ni/N-C catalysts. However, the presence of Ni was found in the same area of the sample through energy dispersive X-ray (EDX) spectroscopy. On the other hand, in the case of Ni/C catalyst, ~3.9 nm-sized Ni NPs were formed. The EXAFS study of the Ni K edge for the Ni/N-C catalyst revealed the existence of individual Ni atoms without any Ni–Ni interactions. The spectrum of Ni–N-C resembled the spectra of Ni phthalocyanine, where four N atoms coordinate every Ni atom. For Ni–C, the Ni K edge spectrum resembled the Ni foil and NiO spectrum and showed a clear Ni–Ni bond. According to the Ni 2p XPS measurement, Ni is present in Ni^{2+} in both Ni–N-C and Ni–C catalysts. But the line corresponding to Ni^{2+} species was negatively shifted by 0.8 eV in Ni–N-C (855.5 eV) as compared to the Ni–C (856.3 eV), indicating the different electronic environment in both the catalysts. Ni atoms in phthalocyanine [63] and Ni single atoms distributed on Ni–N-C also give similar spectra, where Ni is coordinated to 4 N atoms, as reported earlier. Therefore, the authors concluded that Ni single atom in the Ni–N-C catalyst was coordinated by four pyridinic N species, thus resulting in an Ni-N_4 site [64, 65].

The Ni–N-C SAC exhibited more Ni mass-based catalytic activity toward the FA decomposition at high temperatures than the Ni–C catalyst. The activation energies for decomposition of FA over the Ni–N-C were found to be about 100 ± 9 kJ mol^{-1}. The single-atom Ni–N-C catalysts showed outstanding stability and selectivity of the hydrogen production (97%) in the FA decomposition reaction.

4.4.2.2 Co SACs

Co metal SACs are widely studied for the decomposition of FA. Beller and coworkers synthesized a Co catalyst by pyrolysis of Co acetate complex with phenanthroline (Phen) ligands on a Vulcan XC72R carbon at 800°C [66]. They showed that the content of phenanthroline in the initial mixture has a vital role in the nature of Co species formed and in their activity for FA dehydrogenation. They prepared catalysts with different Co/phen ratios (1:3, 1:7, and 1:10). The XRD pattern of all the catalysts revealed that the diffraction pattern of Co/phen (1:7) and Co/phen (1:10) were not similar to Co/phen (1:3). The HAADF imaging technique showed that, at low ratios of Co/phen (1:7 and 1:10), single atoms were formed (visible, but their concentration was found to be too low), whereas at high ratios of Co/phen (1:3), NPs were formed. XPS measurement revealed the presence of mainly Co^{2+} species and formation of Co–N bond mainly with pyridinic N species.

A liquid phase FA decomposition was performed in propylene carbonate (PC) at ~100°C. The single atom Co catalyst, Co–phen (1:7), exhibits maximum catalytic activity of 423.3 mL $g_{cat}{}^{-1}h^{-1}$ (19.32 L $g_{Co}{}^{-1}h^{-1}$) and selectivity (99.8%) as compared to the catalyst with NPs, Co–phen (1:3). The activity of the Co–phen (1:10) catalyst was less because of the lower Co content. The presence of CoN_x species as single active sites for this reaction was confirmed by thiocyanate ion (SCN^-) poisoning and acid leaching reactions. For the FA dehydrogenation process, the catalyst exhibited superior recyclability and stability under the reaction conditions within 6 days.

In another report, Beller and coworkers performed a comparative study of metal organic framework-based Co-N-C catalysts having Co NPs and single Co atoms for FA dehydrogenation under similar reaction conditions [67]. They synthesized a Co single-atom catalyst (Co-N-C(SACs)) by pyrolysis of ZnCo zeolite imidazolate frameworks (ZIFs) at 1000°C and Co NPs (Co-N-C(NPs)) by pyrolysis of Co-ZIFs at 800°C, as shown in Figure 4.6a. The atomically distributed Co species on carbon were discovered in an ac HAADF-STEM image of single atom Co-N-C (Figure 4.6b). Electron paramagnetic resonance (EPR) and XPS measurements of Co-N-C(NPs) and Co-N-C(SACs) revealed that Co is in the +2 state, and only Co^{2+} species are present in SACs, whereas Co^{3+}, Co^{2+}, and Co are present in NPs (Figure 4.6c). Pyridinic N atoms act as major anchor sites for atomically dispersed Co. XANES spectrum also confirmed that Co is in a +2 state in Co-N-C(SACs) (Figure 4.6d). The single-atom catalyst spectrum resembled the spectrum of Co phthalocyanine (CoPC), suggesting that single Co^{2+} cations coordinated to N species [68]. The Co-N coordination stabilizes the Co single atoms, and there is no Co-Co coordination as implied by EXAFS analysis (Figure 4.6e).

The catalytic activity of Co-N-C(NPs) and Co-N-C(SACs) was checked for FA dehydrogenation in PC, at 98°C. The Co SACs exhibited considerably greater mass-based activities than those for the catalysts with NPs. This is because of highly dispersed $Co^{II}N_x$ sites on Co-N-C(SACs), which are responsible for the dehydrogenation of FA. Pyrolysis temperature determines the catalytic activity since at different pyrolysis temperatures, atomically distributed Co catalysts with Co-N_2, Co-N_3, and Co-N_4 sites can be obtained [69]. The surface energy can be enhanced by varying the Co-N coordination from Co-N_4 and Co-N_3 to Co-N_2, which may further improve the activity of Co SACs. In Figure 4.6f, the catalytic activity of different catalysts synthesized at different pyrolysis temperatures is compared. Increase in pyrolysis temperature decreased the reactivity of Co-N-C(NPs) catalysts because Co NPs agglomerated at higher pyrolysis temperatures. The highest activity was observed for the single-atom Co catalyst synthesized at 1000°C, followed by the catalysts synthesized at 900 and 800°C. According to the authors, the higher activity of the catalyst synthesized at 1000°C is due to the higher concentration of Co-N_2 sites. In the catalysts obtained by pyrolysis at lower temperatures, Co-N_3 and Co-N_4 sites are mainly present.

Theoretical studies of SACs for the decomposition of FA are also reported. The DFT studies provide knowledge about the metal atom–support interaction and transformation of reactants into products. The decomposition of the FA molecule to H_2 and CO_2 may proceed through two pathways. If it proceeds through the breaking of the O–H bond, it gives a formate intermediate (HCOO), or if it proceeds through the breaking of the C–H bond, it results in a carboxyl intermediate (COOH). Since the metal atoms in SACs are coordinated with the heteroatoms of the support for stability, the coordination number of metal atoms is significant for determining their catalytic activity. The lower the coordination number, the higher the reactivity to the FA molecules. SACs of Pt, Pd, Ru, and Ni coordinated with two N atoms of the support are highly reactive and can break the C–H bond of FA very easily; as a result, a carboxyl and an adsorbed hydrogen atom are formed on the single atom active site [52, 53, 56]. For the Pt and Pd single atom sites, the adsorbed hydrogen atom of

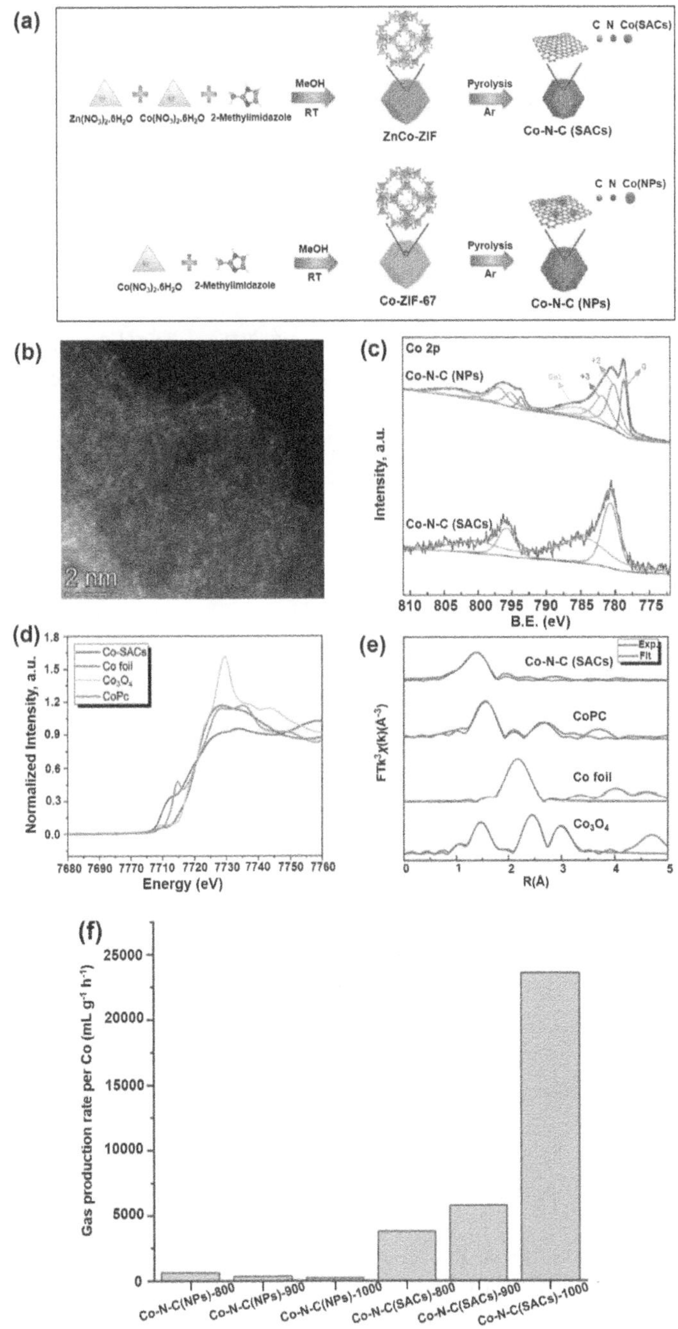

Figure 4.6 (a) Synthetic route of Co-SACs and Co-NPs; (b) ac HAADF-STEM image of Co-SACs (magnified); (c) XPS spectra of Co 2p; (d) XANES; (e) EXAFS of Co-SACs; and (f) activity of catalysts synthesized at different pyrolysis temperature. *Source*: Li et al. (2020), Reproduced with permission of John Wiley & Sons.

carboxyl and the adsorbed hydrogen atom come closer to releasing H_2 and CO_2. The presence of N species in the SACs is important for the higher activity of FA decomposition, since the formation of adsorbed hydrogen and carboxyl is not easy for the catalysts with O and C atoms.

Jeyakumar et al. performed a DFT study of single iron (Fe), ruthenium (Ru), and osmium (Os) embedded N-doped graphene for dehydrogenation of FA [70]. They observed that the Os forms a very strong bond with the H atom from FA, and this was not easily broken to form hydrogen. Thus, a single Os atom does not act as a catalyst. Single Fe and Ru atoms can function as catalysts, but the reaction requires higher energy to cleave C–H and O–H bonds.

When Au or Cu single atoms are coordinated to a pair of pyridinic N atoms, dehydrogenation of FA occurs via breaking of the O–H bond. The pyridinic N atom near the single Au atom activates the FA through protonation of pyridinic N, forming pyridinium formate. Then the CO_2 molecule is released by breaking of the C–H bond and forming Au-H. Bulushev et al. proposed the following possible steps for the reaction [61]:

$$\begin{aligned} &> \mathrm{N} + \mathrm{HCOOH} \rightarrow \; > \mathrm{NH^{+}HCOO^{-}} \\ &> \mathrm{NH^{+}HCOO^{-}} + \mathrm{Au} \rightarrow \; > \mathrm{NH} + \mathrm{AuH} + \mathrm{CO_2} \\ &> \mathrm{NH} + \mathrm{AuH} \rightarrow \; > \mathrm{N} + \mathrm{Au} + \mathrm{H_2} \end{aligned} \tag{4.5}$$

With the help of DFT, Zhong et al. demonstrated that the FA dehydrogenation energy barrier for C_2N supported single Co^{2+}, Cu^{2+}, and Ni^{2+} ions were found to be lower as compared to pure Pd and Pt catalysts [71]. The increased catalytic activity of these catalysts is because of the formation of a dual-active center containing metal cations and the surrounding N atom of C_2N.

Liu and coworkers designed a class of SACs using 12 transition metals (Mn, Fe, Ni, Co, Cu, Ru, Pd, Rh, Pt, Ir, Au, and Ag) anchored over a unique 2D hole-free carbon-nitrogen material, C_3N [72]. Using DFT calculations, they employed these catalysts (TM@ C_3N) for the dehydrogenation of FA. By calculating the binding energies of transition metal atoms, the absorption stability of FA, and the H_2 generating efficiency of H species from an individual transition metals site, they found that Ni@C_3N, Pd@C_3N, and Pt@C_3N performed better for FA dehydrogenation. The calculations of different reaction mechanisms suggested that these three SACs prefer the formation of a formate intermediate pathway rather than carboxyl intermediate for FA dehydrogenation. Pd@C_3N shows the best dehydrogenation efficiency compared to the Ni@C_3N, Pt@C_3N, and well-defined Pd (111) surfaces. So, Pd@C_3N possesses great scope of application as a superior and cost-effective SAC in the dehydrogenation of FA.

Recently, Pei and coworkers studied the catalytic activity of metal SACs anchored on g-C_3N_4 (M@g-C_3N_4) for dehydrogenation of FA by using the DFT method and machine learning [73]. According to DFT calculation results, among Ru, Sc, Ti, Cr, Fe, Mn, Ni, Co, Zr, Mo, V, Rh, Pd, Pt, Os, Ag, Au, W, and Ir containing SACs, the SACs of Pd-, Rh-, and Pt@g-C_3N_4 exhibited good activity. Machine learning revealed that the electronegativity and metal's d-band center influence the adsorption strength of FA on the catalyst, which further determines the reaction energy of the rate-limiting step.

SAAs are a class of SACs wherein catalytically active single metal atoms are deposited on the surface of other less reactive metal support by alloying [74]. Sykes and coworkers synthesized Pt_1Cu-SAA having reactive Pt single metal atoms incorporated in the Cu lattice [75]. They discovered that SAA effectively activates O–H bonds of FA, resulting in the generation of formate. The generated formate undergoes spill-over from the active Pt site to Cu that increases the formation of CO_2 and hydrogen by 6-fold as compared to that of Cu NPs.

4.5 Conclusion and Future Outlook

The rising need for energy, along with the adverse environmental effects of fossil fuels, necessitates the development of an environmentally beneficial and sustainable energy source. Because of its high energy density, hydrogen has a lot of potential as a first-choice clean alternative. But difficulty related to its efficient and safe storage has created a bottleneck in its widespread applications. Among all hydrogen storage materials, chemical hydrogen storage materials attract significant research interest because of their high stability, hydrogen density, and safe storability. AB and FA are two widely studied chemical hydrogen storage compounds with high hydrogen content, 19.6 and 4.4 wt%, respectively. In this chapter, we discussed the hydrogen generation from AB and FA by using SACs.

The size of a catalyst's active sites is crucial in determining its catalytic activity. Because of the effect of quantum size, enhanced metal-support interactions, and a low-coordination environment, the catalytic efficacy of typical heterogeneous catalysts improves when the size of the metal NPs is lowered from nanometer to atomic scale in SACs. SACs are believed to be the bridge between homogeneous and heterogeneous catalysts. SACs offer homogeneity of single active sites at the support surface, which gives selectivity to a specific product. SACs maximize the atom utilization efficiency with fully exposed active sites, providing an efficient method for reducing expensive metal consumption. Single-atom catalysis opens up a very broad avenue for the design of cost-effective catalysts, and it is expected that the SACs for commercial applications will be developed in the near future.

Although SACs have exhibited great performances, there exist some challenges associated with them, such as: i) bulk scale synthesis of SACs with improvement in the metal loading content; ii) strong stabilization of the SACs on the support by enhancing metal–support interactions between atomically dispersed metal atoms and surrounding coordinating atoms; and iii) atomic-level designing of active sites in SAC at desired positions.

We have discussed Rh and Pt metal containing catalysts for dehydrogenation of AB. Very little research has been done in this area, and there is still ample scope for progress. For FA decomposition to hydrogen and CO_2, we have discussed noble, non-noble SACs, and SAAs catalysts that are much more resistant to CO than nanocatalysts. We have also discussed the DFT calculations to gain a theoretical understanding of the metal–support interaction and interaction of reactants with the active site of SACs. Decomposition of FA upon coordination with SACs via formate or carboxyl intermediate is discussed in detail. The work highlighted in this chapter will instigate further design and development of SACs for hydrogen production and allied research areas.

Acknowledgements

We thank the Department of Science and Technology (DST), New Delhi, India, for the DST INSPIRE faculty award to AAV. RKB thanks the Council of Scientific and Industrial Research (CSIR) for the research fellowship. CSIR-CLRI communication number 1639.

Note

The authors declare no conflict of interests and no financial interests.

References

1 Dresselhaus, M.S. and Thomas, I.L. (2001). Alternative energy technologies. *Nature* 414(6861): 332–337.
2 Rosen, M.A. and Koohi-Fayegh, S. (2016). The prospects for hydrogen as an energy carrier: An overview of hydrogen energy and hydrogen energy systems. *Energy, Ecol Environ.* 1(1): 10–29.
3 Nejat, P., Jomehzadeh, F., Taheri, M.M. et al. (2015). A global review of energy consumption, CO_2 emissions and policy in the residential sector (with an overview of the top ten CO_2 emitting countries). *Renew Sustain Energy Rev.* 43: 843–862.
4 Schlapbach, L. and Züttel, A. (2001). Hydrogen-storage materials for mobile applications. *Nature* 414(6861): 353–358.
5 Edwards, P.P., Kuznetsov, V.L., David, W.I.F. et al. (2008). Hydrogen and fuel cells: Towards a sustainable energy future. *Energy Policy* 36(12): 4356–4362.
6 Lang, C., Jia, Y., and Yao, X. (2020). Recent advances in liquid-phase chemical hydrogen storage. *Energy Storage Mater.* 26: 290–312.
7 Sinigaglia, T., Lewiski, F., Santos Martins, M.E. et al. (2017). Production, storage, fuel stations of hydrogen and its utilization in automotive applications: A review. *Int J Hydrogen Energy* 42(39): 24597–24611.
8 Navlani-García, M., Mori, K., Kuwahara, Y. et al. (2018). Recent strategies targeting efficient hydrogen production from chemical hydrogen storage materials over carbon-supported catalysts. *NPG Asia Mater.* 10(4): 277–292.
9 Dalebrook, A.F., Gan, W., Grasemann, M. et al. (2013). Hydrogen storage: Beyond conventional methods. *Chem Commun.* 49(78): 8735–8751.
10 Eberle, U., Felderhoff, M., and Schüth, F. (2009). Chemical and physical solutions for hydrogen storage. *Angew Chemie Int Ed.* 48(36): 6608–6630.
11 Zhu, Q-L. and Xu, Q. (2015). Liquid organic and inorganic chemical hydrides for high-capacity hydrogen storage. *Energy Environ Sci.* 8(2): 478–512.
12 Zhan, n W-W., Zhu, Q-L., and Xu, Q. (2016). Dehydrogenation of ammonia borane by metal nanoparticle catalysts. *ACS Catal.* 6(10): 6892–6905.
13 Li, Z. and Xu, Q. (2017). Metal-nanoparticle-catalyzed hydrogen generation from formic acid. *Acc Chem Res.* 50(6): 1449–1458.
14 Sun, Q., Wang, N., Xu, Q. et al. (2020). Nanopore-supported metal nanocatalysts for efficient hydrogen generation from liquid-phase chemical hydrogen storage materials. *Adv Mater.* 32(44): 2001818.

15 Zhang, H., Liu, G., Shi, L. et al. (2018). Single-atom catalysts: Emerging multifunctional materials in heterogeneous catalysis. *Adv Energy Mater.* 8(1): 1701343.

16 Yang, X-F., Wang, A., Qiao, B. et al. (2013). Single-atom catalysts: A new frontier in heterogeneous catalysis. *Acc Chem Res.* 46(8): 1740–1748.

17 Wang, A., Li, J., and Zhang, T. (2018). Heterogeneous single-atom catalysis. *Nat Rev Chem.* 2(6): 65–81.

18 Qi, F., Howard, S., and Maria, F-S. (2003). Active nonmetallic Au and Pt species on ceria-based water-gas shift catalysts. *Science* 301(5635): 935–938.

19 Bashyam, R. and Zelenay, P. (2006). A class of non-precious metal composite catalysts for fuel cells. *Nature* 443(7107): 63–66.

20 Qiao, B., Wang, A., Yang, X. et al. (2011). Single-atom catalysis of CO oxidation using Pt_1/FeO_x. *Nat Chem.* 3(8): 634–641.

21 Van Leeuwen, P.W.N.M. (2006). *Homogeneous Catalysis: Understanding the Art.* Springer Science & Business Media.

22 Ertl, G., Knözinger, H. and Weitkamp, J. (1997). *Handbook of Heterogeneous Catalysis*, vol. 2. VCH Weinheim.

23 Chen, F., Jiang, X., Zhang, L. et al. (2018). Single-atom catalysis: Bridging the homo- and heterogeneous catalysis. *Chinese J Catal.* 39(5): 893–898.

24 Liu, L. and Corma, A. (2018). Metal catalysts for heterogeneous catalysis: From single atoms to nanoclusters and nanoparticles. *Chem Rev.* 118(10): 4981–5079.

25 Samantaray, M.K., D'Elia, V., Pump, E. et al. (2020). The comparison between single atom catalysis and surface organometallic catalysis. *Chem Rev.* 120(2): 734–813.

26 Liu, J-C., Tang, Y., Wang, Y-G. et al. (2018). Theoretical understanding of the stability of single-atom catalysts. *Natl Sci Rev.* 5(5): 638–641.

27 Wang, L., Li, H., Zhang, W. et al. (2017). Supported rhodium catalysts for ammonia-borane hydrolysis: Dependence of the catalytic activity on the highest occupied state of the single rhodium atoms. *Angew Chemie Int Ed.* 56(17): 4712–4718.

28 Ji, S., Chen, Y., Wang, X. et al. (2020). Chemical synthesis of single atomic site catalysts. *Chem Rev.* 120(21): 11900–11955.

29 Krivanek, O.L., Chisholm, M.F., Nicolosi, V. et al. (2010). Atom-by-atom structural and chemical analysis by annular dark-field electron microscopy. *Nature* 464(7288): 571–574.

30 Nellist, P.D. and Pennycook, S.J. (2000). The principles and interpretation of annular dark-field Z-contrast imaging. *Adv Imaging Electron Phys.* 113: 147–203.

31 Korshin, G.V., Frenkel, A.I., and Stern, E.A. (1998). EXAFS study of the inner shell structure in Copper(II) complexes with humic substances. *Environ Sci Technol.* 32(18): 2699–2705.

32 Rossin, A. and Peruzzini, M. (2016). Ammonia–borane and amine–borane dehydrogenation mediated by complex metal hydrides. *Chem Rev.* 116(15) :8848–8872.

33 Nagyházi, M., Turczel, G., Anastas, P.T. et al. (2020). Highly efficient ammonia borane hydrolytic dehydrogenation in neat water using phase-labeled CAAC-Ru catalysts. *ACS Sustain Chem Eng.* 8(43): 16097–16103.

34 Zylbersztejn, A. and Mott, N.F. (1975). Metal-insulator transition in vanadium dioxide. *Phys Rev B* 11(11): 4383–4395.

35 Wu, C., Zhang, X., Dai, J. et al. (2011). Direct hydrothermal synthesis of monoclinic VO_2(M) single-domain nanorods on large scale displaying magnetocaloric effect. *J Mater Chem.* 21(12): 4509–4517.

36 Sun, Q., Wang, N., Zhang, T. et al. (2019). Zeolite-encaged single-atom rhodium catalysts: Highly-efficient hydrogen generation and shape-selective tandem hydrogenation of nitroarenes. *Angew Chemie Int Ed.* 58(51): 18570–18576.

37 Sun, Q., Wang, N., Bai, R. et al. (2019). Synergetic effect of ultrasmall metal clusters and zeolites promoting hydrogen generation. *Adv Sci.* 6(10): 1802350.

38 Li, Z., He, T., Matsumura, D. et al. (2017). Atomically dispersed Pt on the surface of Ni particles: Synthesis and catalytic function in hydrogen generation from aqueous ammonia–borane. *ACS Catal.* 7(10): 6762–6769.

39 Chen, W., Ji, J., Duan, X. et al. (2014). Unique reactivity in Pt/CNT catalyzed hydrolytic dehydrogenation of ammonia borane. *Chem Commun.* 50(17): 2142–2144.

40 Li, J., Guan, Q., Wu, H. et al. (2019). Highly active and stable metal single-atom catalysts achieved by strong electronic metal–support interactions. *J Am Chem Soc.* 141(37): 14515–14519.

41 O'Connor, N.J., Jonayat, A.S.M., Janik, M.J. et al. (2018). Interaction trends between single metal atoms and oxide supports identified with density functional theory and statistical learning. *Nat Catal.* 1(7): 531–539.

42 Lytle, F.W. (1976). Determination of d-band occupancy in pure metals and supported catalysts by measurement of the LIII X-ray absorption threshold. *J Catal.* 43(1): 376–379.

43 Jin, F. and Enomoto, H. (2011). Rapid and highly selective conversion of biomass into value-added products in hydrothermal conditions: Chemistry of acid/base-catalysed and oxidation reactions. *Energy Environ Sci.* 4(2): 382–397.

44 Joó, F. (2008). Breakthroughs in hydrogen storage: Formic acid as a sustainable storage material for hydrogen. *ChemSusChem.* 1(10): 805–808.

45 Williams, R., Crandall, R.S., and Bloom, A. (1978). Use of carbon dioxide in energy storage. *Appl Phys Lett.* 33(5): 381–383.

46 Bulushev, D.A. and Ross, J.R.H. (2018). Heterogeneous catalysts for hydrogenation of CO_2 and bicarbonates to formic acid and formates. *Catal Rev.* 60(4): 566–593.

47 Sordakis, K., Tang, C., Vogt, L.K. et al. (2018). Homogeneous catalysis for sustainable hydrogen storage in formic acid and alcohols. *Chem Rev.* 118(2): 372–433.

48 Rivera-Cárcamo, C. and Serp, P. (2018). Single atom catalysts on carbon-based materials. *ChemCatChem.* 10(22): 5058–5091.

49 Zhang, L., Wang, Y., and Niu, Z. (2019). Single atoms on graphene for energy storage and conversion. *Small Methods* 3(9): 1800443.

50 Bulushev, D.A. and Bulusheva, L.G. (2021). Catalysts with single metal atoms for the hydrogen production from formic acid. *Catal Rev.* 1–40. doi: 10.1080/01614940.2020.1864860.

51 Podyacheva, O.Y., Bulushev, D.A., Suboch, A.N. et al. (2018). Highly stable single-atom catalyst with ionic Pd active sites supported on N-doped carbon nanotubes for formic acid decomposition. *ChemSusChem.* 11(21): 3724–3727.

52 Bulushev, D.A., Zacharska, M., Shlyakhova, E.V. et al. (2016). Single isolated Pd^{2+} cations supported on N-doped carbon as active sites for hydrogen production from formic acid decomposition. *ACS Catal.* 6(2): 681–691.

53 Zacharska, M., Bulusheva, L.G., Lisitsyn, A.S. et al. (2017). Factors influencing the performance of Pd/C catalysts in the green production of hydrogen from formic acid. *ChemSusChem.* 10(4): 720–730.

54 Arrigo, R., Schuster, M.E., Xie, Z. et al. (2015). Nature of the N–Pd interaction in nitrogen-doped carbon nanotube catalysts. *ACS Catal.* 5(5): 2740–2753.

55 Podyacheva, O., Lisitsyn, A., Kibis, L. et al. (2019). Nitrogen-doped carbon nanotubes and nanofibers for green hydrogen production: Similarities in the nature of nitrogen species, metal–nitrogen interaction, and catalytic properties. *Energies* 12(20): 3976.
56 Bulushev, D.A., Zacharska, M., Lisitsyn, A.S. et al. (2016). Single atoms of Pt-group metals stabilized by N-doped carbon nanofibers for efficient hydrogen production from formic acid. *ACS Catal.* 6(6): 3442–3451.
57 Jia, L., Bulushev, D.A., Podyacheva, O.Y. et al. (2013). Pt nanoclusters stabilized by N-doped carbon nanofibers for hydrogen production from formic acid. *J Catal.* 307: 94–102.
58 Yi, N., Saltsburg, H., and Flytzani-Stephanopoulos, M. (2013). Hydrogen production by dehydrogenation of formic acid on atomically dispersed gold on ceria. *ChemSusChem.* 6(5): 816–819.
59 Bulushev, D.A., Chuvilin, A.L., Sobolev, V.I. et al. (2019). Single Au atoms on the surface of N-free and N-doped carbon: Interaction with formic acid and methanol molecules. *Top Catal.* 62(5): 508–517.
60 Bulushev, D.A., Yuranov, I., Suvorova, E.I. et al. (2004). Highly dispersed gold on activated carbon fibers for low-temperature CO oxidation. *J Catal.* 224(1): 8–17.
61 Bulushev, D.A., Sobolev, V.I., Pirutko, L.V. et al. (2019). Hydrogen production from formic acid over Au catalysts supported on carbon: Comparison with Au catalysts supported on SiO_2 and Al_2O_3. *Catalysts* 9: 376.
62 Nishchakova, A.D., Bulushev, D.A., Stonkus, O.A. et al. (2019). Effects of the carbon support doping with nitrogen for the hydrogen production from formic acid over Ni catalysts. *Energies* 12(21): 4111.
63 Berríos, C., Cárdenas-Jirón, G.I., Marco, J.F. et al. (2007). Theoretical and spectroscopic study of Nickel(II) Porphyrin derivatives. *J Phys Chem A* 111(14): 2706–2714.
64 Liu, W., Chen, Y., Qi, H. et al. (2018). A durable nickel single-atom catalyst for hydrogenation reactions and cellulose valorization under harsh conditions. *Angew Chemie Int Ed.* 57(24): 7071–7075.
65 Yang, H., Shang, L., Zhang, Q. et al. (2019). A universal ligand mediated method for large scale synthesis of transition metal single atom catalysts. *Nat Commun.* 10(1): 4585.
66 Tang, C., Surkus, A-E., Chen, F. et al. (2017). A stable nanocobalt catalyst with highly dispersed CoN_x active sites for the selective dehydrogenation of formic acid. *Angew Chemie Int Ed.* 56(52): 16616–16620.
67 Li, X., Surkus, A-E., Rabeah, J. et al. (2020). Cobalt single-atom catalysts with high stability for selective dehydrogenation of formic acid. *Angew Chemie Int Ed.* 59(37): 15849–15854.
68 Liu, W., Zhang, L., Yan, W. et al. (2016). Single-atom dispersed Co–N–C catalyst: Structure identification and performance for hydrogenative coupling of nitroarenes. *Chem Sci.* 7(9): 5758–5764.
69 Wang, X., Chen, Z., Zhao, X. et al. (2018). Regulation of coordination number over single Co sites: Triggering the efficient electroreduction of CO_2. *Angew Chemie.* 130(7): 1962–1966.
70 Christopher Jeyakumar, T. and Sivasankar, C. (2019). Dehydrogenation of formic acid catalysed by M-embedded nitrogen-doped graphene (M = Fe, Ru, Os): A DFT study. *New J Chem.* 43(3): 1440–1448.
71 Zhong, W., Liu, Y., Deng, M. et al. (2018). C_2N-supported single metal ion catalysts for HCOOH dehydrogenation. *J Mater Chem A* 6(24): 11105–11112.

72 Bing, Q. and Liu, J. (2021). Transition metal single atom anchored C_3N for highly efficient formic acid dehydrogenation: A DFT study. *Appl Surf Sci.* 562: 150186.

73 Zhao, X., Wang, L., and Pei, Y. (2021). Single metal atom catalyst supported on g-C_3N_4 for formic acid dehydrogenation: A combining density functional theory and machine learning study. *J Phys Chem C* 125(41): 22513–22521.

74 Hannagan, R.T., Giannakakis, G., Flytzani-Stephanopoulos, M. et al. (2020). Single-atom alloy catalysis. *Chem Rev.* 120(21): 12044–12088.

75 Marcinkowski, M.D., Liu, J., Murphy, C.J. et al. (2017). Selective formic acid dehydrogenation on Pt-Cu single-atom alloys. *ACS Catal.* 7(1): 413–420.

5

Non-Noble Metal Ion-Based Metal-Organic Framework Electrocatalyst for Electrochemical Hydrogen Generation

Satya Prakash[1,2], Kamlesh[1,2], Deepika Tanwar[1,2], Pankaj Raizda[3], Pardeep Singh[3], Manish Mudgal[1,2], A.K. Srivastava[1,2], and Archana Singh[1,2]

[1] *Academy of Scientific and Innovative Research (AcSIR), Ghaziabad 201002, India*
[2] *CSIR – Advanced Material and Processes Research Institute, Bhopal 462026, India*
[3] *School of Chemistry, Shoolini University, Himachal Pradesh 173229, India*

5.1 Introduction

Energy is a vital component of our daily lives as it is essential for almost all human activity. Nature has provided humanity with vast energy reserves such as coal, petroleum, gases, and other fossil fuels beneath the earth's surface, as well as sunlight, wind, and other sources of energy above the surface. About 80% of global energy demand is fulfilled by fossil fuels [1]. Unfortunately, fossil fuels are exhaustible sources of energy. With the current pace of consumption (624×10^{18} J), these fuels are likely to remain expensive, and if precautions are not taken, they may run out shortly [2]. There is no ray of hope in the future that the consumption rate will decrease because the world's current population is estimated to be 786 million people, with an annual growth rate of 1.05%. The global population is expected to expand by 81 million people per year. The rapid extinction of fossil fuels is a matter of concern, but the pollution caused by their burning is an even more worrying issue. The combustion of fossil fuels emits greenhouse gases which lead to planet-warming. There are lots of renewable forms of energy in nature but we are unable to harness all of them because of the limiting factors such as intermittent availability, requirement of high capital cost, and regional and seasonal specificity. As our dependence on renewable energy continuously grows day by day, this energy must be stored for the time when there is not a ray of sunshine [1], or if the wind has stopped blowing [3]. The best way is to use renewable energy to get hydrogen and oxygen by splitting water by following Equation 5.1 [4].

$$2H_2O \rightarrow O_2 + 2H_2 \qquad \Delta H_{298K} = 286 \text{ kJ mol}^{-1} \tag{5.1}$$

The major advantage of storing this energy in the form of hydrogen is that when it becomes oxidized it again forms water and there is no production of carbon-containing by-products like CO, CO_2, etc. As with any redox reaction consisting of two half-cell reactions, so it is also with water splitting; one is known as the hydrogen evolution reaction (HER) at the cathode, whereas the other is the oxygen evolution reaction (OER) at the anode. The hypothetical value of cell potential for water splitting calculated based on thermodynamic parameters is 1.23 V [5] but the experimental value is always higher than the ideal or calculated value (1.23 V) due to the kinetic barriers. To deal with this problem electrocatalysts are used. In this chapter, our main focus is on the electrocatalyst based on MOFs and their derivatives used for water splitting and related challenges [5].

Green Energy Harvesting: Materials for Hydrogen Generation and Carbon Dioxide Reduction,
First Edition. Edited by Pooja Devi.

5.2 Need for Hydrogen

For the rapid growth of a nation, a sustainable and stable energy source is required, so it is necessary to use renewable energy more effectively and consistently. Solar and wind are ecologically-beneficial primary renewable energy sources but they are unstable and non-transportable [6, 7]. To avoid these issues, water electrolysis can convert such types of unstable energy sources into stable and transportable hydrogen energy. Presently, most global countries have focused on the development of new hydrogen technologies as a plausible solution. Moreover, hydrogen as a fuel has a superior energy storage capability with a calorific value of 1.43×10^8 Jkg^{-1}, which implies that 1 kg of hydrogen is released when completely burnt [8], whereas only one-third of the heat is released when burning the same amount of gasoline [6]. Most importantly, the total combustion of hydrogen produces water as a by-product, and there are no released greenhouse gases or polluting products.

5.3 Hydrogen Production

Despite being one of the most abundant atoms in the universe, hydrogen never exists independently because of its low molecular weight and interacts readily with other elements. As a result, the concept of hydrogen generation is dependent on the removal of other hydrogen-rich molecules. Presently various available methods are represented in Figure 5.1, including 46% steam reforming of natural gases, 30% oil reforming, 18% coal gasification, and 4–5% water electrolysis, as well as other negligible sources currently used methods for hydrogen production [9]. Water electrolysis as a technique of hydrogen production has the benefit of being a simple, huge production that does not require the use of nonrenewable sources such as fossil fuels and also produces no greenhouse emissions.

5.4 Water Splitting

The water-splitting reaction ($2H_2O \rightarrow 2H_2 + O_2$) is the combined form of two half-reactions called the oxygen evolution reaction (OER) and the hydrogen evolution reaction (HER) [7, 10]. The water-splitting reaction is a thermodynamically non-spontaneous reaction and completion of it requires extra energy. Based on applied energy sources like heat, light, and

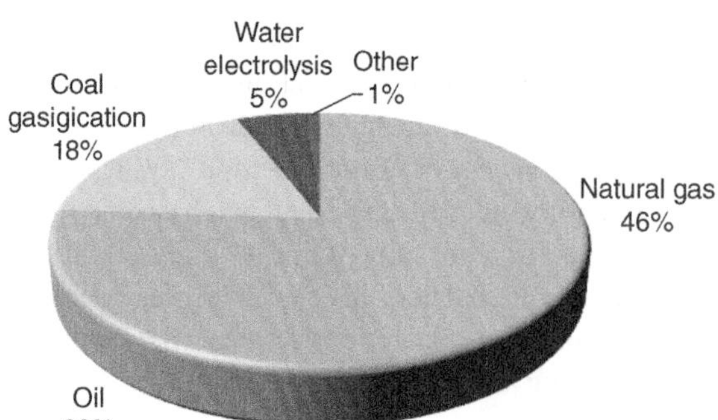

Figure 5.1 Hydrogen production percentage from different sources.

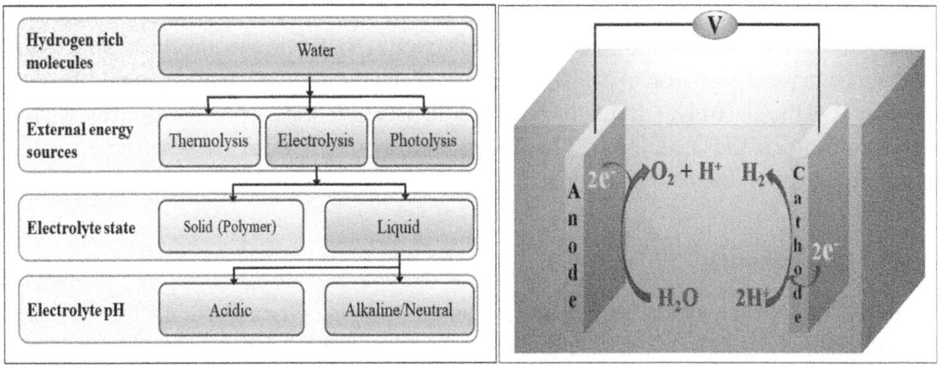

Figure 5.2 (a) Types of water-splitting methods; and (b) schematic of overall water electrolysis using two electrode system.

electricity, water-splitting reactions are known as thermolysis, photolysis, and electrolysis respectively and are shown in Figure 5.2a.

Typical electrolysis, as illustrated in Figure 5.2b, is made up of two parts; an electrolyte and an electrode (an anode and a cathode). Based on the physical state of the electrolyte, the water electrolysis reaction can be divided into solid (polymer) electrolyte membrane (PEM) water electrolysis and liquid electrolyte water electrolysis, as described in Figure 5.2b. The PEM water electrolysis process operated in an acidic environment can produce a higher current density (2A cm^{-2}) but it operates at some expensive electrode materials such as Pt-group metals [6]. In addition, PEM electrolyzers show worse long-term stability compared to liquid electrolyte electrochemical cells. Although the liquid electrolyte can only function at a low current density due to the nature of the liquid electrolyte, it allows for cheaper electrode materials instead of noble metals and also provides acceptable chemical and mechanical stability [11, 12]. Based on the liquid electrolyte, water electrolysis reactions are of different types: acidic water electrolysis and alkaline/neutral water electrolysis.

In acidic water electrolyte
At cathode: HER; $2H^+ + e^- \rightarrow H_2$
At anode: OER; $2H_2O \rightarrow O_2 + 4H^+ + e^-$

In neutral/alkaline electrolyte
At cathode: HER; $2H_2O + 2e^- \rightarrow H_2 + 2OH^-$
At anode: OER; $4OH^- \rightarrow 2H_2O + O_2 + 4e^-$

In comparison to alkaline electrolytes, acidic electrolytes allow the substitution of precious metals such as Ir-, Ru-, and Pt-based catalysts with inexpensive transition metals. Further, alkaline electrolytes promote the OER, allowing for significantly higher kinetics compared to acidic electrolytes [13].

5.5 Electrocatalysts

The overall water electrolysis reaction is a combined form of two half-reactions that occur simultaneously called HER and OER at the negative (cathode) and positive (anode), respectively [14, 10, 15]. According to the Nernstian equation at 25°C and 1 atm, the Nernstian potential is described in the normal hydrogen electrode (NHE), decreasing linearly by 59 mV with an increase of each unit of pH [6, 16]. The water electrolysis mechanism highly depends on the pH of the electrolyte. In highly basic/acidic pH conditions, a greater number of available ions facilitate the OER and HER pathways compared to neutral pH, where extra energy is required to generate the ions by breaking up the water molecules [16].

5.5.1 Overpotential

As mentioned earlier, water electrolysis is a combined form of the HER and OER processes. For completion of this two half-reaction, some theoretically applied potentials, 0 V for HER and 1.23 V vs. NHE (at 25°C; 1 atm) for OER are needed, which are considered as the thermodynamic equilibrium potentials. However, practically applied potentials are greater than the thermodynamical equilibrium potentials for both half-reactions [16]. During the electrolysis of water, a few unfavorable factors of electrodes materials, including activation energy, gas bubbles, and ion diffusion, as well as some device-related factors, in which electrolyte concentration, electrode and wire resistance, solution diffusion stoppage, gas bubbles formation on the surface of electrodes, and temperature rise should be taken into account [17, 18]. All these factors result in some extra amount of potential higher than the standard potential needed to be called overpotential. The overpotential is commonly calculated at a current density of 10 mA cm^{-2} to estimate the electrocatalytic activity of a catalyst. However, a major aspect is to choose suitable electrode material which may significantly reduce the overpotential and as a result increase the reaction kinetics and total electrochemical cell efficiency. Sometimes, overpotential at the current density of 10 mA cm^{-2} cannot be used to assess the activity of the electrocatalyst, therefore some other parameters are also needed due to different loading amounts of the catalyst.

5.5.2 Tafel Slope

The Tafel slope is a fundamental parameter for studying the reaction mechanism and kinetics of various electrocatalysts. The Tafel slope represents the rate at which current density changes as the overpotential changes. The Tafel slope is derived from the Tafel plot using the Tafel Equation 5.2 via linear fitting.

$$\eta = a + b \log j \tag{5.2}$$

where b is represented by the Tafel slope between the overpotential (η) and current density (j_o). One other catalytic activity parameter is known as the exchange current density which can be calculated from the Tafel diagram. Generally, the electrocatalyst, which has a lower Tafel slope (b) and higher exchange current density (j_o), means faster kinetics of the reaction and higher activity.

5.5.3 Turnover Frequency (TOF)

TOF is the number of reactants, i.e., H_2O transformed to the desired product (H_2 and O_2) on each catalyst active site per unit time. The value of TOF (in per seconds) can be calculated using Equation 5.3.

$$\mathrm{TOF} = \frac{jS_a}{\alpha Fn} \tag{5.3}$$

where j (mA cm^{-2}) is the current density at a particular overpotential, S_a (cm^{-2}) is the surface area of the working electrode, and α (per mol) is a number of electrons involved in a particular half-reaction for HER and OER, with values of 2 and 4, respectively. F ($Cmol^{-1}$) is the Faraday constant and n (mol) is the catalytic active site.

However, in both half-reactions, OER in the more sluggish and kinetically hindered process because of their $4e^-/4H^+$ transfer process paired with the dissociation of the O–H bond and the formation of the O–O double bond.

5.5.4 Stability

Stability is another significant factor that has a major impact on the practical application of electrocatalysts. The long-term stability of the electrocatalyst is generally studied using two methods. One is by the chronoamperometry curve at a constant potential, where a change in current density was measured over time. The second method is the cyclic voltammetry technique where often 1000 cycles were measured. The catalyst's higher stability shown in the overpotential remains close to the initial polarization curves of cyclic voltammetry. In some cases, the stability in terms of structure, morphology, and composition also changes during the electrochemical process. In the last few years, a range of techniques, including X-ray diffraction (XRD), X-ray photoelectron microscopy (XPS), transmission electron microscopy (TEM), and scanning electron microscopy (SEM), have been used to determine the morphology and composition stability of the catalyst.

5.6 MOF-Based Electrocatalyst for Water Oxidation Reaction

As already discussed, the major challenge in water oxidation is overpotential. Some electrocatalysts are needed to deal with this problem. To date, the best candidate for HER is platinum and for OER are the oxides of iridium and ruthenium [19], but as they are noble metals, issues are associated with them in terms of low abundance and high cost. Earth-abundant elements-based electrocatalysts are the better alternative, as they are less costly and highly abundant [4, 5, 20]. Metal-organic frameworks based on non-precious elements are receiving a lot of attention from researchers because of their astonishing properties, such as large surface areas, high porosity, tunable pore size, facile composition modification, a wide range of morphological forms, and ability to serve as a precursor in synthesizing different composites [5]. The metal-organic framework is defined as the hybrid of crystalline inorganic and organic components linked together by the coordination bond between the metal ions or metal cluster and organic linker [20]. The major advantage of the porosity and high surface area is that the porous framework acts as the microreactor for different types of chemical reactions in which one important reaction is water splitting by using an electrocatalyst. This microreactor not only holds the gas molecules but also provides the catalytic sites in the form of metal nodes and altered ligands after post-synthetic approaches. MOFs can be synthesized using a variety of processes including hydrothermal, ultrasonic, microwave, electrochemical, and mechanochemical methods [5]. Most of the MOFs are not stable in the electrolyte since they generally contain acid or alkali. In acid, the organic linker part of MOFs becomes protonated and in alkali, the metal ions from the metal become hydroxidized [21]. As the framework of MOFs mainly contains carbon, they have low conductivity. In this chapter, we discuss how all the above-mentioned problems have been addressed. We will further focus on the MOFs based on cheap and abundant first-row transition series metal ions, which have been used in the electrocatalytic water-splitting reaction.

5.6.1 Strategies to Design an MOFs-Based Electrocatalyst

The property which must be kept in mind while designing MOFs, which are used as electrocatalysts for water splitting, is their long-term stability, high catalytic activities, and conductivity. As mentioned earlier, most MOFs are naturally insulated which means that their electronic uses are hampered severely [22]. The design techniques of MOF-based electrocatalysts used for water splitting are summarized in Figure 5.3. The presence of vacant coordination sites is a vital strategy in catalyst design. These are induced in the pristine MOFs by post-synthetic approaches like electrochemical oxidation, gas plasma etching, etc. [19]. As a result of the post-synthetic approach, the coordinated unsaturated ions or clusters are formed, which act as sites for catalysis. Heteroatoms such as N, O, S, Se, etc., are introduced into the organic linkers to provide additional catalytic sites. Since the MOFs have a large vacant space they can be incorporated with the guest species like metal ions, polymer chains, etc., and also doped with graphene, carbon nanotubes, MoS_2, etc. to create MOF composites, hence enhancing the conductive property [22]. They can be employed as precursors or templates to change into conductive frameworks via hydrothermal treatment or pyrolysis at high temperatures in an inert atmosphere due to their low conductivity. As the HER and OER proceed on their respective electrodes, the gases evolve to cover the surface of the electrode or reside in the pores, which leads to the blocking of contact between the electrode surface and the reactants. To overcome this problem, super aerophobic functional groups are introduced in as-synthesized MOFs [19].

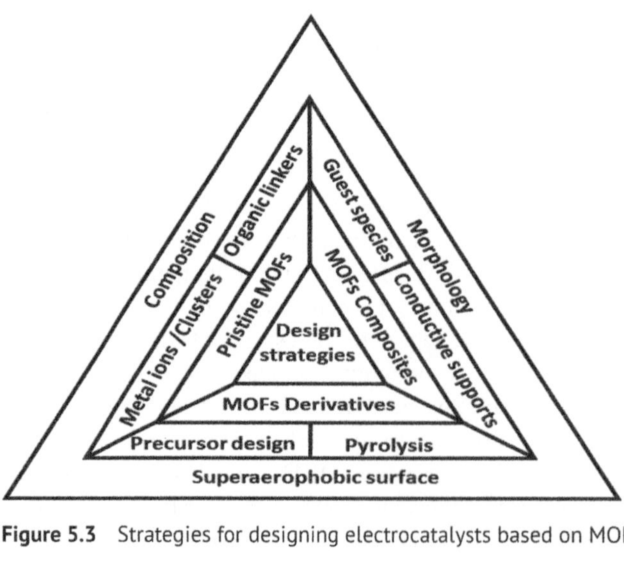

Figure 5.3 Strategies for designing electrocatalysts based on MOF for water splitting.

5.6.2 Pristine MOFs and Their Composites

To date, lots of molecular catalysts are synthesized, which are used in the process of hydrogen production, but the significant challenges associated with them are their low recoverability and recyclability. To deal with these problems, active catalytic sites of molecular catalysts are coordinated with appropriate organic linkers to form MOFs. First-row transitions metal such as Ni, Fe, Mn, etc., mostly behave as the catalytic sites in MOF-based electrocatalysts. Table 5.1 lists the pristine MOFs and their derivatives used as electrocatalysts for water splitting with the crucial electrochemical parameters published in recent years. In 2011, Nohra et al. reported POMOF (Polyoxometalate-based metal-organic frameworks), which were the first pristine MOFs for HER [23]. POMs have certain significant disadvantages of high solubility and poor surface area. $(TBA)_3[PMo_8Mo_4O_{36}(OH)_4Zn_4][C_6H_3(COO)_3]_{4/3}.6H_2O(\varepsilon(trim)_{4/3})$ is a three-dimensional (3D) open framework comprised of keggin units linked by trimesate linker and tetrabutylammonium (TBA) to function as counter ions inhabiting the channels, and showed exceptional catalytic property for HER in highly acidic ($pH = 1$) medium and outperformed platinum, but electrochemical studies were insufficient to demonstrate its catalytic activity and efficiency. Following Nohra, Qin et al. in 2015 synthesized two novel POMOFs-based electrocatalysts, NENU-500 and NENU-501, with ultra-stability in acidic conditions for HER [24]. NENU-500 shows the least onset potential of 180 mV with the Tafel slope of 96 mV dec^{-1} and 237 mV overpotential at 10 mA cm^{-2} current density among POM-based MOFs. NENU-500, 501 maintains selectrocatalytic activity up to 2000 cycles.

The more the exposed catalytic sites, the more will be the catalytic activity, therefore the design of nano two- or three-dimensional MOFs result in a large surface area for HER. Marinescu et al. effectively synthesized two MOFs having two dimensions (2D), THTCo and

Table 5.1 Representation of important parameters of pristine MOF and their derivatives.

Sample ID	Application	Electrolyte	Overpotential (mV) at current density 10 mA cm^{-2})	Tafel slope (mV dec^{-1})	References
AB and CTGU-	HER	0.5 M H_2SO_4	44	45	[32]
AB and CTGU-9	HER	0.5 M H_2SO_4	128	87	[33]
HUST-200	HER	0.5 M H_2SO_4	131	51	[34]
HUST-201	HER	0.5 M H_2SO_4	192	79	[34]
UiO-66-NH_2-Mo-5	HER	0.5 M H_2SO_4	200	59	[35]
NENU-500	HER	0.5 M H_2SO_4	237	96	[24]
NU-1000-Ni-S	HER	0.1 M HCl	238	120	[27]
THTA-Co	HER	0.5 M H_2SO_4	283	71	[36]
THTNi 2DSP	HER	0.5 M H_2SO_4	333	81	[26]
NENU-501	HER	0.5 M H_2SO_4	392	137	[24]
BHT-Co	HER	0.5 M H_2SO_4	340	108	[25]
BTSe-Co	HER	0.5 M H_2SO_4	343	97	[37]

(Continued)

Table 5.1 (Continued)

Sample ID	Application	Electrolyte	Overpotential (mV) at current density 10 mA cm^{-2})	Tafel slope (mV dec^{-1})	References
THTA-Ni	HER	0.5 M H_2SO_4	315	76	[36]
MAF-X27-OH	OER	1 M KOH	292	88	[38]
Fe/Ni-BTC	OER	0.1 M KOH	270	47	[39]
NiCo-UMOFNs	OER	1 M KOH	250	42	[29]
Fe3-Co_2	OER	0.1 M KOH	225	48	[40]
NiFe-UMNs	OER	1 M KOH	260	30	[41]
Fe: 2D-Co-NS	OER	0.1 M KOH	211	46	[42]
NiPc-MOF	OER	1 M KOH	250	74	[43]
Fe/$Ni_{2.4}$/$Mn_{0.4}$-MIL-53	OER	1 M KOH	219	52	[44]
Ni-Fe-MOFNSs	OER	1 M KOH	221	56	[45]
Ni-MOF@Fe-MOF	OER	1 M KOH	265	82	[46]
CTGU-10C_2	OER	0.1 M KOH	240	58	[15]
NiFe-MOF-74	OER	1 M KOH	223	72	[47]
$Co_{0.6}$Fe-MOF-74	OER	1 M KOH	280	56	[48]
NiFe-NFF	OER	1 M KOH	227	39	[49]
CoBDC-Fc-NF	OER	1 M KOH	178	51	[50]
HE-MOF-RT	OER	1 M KOH	245	54	[51]
CoFe-LM-160-12(BDC)	OER	1 M KOH	274	47	[52]

BHTCo, with pore size and thickness of almost 2 nm and 360 nm, respectively, by mixing Co(II) with Benzene-1,2,3,4,5,6-hexathiol (BHT) and Triphenylene-2,3,6,7,10,11-hexathiolate (THT) [25]. Two-dimensional BHTCo and THTCo demonstrate overpotentials of 340 and 530 mV, respectively, in an acidic environment (pH = 1.3) to attain the qualifying current density of 10 mA cm^{-2} with a Tafel slope of 149 and 189 mV dec^{-1} in less acidic conditions (pH = 4.2). Both BHTCo and THTCo exhibit specific electrocatalytic HER performance attributed to different active catalytic sites or centers. However, the major flaw linked with these 2D MOFs is their finite lateral dimension (<5 mm) and large layer thickness. Following Marinescu's findings, Feng et al. used the Langmuir Blodgett (LB) method to synthesize 2D THTNi MOF, which has a large surface area and 0.7–0.9 mm thickness [26]. The HER activity was tested in 0.5 H_2SO_4 solutions giving an overpotential of 333 mV for 10 mA cm^{-2} current density having a Tafel slope of 81 mV dec^{-1}. According to the author, a large surface area and single-layered structure were responsible for the high electrocatalytic HER activity, because it led to exposing most of the active sites for the HER process.

A useful technique for achieving a notable improvement in performance, in addition to changing the electrocatalyst composition, is where a porous supportive framework is used to create a composite electrode. In Figure 5.4a, the NU-1000 crystal structure is shown with its nodes and linker. Hexagonal and triangular 1D channels with 31 and 10 Å diameters, respectively lead to high porosity with 71% void volume. In Figure 5.4b, deposition of Ni-S on FTO is represented. On mixing the MOF and inorganic substance (Ni-S), a hybrid electrocatalyst containing the Ni-S flat layer on the (FTO and below the Zr-based MOF (NU-1000) rods on the FTO parts was created [27]. In contrast to expectation, MOF did not act as a physical template but it altered the chemical environment near the sulfide catalyst. As a result, local proton delivery and transportation have improved. This improvement is responsible for its high catalytic activity, which is 238 mV overpotential for 10 mA cm^{-2} with a 120 mV dec^{-1} Tafel slope value. Figure 5.4c shows the current density vs. potential plot of different substances in 0.1 HCl and Figure 5.4d shows Tafel plots of four different electrodes.

In comparison to HER, OER has sluggish kinetics due to the involvement of 4 electrons during the process. It is widely known that transition metals show great activity toward water oxidation. MOF based on the transition metals like Ni, Fe, Co, etc. and their composites were mostly synthesized.

Wang et al. synthesized an MOF based on a zeolitie imidazolate framework (ZIF) Co-ZIF-9, which carries out water oxidation over a wide pH range [28]. By DFT calculation, it was depicted that water molecules were activated on binding to the OH- group with metal sites, while the proton that had been removed was absorbed by a neighboring benzimidazolate in the framework. But the better way is to create the coordinatively unsaturated metal sites (CUMS) to boost the OER performance. Li et al. succeeded

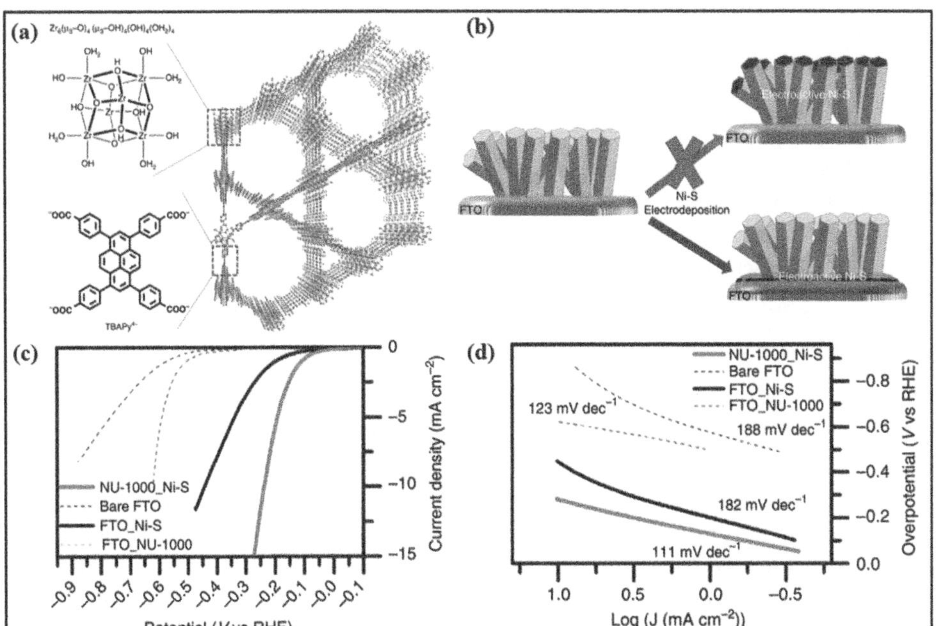

Figure 5.4 (a) Crystal structure of NU-1000; (b) electrodeposition of Ni-S on FTO (conductive surface) against expectation; and (c, d) electrocatalytic HER performance was compared. M. D., PETERS et al., 2015 / Springer Nature / CC BY-4.0.

in producing the CUMS by dielectric barrier discharge (DBD) plasma etching in ZIF-67. ZIF-67 with CUMS shows an overpotential of 320 mV for 10 mA cm^{-2}. The reason for the high OER activity is the knocking down of the ligand from the metal(Co) sites, which carry the high positive charges. They also have significant interaction with oxygen, allowing the adsorption of O_2, O, OH, and OOH species as well as the creation of active sites for OER.

Zhao et al. reported the creation of a 2D nanostructured OER electrocatalyst based on MOFs to enhance the exposure of the active sites and regulate them with other metal ions [29]. Ultrathin Ni-Co MOF nanosheets (NiCo-UMOFN) showed the overpotential value of 250 mV at 10 mA cm^{-2}, which was significantly lower in comparison to bulky Ni/Co-UMOFNs and commercial RuO_2, and they had long-term catalytic durability. According to the author, the exposure of the dominant active centers, coordinative unsaturation of the active metal sites, and synergistic effect between cobalt and nickel improved the OER performance. With the help of powder X-ray diffraction (PXRD) and SEM, the crystal structure of NiCo-UMOFNs was obtained. Ni and Co were octahedrally linked by oxygen atoms and these octahedral units were joined together in the (200) plane along the [010]/[001] direction to generate a 2D bimetal layer separated by BDC linkers. (200) crystallographic planes in (NiCo-UMOFNs) were the most exposed plane. The potential region from 1.25 to 1.38 V is endowed with the oxidation of Ni^{2+} to Ni^{3+}. NiCo-UMOFNs have the least value overpotential (250 mV @ 10 mA cm^{-2}) and Tafel slope of just 42 mV dec^{-1} among other MOFs.

Sun et al. grew NiFe-based MOF nanosheets on the NF (nickel foam) by the solvothermal technique [30]. The advantage of as-synthesized MIL-53(NiFe)/NF is that it can be utilized as an electrode without using a binder. It displays outstanding stability and remarkable OER activity having an overpotential 233 mV for 50 mA cm^{-2} current density with a Tafel value of just 31.3 mV dec^{-1} in 1 M KOH solution. The high OER activity was attributed to the increase in ECSA, reaction sites, enhancing electron transport capabilities, and altering the electronic structures by introducing the Fe atoms.

MOFs with a large vacant space are often good hosts, able to encapsulate the guest species. Manna et al. were able to encapsulate the Co(II) complex $[Co^{II}(H2O)_4(DMF)_2]^{2+}$ in the void of MOF $[\{Co_3(\mu_3\text{-}OH)(BTB)_2(dpe)_2\}^{1-}]$ where BTB is benzene 1,3,5 tribenzoic acid and dpe is 1,2-di(4-pyridyl)ethylene [31]. The hybrid formed had the overpotential of 390 mV for 1 mA cm^{-2} in 0.1 KOH solution.

5.6.3 MOF Derivatives

We see that MOFs have great properties like high surface area and tailorable pore size, easily functionalized by doping with heteroatoms but major drawbacks associated with the pristine MOFs are low conductivity and less stability, which limit their use as electrocatalysts for water splitting. Due to these drawbacks, MOFs are used as precursors or templates to transform them into MOF derivatives that have a similar fundamental morphology. Further, the direct growth of distinctive morphologies (nanorods, nanosheets, etc.) on conductive substrates (carbon cloths, carbon paper, Cu foam, Ni foam, Ti foil) can dramatically improve conductivity as well as charge/mass transfer efficiency, thereby improving water-splitting output [5]. A list of MOF-derived materials with their crucial parameters such as overpotential and Tafel slope is presented in Table 5.2.

Table 5.2 Representation of electrocatalytic parameters of MOF and their derivative.

Sample ID	Application	Electrolyte	Overpotential @ current density (mV@mA cm^{-2})	Tafel (mV dec^{-1})	References
$Co_{0.59}Fe_{0.41}P$	HER	0.5 M H_2SO_4	72	52	[58]
CoP@BCN-1	HER	0.5 M H_2SO_4	87	46	[59]
A-Ni-C	HER	0.5 M H_2SO_4	34	41	[60]
MoC@GS(700)	HER	0.5 M H_2SO_4	124	43	[61]
Co@BCN	HER	0.5 M H_2SO_4	96	64	[62]
N-C/Ni-C/Ni-F	HER	1.0 M KOH	8	24	[63]
WC@NPC	HER	0.5 M H_2SO_4	51	49	[53]
Co@N-CNTs@ rGO	HER	0.5 M H_2SO_4	87	52	[64]
IrCo@NC-500	HER	0.5 M H_2SO_4	24	23	[65]
Cu_3P@NPPC-650	HER	0.5 M H_2SO_4	89	76	[66]
Fe_3C/Mo_2C@ NPGC	HER	0.5 M H_2SO_4	98	45	[67]
Co_3O_4C-NA	OER	1.0 KOH	240	61	[68]
Co_3O_4/N-PC	OER	0.1 M KOH	290	72	[69]
NF@ NC-$CoFe_2O_4$/C	OER	1.0 M KOH	240	45	[70]
Co_3O_4/NSC-50	OER	1.0 M KOH	260	70	[71]
NC@CoO/CuO	OER	0.5 M KOH	270	67	[72]
Co_3O_4/Co-Fe Oxide	OER	1.0 M KOH	297	61	[73]
NiCo/Fe_3O_4/ MOF-74	OER	1 M KOH	238	29	[74]
Ni-Fe-Se cages	OER	1 M KOH	240	24	[75]
$NiCoS_2$/CFP	OER	1 M KOH	270	79	[76]
NiFe@NCX	OER	0.1 M KOH	230	61	[77]
H_3LCoCN800	OER	1 M KOH	215	64	[78]
ZnCo-Fe-20	OER	1 M KOH	176	68	[79]
MoO_2@PC-RGO	HER	0.5 M H_2SO_4	64	41	[80]
HC800	HER	1 M KOH	123	84	[81]
MoCN-3D	HER	0.5 M H_2SO_4	89	51.4	[82]
WC@NPC	HER	0.5 M H_2SO_4	51	49	[53]
NiCoP/C	OER	1 M KOH	330	96	[83]
NCNTFs	OER	1 M KOH	370	93	[57]
CC@$NiCo_2O_4$	OER	1 M KOH	340	72	[84]
MoP@NPC/rGO	HER	0.5 M H_2SO_4	218	57	[85]
Mo_2C/C	HER	1 M KOH	165	66.41	[86]

(Continued)

Table 5.2 (Continued)

Sample ID	Application	Electrolyte	Overpotential @ current density (mV@mA cm^{-2})	Tafel (mV dec^{-1})	References
Ni_2P@C/G	OER	1 M KOH	285	44	[87]
$NiCo_2O_4$/NiO-rGO	OER	1 M KOH	340	66	[88]
CoP@GC	OER	1 M KOH	345	56	[89]
Co@N-CNTs@rGO	HER	0.5 M H_2SO_4	87	52	[64]
Zn -doped $CoSe_2$	OER	1 M KOH	356	88	[90]
CeO_x/CoS	OER	1 M KOH	269	50	[91]
CeO_x/CoS/CC	OER	1 M KOH	232	-	[91]
Co-Fe-P-1.7	OER	1 M KOH	260	58	[92]
CNT-NC-CoP	OER	1 M KOH	251	82.1	[93]
CoP/Mo_2C-NC	HER	0.5 M H_2SO_4	55.7	49	[94]
CoPS/N-C	HER	0.5 M H_2SO_4	80	68	[95]
Co_3O_4@CoNCNT	HER	0.5 M H_2SO_4	171	121	[96]

By using a cage confinement pyrolysis approach, Xu et al. synthesized ultrasmall WC nanoparticles (2 nm) shielded by or embedded in N-doped nanoporous carbon (denoted as WC@NPC) from a zeolitic metal azolate framework MAF-6, which served as an outstanding electrocatalyst for HER [53]. It attained a 10 mA cm^{-2} current density in a 0.5 M sulfuric acid (H_2SO_4) solution with just an overpotential of 51 mV and a Tafel slope of 49 mV dec^{-1}. Further, throughout the long-term electrolytic procedure, it showed excellent durability. According to the DFT simulation, the enhanced HER performance was due to the presence of ultrasmall WC NPs and the synergetic effect of graphene shells.

Similarly, a nickel carbon-based catalyst (A-Ni-C) was synthesized by the carbonization of nickel-based MOF which was followed by HCl leaching, CV, and constant potential treatments [54]. The A-Ni-C appeared as hollow onion-like nanoshells and the Ni metal was enclosed in the graphene sheet. It shows the overpotential of 34 mV for 10 mA cm^{-2} with a Tafel slope value of 41 mV dec^{-1} in 0.5 H_2SO_4 solutions with stability of about 25 h

Later, Yang et al. proposed an MOF-assisted approach including the use of Zr-UIO-66 MOF as the host, followed by chemical vapour deposition (CVD) of MoO_3 into the structure to generate ultrafine MoO_2 (~3 nm) coated in porous carbon (MoO_2@PC) after pyrolysis and acid processing [55]. Further, MoO_2@PC was used as a precursor to obtain Mo_2C@PC, MoN@PC, and MoP@PC. MoP@PC has the overpotential of 153 mV for 10 mA cm^{-2} with a Tafel slope of 66 mV dec^{-1} and high durability. The extraordinary HER activity was due to the high accessibility of the surface area and the presence of small nanoparticles.

Aijaz et al. synthesized core-shell Co@Co_3O_4 nanoparticles enclosed in CNT-grafted N-doped carbon polyhedrons as an improved bifunctional catalyst (for OER and ORR)

[56]. This electrocatalyst was made by pyrolyzing ZIF-67 in a reduced atmosphere and then oxidative calcining it under controlled conditions. It has the overpotential of 420 mV for 10 mA cm^{-2} current density surpassing commercial RuO_2, and exceptionally long durability for OER in 0.1 M KOH.

Xia et al. gave a general way for synthesizing hollow frameworks of N-doped CNT obtained by pyrolysis and acid processing of a Co-based ZIF-67 framework. ZIF-67 particles were the source of nitrogen and carbon in the formation of Nitrogen-doped carbon nanotubes (NCNTs) [57]. It had an overpotential of 380 mV for 10 mA cm^{-2} existing density with a small value of Tafel slope of 93 mV dec^{-1} and possessed high stability. The exceptional OER capabilities were mostly due to the synergistic effect of elemental composition and the tough hollow-like structure of the linked crystalline N-doped CNTs.

5.7 Challenges and Future Prospects

The major challenges and future prospects of MOFs used as electrocatalysts for water splitting are described as follows:

1) The density of active sites and the shape of the electrocatalyst have a significant impact on electrocatalytic performance. As a result, the composition and specific surface area of the formulated non-precious nanoparticles should be given additional attention [97].
2) Many factors can affect the electrocatalytic OER performance, including electrolytes, catalyst loadings, material morphology, supporting substrates, and electrochemical measuring methodologies. An objective third-party or a benchmarking approach should be used to objectively evaluate the electrocatalysts' OER performance [98, 99, 100].
3) Even though multiple non-noble-metal-based OER electrocatalysts have been created using various ways for several years, there is still a long way to go before OER electrocatalysts are mass marketed as practical devices. Shortly, the most important focus will be on developing novel materials as non-noble-metal-based OER electrocatalysts [99].
4) *A better understanding of the OER/HER reaction mechanism*: Targeted efforts for designing efficient OER/HER catalysts would require a better knowledge of the electrode-electrolyte contact and the associated dynamics [101].
5) *Identify the active site of the OER catalyst*: To boost the activity of an electrocatalyst, one can either increase the number of active sites on the electrode (e.g., by nanostructuring to manifestmore active sites) or increase the intrinsic activity of each active site.
6) *Mass production of the electrocatalyst*: There must be a continuum from academic research to industrial fabrication. There is a need for a scalable technology that can produce low-cost and efficient catalytic materials on a wide scale [101].
7) Currently, non-precious metal-based electrocatalysts operate in alkaline conditions. Protecting OER active centers with structurally stable components (e.g., perovskites and Mn oxides) can allow for the usage of earth-abundant OER electrocatalysts in acidic conditions [102].

8) The multi-electron transfer is involved in the electrochemical OER process. Therefore, the conductivity of the materials has a significant impact on OER performance. In reality, catalytic materials with a high number of active sites but limited conductivity may perform poorly overall. Incorporating the metal framework and the carbon materials as well as a strong interaction between the reactive species and the conductive host, are efficient ways to boost electrocatalyst conductivity [103, 104].
9) Water electrolysis is severely hampered by the unavailability of efficient bifunctional electrocatalysts. The development of highly effective bifunctional ORR/OER electrocatalysis is required to be pushed forward by a mix of theoretical and experimental studies, as well as the application of *in situ/in operando* characterization approaches [101, 99].
10) The temperature and conditions of pyrolysis are also critical factors in defining the form and composition of the carbon nanocomposite derived from the MOFs. The development of next-generation substances with improved activity for the marketing of energy devices such as fuel cells and metal-air batteries could be aided by the controlled fabrication of customized MOFs [105].

5.8 Conclusions

Although there are many flaws and challenges associated with MOF-based electrocatalysts, there is no ambiguity about their potential to be used in the production of green hydrogen without emission of carbon. Although many pristine MOFs have low electronic conductivity and stability, by doping or synthesizing the MOF composites/derived materials, this problem can also be solved. By doping with heteroatoms or with other metals, the catalytic active sites increase, and more exposure leads to improved performance in the water-splitting process. This chapter provides the basic understanding of water splitting, MOFs, and their derivatives as electrocatalysts for OER and HER processes.

References

1 Abe, J.O., Popoola, A., Ajenifuja, E. et al. (2019). Hydrogen energy, economy and storage: Review and recommendation. *International Journal of Hydrogen Energy* 44: 15072–15086.
2 Afgan, N. and Veziroglu, A. (2012). Sustainable resilience of hydrogen energy system. *International Journal of Hydrogen Energy* 37: 5461–5467.
3 Dunn, B., Kamath, H., and Tarascon, J.-M. (2011). Electrical energy storage for the grid: A battery of choices. *Science* 334: 928–935.
4 Roger, I. and Symes, M.D. (2016). First row transition metal catalysts for solar-driven water oxidation produced by electrodeposition. *Journal of Materials Chemistry A* 4: 6724–6741.
5 Tan, J.-B. and Li, G.-R. (2020). Recent progress on metal–organic frameworks and their derived materials for electrocatalytic water splitting. *Journal of Materials Chemistry A* 8: 14326–14355.
6 Li, X., Hao, X., Abudula, A. et al. (2016b). Nanostructured catalysts for electrochemical water splitting: Current state and prospects. *Journal of Materials Chemistry A* 4: 11973–12000.

7 Song, F., Bai, L., Moysiadou, A. et al. (2018). Transition metal oxides as electrocatalysts for the oxygen evolution reaction in alkaline solutions: An application-inspired renaissance. *Journal of the American Chemical Society* 140: 7748–7759.

8 Du, X., Fu, J., and Zhang, X. (2018). $NiCo_2O_4$@$NiMoO_4$ supported on nickel foam for electrocatalytic water splitting. *ChemCatChem* 10: 5533–5540.

9 Sljukic Paunkovic, B., Santos, D., Vujković, M. et al. (2016). Molybdenum carbide nanoparticles on carbon nanotubes and carbon xerogel: Low-cost cathodes for hydrogen production by alkaline water electrolysis. *ChemSusChem* 9: 1200–1208.

10 Yano, J. and Yachandra, V. (2014). Mn_4Ca cluster in photosynthesis: Where and how water is oxidized to dioxygen. *Chemical Reviews* 114: 4175–4205.

11 Chowdhury, D.R., Spiccia, L., Amritphale, S.S. et al. (2016). A robust iron oxyhydroxide water oxidation catalyst operating under near neutral and alkaline conditions. *Journal of Materials Chemistry A* 4: 3655–3660.

12 Garcés-pineda, F.A., Blasco-Ahicart, M., Nieto-Castro, D. et al. (2019). Direct magnetic enhancement of electrocatalytic water oxidation in alkaline media. *Nature Energy* 4: 519–525.

13 Yan, Y., Xia, B.Y., Zhao, B. et al. (2016). A review on noble-metal-free bifunctional heterogeneous catalysts for overall electrochemical water splitting. *Journal of Materials Chemistry A* 4: 17587–17603.

14 Chen, M., Wu, Y., Han, Y. et al. (2015). An iron-based film for highly efficient electrocatalytic oxygen evolution from neutral aqueous solution. *ACS Applied Materials & Interfaces* 7: 21852–21859.

15 Zhou, W., Huang, D.D., Wu, Y.P. et al. (2019). Stable hierarchical bimetal–organic nanostructures as high performance electrocatalysts for the oxygen evolution reaction. *Angewandte Chemie International Edition* 58: 4227–4231.

16 Peng, J., Dong, W., Wang, Z. et al. (2020). Recent advances in 2D transition metal compounds for electrocatalytic full water splitting in neutral media. *Materials Today Advances* 8: 100081.

17 Anantharaj, S., Ede, S.R., Karthick, K. et al. (2018). Precision and correctness in the evaluation of electrocatalytic water splitting: Revisiting activity parameters with a critical assessment. *Energy and Environmental Science* 11: 744–771.

18 Anantharaj, S., Ede, S.R., Sakthikumar, K. et al. (2016). Recent trends and perspectives in electrochemical water splitting with an emphasis on sulfide, selenide, and phosphide catalysts of Fe, Co, and Ni: A review. *ACS Catalysis* 6: 8069–8097.

19 Yan, Y., He, T., Zhao, B. et al. (2018). Metal/covalent–organic frameworks-based electrocatalysts for water splitting. *Journal of Materials Chemistry A* 6: 15905–15926.

20 Lee, Y.-R., Kim, J., and Ahn, W.-S. (2013). Synthesis of metal-organic frameworks: A mini review. *Korean Journal of Chemical Engineering* 30: 1667–1680.

21 Mahmood, A., Guo, W., Tabassum, H. et al. (2016). Metal-organic framework-based nanomaterials for electrocatalysis. *Advanced Energy Materials* 6: 1600423.

22 Meng, H., Han, Y., Zhou, C. et al. (2020). Conductive metal–organic frameworks: Design, synthesis, and applications. *Small Methods* 4: 2000396.

23 Nohra, B., El Moll, H., Rodriguez Albelo, L.M. et al. (2011). Polyoxometalate-based metal organic frameworks (POMOFs): Structural trends, energetics, and high electrocatalytic efficiency for hydrogen evolution reaction. *Journal of the American Chemical Society* 133: 13363–13374.

24 Qin, J.-S., Du, D.-Y., Guan, W. et al. (2015). Ultrastable polymolybdate-based metal–organic frameworks as highly active electrocatalysts for hydrogen generation from water. *Journal of the American Chemical Society* 137: 7169–7177.

25 Clough, A.J., Yoo, J.W., Mecklenburg, M.H. et al. (2015). Two-dimensional metal–organic surfaces for efficient hydrogen evolution from water. *Journal of the American Chemical Society* 137: 118–121.

26 Dong, R., Pfeffermann, M., Liang, H. et al. (2015). Large-area, free-standing, two-dimensional supramolecular polymer single-layer sheets for highly efficient electrocatalytic hydrogen evolution. *Angewandte Chemie International Edition* 54: 12058–12063.

27 Hod, I., Deria, P., Bury, W. et al. (2015). A porous proton-relaying metal-organic framework material that accelerates electrochemical hydrogen evolution. *Nature Communications* 6: 1–9.

28 Wang, S., Hou, Y., Lin, S. et al. (2014). Water oxidation electrocatalysis by a zeolitic imidazolate framework. *Nanoscale* 6: 9930–9934.

29 Zhao, S., Wang, Y., Dong, J. et al. (2016). Ultrathin metal–organic framework nanosheets for electrocatalytic oxygen evolution. *Nature Energy* 1: 1–10.

30 Sun, F., Wang, G., Ding, Y. et al. (2018). NiFe-based metal–organic framework nanosheets directly supported on nickel foam acting as robust electrodes for electrochemical oxygen evolution reaction. *Advanced Energy Materials* 8: 1800584.

31 Manna, P., Debgupta, J., Bose, S. et al. (2016). A mononuclear coii coordination complex locked in a confined space and acting as an electrochemical water-oxidation catalyst: A "ship-in-a-bottle" approach. *Angewandte Chemie* 128: 2471–2476.

32 Wu, Y., Zhou, W., Zhao, J. et al. (2017). Surfactant-assisted phase-selective synthesis of new cobalt MOFs and their efficient electrocatalytic hydrogen evolution reaction. *Angew. Chemie-Int. Ed.* 56: 13001–13005. *Angewandte Chemie* 129: 13181–13185.

33 Zhou, W., Wu, Y.-P., Wang, X. et al. (2018). Improved conductivity of a new Co(II)-MOF by assembled acetylene black for efficient hydrogen evolution reaction. *CrystEngComm* 20: 4804–4809.

34 Zhang, L., Li, S., Gomez-garcia, C.J. et al. (2018b). Two novel polyoxometalate-encapsulated metal–organic nanotube frameworks as stable and highly efficient electrocatalysts for hydrogen evolution reaction. *ACS Applied Materials and Interfaces* 10: 31498–31504.

35 Dai, X., Liu, M., Li, Z. et al. (2016). Molybdenum polysulfide anchored on porous Zr-metal organic framework to enhance the performance of hydrogen evolution reaction. *The Journal of Physical Chemistry C* 120: 12539–12548.

36 Dong, R., Zheng, Z., Tranca, D.C. et al. (2017). Immobilizing molecular metal dithiolene–diamine complexes on 2D metal–organic frameworks for electrocatalytic H_2 production. *Chemistry – A European Journal* 23: 2255–2260.

37 Jiang, J., Huang, L., Liu, X. et al. (2017). Bioinspired cobalt–citrate metal–organic framework as an efficient electrocatalyst for water oxidation. *ACS Applied Materials & Interfaces* 9: 7193–7201.

38 Lu, X.-F., Liao, P.-Q., Wang, J.-W. et al. (2016). An alkaline-stable, metal hydroxide mimicking metal–organic framework for efficient electrocatalytic oxygen evolution. *Journal of the American Chemical Society* 138: 8336–8339.

39 Wang, L., Wu, Y., Cao, R. et al. (2016). Fe/Ni metal–organic frameworks and their binder-free thin films for efficient oxygen evolution with low overpotential. *ACS Applied Materials and Interfaces* 8: 16736–16743.

40 Shen, J.-Q., Liao, P.-Q., Zhou, D.-D. et al. (2017). Modular and stepwise synthesis of a hybrid metal-organic framework for efficient electrocatalytic oxygen evolution. *Journal of the American Chemical Society* 139: 1778–1781.

41 Hai, G., Jia, X., Zhang, K. et al. (2018). High-performance oxygen evolution catalyst using two-dimensional ultrathin metal-organic frameworks nanosheets. *Nano Energy* 44: 345–352.

42 Huang, J., LI, Y., Huang, R.K. et al. (2018). Electrochemical exfoliation of pillared-layer metal–organic framework to boost the oxygen evolution reaction. *Angewandte Chemie* 130: 4722–4726.

43 Jia, H., Yao, Y., Zhao, J. et al. (2018). A novel two-dimensional nickel phthalocyanine-based metal–organic framework for highly efficient water oxidation catalysis. *Journal of Materials Chemistry A* 6: 1188–1195.

44 Li, F.L., Shao, Q., Huang, X. et al. (2018a). Nanoscale trimetallic metal–organic frameworks enable efficient oxygen evolution electrocatalysis. *Angewandte Chemie* 130: 1906–1910.

45 Li, F.L., Wang, P., Huang, X. et al. (2019). Large-scale, bottom-up synthesis of binary metal–organic framework nanosheets for efficient water oxidation. *Angewandte Chemie* 131: 7125–7130.

46 Rui, K., Zhao, G., Chen, Y. et al. (2018). Hybrid 2D dual-metal–organic frameworks for enhanced water oxidation catalysis. *Advanced Functional Materials* 28: 1801554.

47 Xing, J., Guo, K., Zou, Z. et al. (2018). *In situ* growth of well-ordered NiFe-MOF-74 on Ni foam by Fe_{2+} induction as an efficient and stable electrocatalyst for water oxidation. *Chemical Communications* 54: 7046–7049.

48 Zhao, X., Pattengale, B., Fan, D. et al. (2018). Mixed-node metal–organic frameworks as efficient electrocatalysts for oxygen evolution reaction. *ACS Energy Letters* 3: 2520–2526.

49 Cao, C., Ma, D.D., Xu, Q. et al. (2019). Semisacrificial template growth of self-supporting MOF nanocomposite electrode for efficient electrocatalytic water oxidation. *Advanced Functional Materials* 29: 1807418.

50 Xue, Z., Liu, K., Liu, Q. et al. (2019). Missing-linker metal-organic frameworks for oxygen evolution reaction. *Nature Communications* 10: 1–8.

51 Zhao, X., Xue, Z., Chen, W. et al. (2019). Ambient fast, large-scale synthesis of entropy-stabilized metal–organic framework nanosheets for electrocatalytic oxygen evolution. *Journal of Materials Chemistry A* 7: 26238–26242.

52 Cai, M., Liu, Q., Xue, Z. et al. (2020). Constructing 2D MOFs from 2D LDHs: A highly efficient and durable electrocatalyst for water oxidation. *Journal of Materials Chemistry A* 8: 190–195.

53 Xu, Y.-T., Xiao, X., Ye, Z.-M. et al. (2017). Cage-confinement pyrolysis route to ultrasmall tungsten carbide nanoparticles for efficient electrocatalytic hydrogen evolution. *Journal of the American Chemical Society* 139: 5285–5288.

54 Fang, L.J., Wang, X.L., Li, Y.H. et al. (2017). Nickel nanoparticles coated with graphene layers as efficient co-catalyst for photocatalytic hydrogen evolution. *Applied Catalysis B: Environmental* 200: 578–584.

55 Yang, J., Zhang, F., Wang, X. et al. (2016). Porous molybdenum phosphide nano-octahedrons derived from confined phosphorization in UIO-66 for efficient hydrogen evolution. *Angewandte Chemie* 128: 13046–13050.

56 Aijaz, A., Masa, J., Rösler, C. et al. (2016). Co@Co_3O_4 encapsulated in carbon nanotube-grafted nitrogen-doped carbon polyhedra as an advanced bifunctional oxygen electrode. *Angewandte Chemie International Edition* 55: 4087–4091.

57 Xia, B.Y., Yan, Y., Li, N. et al. (2016). A metal–organic framework-derived bifunctional oxygen electrocatalyst. *Nature Energy* 1: 1–8.

58 Hao, J., Yang, W., Zhang, Z. et al. (2015). Metal–organic frameworks derived Co_xFe-xP nanocubes for electrochemical hydrogen evolution. *Nanoscale* 7: 11055–11062.

59 Tabassum, H., Guo, W., Meng, W. et al. (2017). Hydrogen evolution: Metal–organic frameworks derived cobalt phosphide architecture encapsulated into B/N Co-doped graphene nanotubes for all pH value electrochemical hydrogen evolution. *Advanced Energy Materials* 7(9): 1601671.

60 Fan, L., Liu, P.F., Yan, X. et al. (2016). Atomically isolated nickel species anchored on graphitized carbon for efficient hydrogen evolution electrocatalysis. *Nature Communications* 7: 1–7.

61 Shi, Z., Wang, Y., Lin, H. et al. (2016). Porous nanoMoC@graphite shell derived from a MOFs-directed strategy: An efficient electrocatalyst for the hydrogen evolution reaction. *Journal of Materials Chemistry A* 4: 6006–6013.

62 Zhang, H., Ma, Z., Duan, J. et al. (2016a). Active sites implanted carbon cages in core–shell architecture: Highly active and durable electrocatalyst for hydrogen evolution reaction. *ACS Nano* 10: 684–694.

63 Lu, X., Tan, X., Wang, D.W. et al. (2017a). Nitrogen-doped carbon nanosheets coupled nickel–carbon pyramid arrays toward efficient evolution of hydrogen. *Advanced Sustainable Systems* 1: 1700032.

64 Chen, Z., Wu, R., Liu, Y. et al. (2018). Ultrafine Co nanoparticles encapsulated in carbon-nanotubes-grafted graphene sheets as advanced electrocatalysts for the hydrogen evolution reaction. *Advanced Materials* 30: 1802011.

65 Jiang, P., Chen, J., Wang, C. et al. (2018). Tuning the activity of carbon for electrocatalytic hydrogen evolution via an iridium-cobalt alloy core encapsulated in nitrogen-doped carbon cages. *Advanced Materials* 30: 1705324.

66 Wang, R., Dong, X.Y., Du, J. et al. (2018a). MOF-derived bifunctional Cu_3P nanoparticles coated by a N, P-codoped carbon shell for hydrogen evolution and oxygen reduction. *Advanced Materials* 30: 1703711.

67 Li, J.-S., Tang, Y.-J., Liu, C.-H. et al. (2016a). Polyoxometalate-based metal–organic framework-derived hybrid electrocatalysts for highly efficient hydrogen evolution reaction. *Journal of Materials Chemistry A* 4: 1202–1207.

68 Ma, T.Y., Dai, S., Jaroniec, M. et al. (2014). Metal–organic framework derived hybrid Co_3O_4-carbon porous nanowire arrays as reversible oxygen evolution electrodes. *Journal of the American Chemical Society* 136: 13925–13931.

69 Hou, Y., Li, J., Wen, Z. et al. (2015). Co_3O_4 nanoparticles embedded in nitrogen-doped porous carbon dodecahedrons with enhanced electrochemical properties for lithium storage and water splitting. *Nano Energy* 12: 1–8.

70 Lu, X.F., Gu, L.F., Wang, J.W. et al. (2017b). Bimetal-organic framework derived $CoFe_2O_4$/C porous hybrid nanorod arrays as high-performance electrocatalysts for oxygen evolution reaction. *Advanced Materials* 29: 1604437.

71 Khalid, M., Honorato, A.M., Ticianelli, E.A. et al. (2018). Uniformly self-decorated Co_3O_4 nanoparticles on N, S co-doped carbon layers derived from a camphor sulfonic acid and metal–organic framework hybrid as an oxygen evolution electrocatalyst. *Journal of Materials Chemistry A* 6: 12106–12114.

72 Kuang, X., Wang, Z., Sun, X. et al. (2018). Metal oxide-and N-codoped carbon nanosheets: Facile synthesis derived from MOF nanofibers and their application in oxygen evolution. *Chemical Communications* 54: 264–267.

73 Wang, X., Yu, L., Guan, B.Y. et al. (2018c). Metal–organic framework hybrid-assisted formation of Co_3O_4/Co-Fe oxide double-shelled nanoboxes for enhanced oxygen evolution. *Advanced Materials* 30: 1801211.

74 Wang, X., Xiao, H., Li, A. et al. (2018b). Constructing NiCo/Fe_3O_4 heteroparticles within MOF-74 for efficient oxygen evolution reactions. *Journal of the American Chemical Society* 140: 15336–15341.

75 Nai, J., Lu, Y., Yu, L. et al. (2017). Formation of Ni–Fe mixed diselenide nanocages as a superior oxygen evolution electrocatalyst. *Advanced Materials* 29: 1703870.

76 Xie, Z., Tang, H., and Wang, Y. (2019). MOF-derived Ni-doped CoS_2 grown on carbon fiber paper for efficient oxygen evolution reaction. *ChemElectroChem* 6: 1206–1212.

77 Zhu, J., Xiao, M., Zhang, Y. et al. (2016). Metal–organic framework-induced synthesis of ultrasmall encased NiFe nanoparticles coupling with graphene as an efficient oxygen electrode for a rechargeable Zn–air battery. *ACS Catalysis* 6: 6335–6342.

78 Zhou, T., Du, Y., Wang, D. et al. (2017). Phosphonate-based metal–organic framework derived Co–P–C hybrid as an efficient electrocatalyst for oxygen evolution reaction. *ACS Catalysis* 7: 6000–6007.

79 Fu, Y., Wang, W., Wang, J. et al. (2019). MOFs-derived ZnCo–Fe core–shell nanocages with remarkable oxygen evolution reaction performance. *Journal of Materials Chemistry A* 7: 17299–17305.

80 Tang, Y.J., Gao, M.R., Liu, C.H. et al. (2015). Porous molybdenum-based hybrid catalysts for highly efficient hydrogen evolution. *Angewandte Chemie* 127: 13120–13124.

81 Xu, X., Nosheen, F., and Wang, X. (2016). Ni-decorated molybdenum carbide hollow structure derived from carbon-coated metal–organic framework for electrocatalytic hydrogen evolution reaction. *Chemistry of Materials* 28: 6313–6320.

82 Zhang, H., Ma, Z., Liu, G. et al. (2016b). Highly active nonprecious metal hydrogen evolution electrocatalyst: Ultrafine molybdenum carbide nanoparticles embedded into a 3D nitrogen-implanted carbon matrix. *NPG Asia Materials* 8: e293–e293.

83 He, P., Yu, X.Y., and Lou, X.W. (2017). Carbon-incorporated nickel–cobalt mixed metal phosphide nanoboxes with enhanced electrocatalytic activity for oxygen evolution. *Angewandte Chemie International Edition* 56: 3897–3900.

84 Guan, C., Liu, X., Ren, W. et al. (2017). Rational design of metal-organic framework derived hollow $NiCo_2O_4$ arrays for flexible supercapacitor and electrocatalysis. *Advanced Energy Materials* 7: 1602391.

85 Li, J.-S., Sha, J.-Q., Du, B. et al. (2017). Highly efficient hydrogen evolution electrocatalysts based on coupled molybdenum phosphide and reduced graphene oxide derived from MOFs. *Chemical Communications* 53: 12576–12579.

86 Qamar, M., Adam, A., Merzougui, B. et al. (2016). Metal–organic framework-guided growth of Mo_2C embedded in mesoporous carbon as a high-performance and stable electrocatalyst for the hydrogen evolution reaction. *Journal of Materials Chemistry A* 4: 16225–16232.

87 Wang, M., Lin, M., Li, J. et al. (2017). Metal–organic framework derived carbon-confined Ni_2P nanocrystals supported on graphene for an efficient oxygen evolution reaction. *Chemical Communications* 53: 8372–8375.

88 Zhang, G., Wang, H., Yang, J. et al. (2018a). Temperature effect on Co-based catalysts in oxygen evolution reaction. *Inorganic Chemistry* 57: 2766–2772.

89 Wu, R., Wang, D.P., Zhou, K. et al. (2016). Porous cobalt phosphide/graphitic carbon polyhedral hybrid composites for efficient oxygen evolution reactions. *Journal of Materials Chemistry A* 4: 13742–13745.

90 Dong, Q., Wang, Q., Dai, Z. et al. (2016). MOF-derived Zn-doped $CoSe_2$ as an efficient and stable free-standing catalyst for oxygen evolution reaction. *ACS Applied Materials & Interfaces* 8: 26902–26907.

91 Xu, H., Cao, J., Shan, C. et al. (2018). MOF-derived hollow CoS decorated with CeO_x nanoparticles for boosting oxygen evolution reaction electrocatalysis. *Angewandte Chemie* 130: 8790–8794.

92 Zhang, T., Du, J., Xi, P. et al. (2017). Hybrids of cobalt/iron phosphides derived from bimetal–organic frameworks as highly efficient electrocatalysts for oxygen evolution reaction. *ACS Applied Materials & Interfaces* 9: 362–370.

93 Wang, X., Ma, Z., Chai, L. et al. (2019). MOF derived N-doped carbon coated CoP particle/carbon nanotube composite for efficient oxygen evolution reaction. *Carbon* 141: 643–651.

94 Luo, X., Zhou, Q., Du, S. et al. (2018). One-dimensional porous hybrid structure of Mo_2C-CoP encapsulated in N-doped carbon derived from MOF: An efficient electrocatalyst for hydrogen evolution reaction over the entire pH range. *ACS Applied Materials and Interfaces* 10: 42335–42347.

95 Li, Y., Niu, S., Rakov, D. et al. (2018b). Metal organic framework-derived CoPS/N-doped carbon for efficient electrocatalytic hydrogen evolution. *Nanoscale* 10: 7291–7297.

96 Singh, T., Das, C., Bothra, N. et al. (2020). MOF derived Co_3O_4@ Co/NCNT nanocomposite for electrochemical hydrogen evolution, flexible zinc-air batteries, and overall water splitting. *Inorganic Chemistry* 59: 3160–3170.

97 Ali, A. and Shen, P.K. (2020). Nonprecious metal's graphene-supported electrocatalysts for hydrogen evolution reaction: Fundamentals to applications. *Carbon Energy* 2: 99–121.

98 Mccrory, C.C., Jung, S., Ferrer, I.M. et al. (2015). Benchmarking hydrogen evolving reaction and oxygen evolving reaction electrocatalysts for solar water splitting devices. *Journal of the American Chemical Society* 137: 4347–4357.

99 Wu, Z.P., Lu, X.F., Zang, S.Q. et al. (2020). Non-noble-metal-based electrocatalysts toward the oxygen evolution reaction. *Advanced Functional Materials* 30: 1910274.

100 Zou, X. and Zhang, Y. (2015). Noble metal-free hydrogen evolution catalysts for water splitting. *Chemical Society Reviews* 44: 5148–5180.

101 Huang, Z.F., Wang, J., Peng, Y. et al. (2017). Design of efficient bifunctional oxygen reduction/evolution electrocatalyst: Recent advances and perspectives. *Advanced Energy Materials* 7: 1700544.

102 Huynh, M., Ozel, T., Liu, C. et al. (2017). Design of template-stabilized active and earth-abundant oxygen evolution catalysts in acid. *Chemical Science* 8: 4779–4794.

103 Chen, P., Xu, K., Zhou, T. et al. (2016). Strong-coupled cobalt borate nanosheets/graphene hybrid as electrocatalyst for water oxidation under both alkaline and neutral conditions. *Angewandte Chemie International Edition* 55: 2488–2492.

104 Huan, T.N., Rousse, G., Zanna, S. et al. (2017). A dendritic nanostructured copper oxide electrocatalyst for the oxygen evolution reaction. *Angewandte Chemie International Edition* 56: 4792–4796.

105 Bhattacharyya, S., Das, C., and Maji, T.K. (2018). MOF derived carbon-based nanocomposite materials as efficient electrocatalysts for oxygen reduction and oxygen and hydrogen evolution reactions. *RSC Advances* 8: 26728–26754.

6

2D Materials for CO_2 Reduction and H_2 Generation

Rameez Ahmad Mir[1,2], Sanjay Upadhyay[3,4], and O.P. Pandey[2,3,]*

[1] *Department of Materials Science Engineering, University of Toronto, Canada*
[2] *Center of Excellence for Emerging Materials (CEEMS)-Virginia Tech (VT), TIET, Patiala, Punjab, India*
[3] *School of Physics and Materials Science (SPMS), Thapar Institute of Engineering and Technology, Patiala 147004, Punjab, India*
[4] *Division of Research & Innovation, Uttaranchal University, Dehradun 248007, India*
[*] *Corresponding Author*

6.1 Introduction

6.1.1 Principles of CO_2 Reduction

In view of the emission of toxic gases, CO_2 causes serious issues of global warming. The reduction of CO_2 into a diverse range of hydrocarbons like CO, CH_4, CH_3OH, HCHO, and HCOOH is a promising approach. Syngas (CO and H_2) formed during the CO_2 reduction reaction are essential products used for energy processing like the Fischer–Tropsch process, ammonia, and methanol formation [1]. The electrocatalytic and photocatalytic approaches are the potential procedures adopted for CO_2 reduction utilizing the electricity generated from renewable energy sources and natural solar radiation [2, 3].

The production of carbon monoxide (CO) via electrocatalysis and photocatalysis is superior to the traditional industrial methods, which rely on high temperature/pressure and specific experimental setups. Extensive efforts have been devoted to search ideal catalysts for CO_2 reduction carried out by the electrocatalysis and photocatalysis processes. The rational design of a suitable photocatalyst with a strong visible light absorption ability and having maximum interface sites improves the photocatalytic CO_2 reduction reaction activity. Among all categories of catalyst for CO_2 reduction, two-dimensional (2D) materials have gained special attention and have shown promising results for high conversion of CO_2 into valuable fuel and chemicals [4–7]. The higher activity is ascribed to the enhanced conductivity, increased number of edge sites, and the abundant surface atoms along 2D conducting channels. Moreover, the synthesis of these 2D materials at the nanoscale further enhances their CO_2 reduction rate due to higher surface area and improved active sites. Different families of 2D materials like metal chalcogenides, metal carbides, metal-free 2D materials, and atomically thin metal/metal-metal oxides have been developed for the rapid transformation of CO_2 into valuable chemicals and fuels [4, 8–12]. Some 2D materials exhibit high activity in electrochemical CO_2 reductions and are less active for photocatalytic reduction reactions and vice versa due to their inherent properties. The same features positively influence H_2 production as a new and clean source of fuel.

6.1.2 Principles of H_2 Generation

Hydrogen is the most readily available element in the universe, in the form of hydrocarbons and water. However, different scientific procedures like thermal, electrolytic, and

Green Energy Harvesting: Materials for Hydrogen Generation and Carbon Dioxide Reduction,
First Edition. Edited by Pooja Devi.

photocatalytic procedures have been adopted to produce hydrogen from fossil fuels, biomass, and water [13–15]. Hydrogen production through water splitting has received special attention due to the easy availability of unlimited water reservoirs in nature and as a renewable source. Hydrogen as a source of energy will not only reduce the depletion of fossil fuels but will also overcome the issues caused by CO_2 emissions. Photocatalytic water splitting and water electrocatalysis are the most efficient and affordable ways to produce hydrogen [16, 17]. In photocatalytic water splitting, the solar energy dissociates the water into molecular hydrogen (H_2) and oxygen in the presence of a suitable catalyst [18]. The bandgap of associated catalyst material plays a key role in water dissociation in the presence of unlimited solar radiation. However, the recombination of electron–hole pairs in the wide bandgap catalyst materials and the high overpotential required for water splitting restricts its practical use. Water electrolysis is a well-developed technology to produce a clean source of hydrogen through hydrogen evolution reaction (HER) and oxygen evolution reactions (OER) for commercial utilization in electrolyzers and also for other industrial purposes. It also overcomes a prominent issue faced in photocatalytic water splitting dependent on solar radiation, affected by weather conditions. On the other hand, based on the type of electrolyte used for electrolysis, the electrolyzers (water electrolysis systems) have been divided into two main types that are alkaline exchange membrane (AEM) and proton exchange membrane (PEM) electrolyzers [19, 20]. PEM and AEM electrolyzers utilize the acidic and alkaline mediums as electrolytes for water electrolysis, respectively. The approach still has no practical viability without a suitable electrocatalyst to make them operational at low overpotentials with enhanced efficiency of H_2 production at a lower cost. Depending upon the reaction pathways of HER/OER for H_2 production in PEM and AEM electrolyzers, the electrocatalyst plays a key role, which lowers the energy barrier and overpotential for water splitting into H_2 and O_2, respectively. At present, only the noble metals are suitable but are not commercialized due to their high cost [21–24]. Different materials have been tested and fabricated to replace costly noble metal electrocatalysts for HER/OER. The key parameter that determines the efficiency and stability of an electrolyte is the adsorption energy of the electrocatalyst. The lower attraction of hydroxyl ions and moderate adsorption energy of the electrocatalyst enhances its HER/OER efficiency and stability in an alkaline medium. However, the hydrogen adsorption (H_{ads}) at the active site of the electrocatalyst determines its HER efficiency in an acidic medium [25]. In view of this and for the replacement of costly noble metal species as electrocatalysts in HER, the non-noble materials, especially the thinner 2D electrocatalysts providing highly exposed surface sites for easy charge transport routes, have been recognized as viable materials for enhanced HER activity [26–29]. The different types and the role of 2D materials in electrocatalytic CO_2 reduction and hydrogen production via electrocatalytic water splitting, in particular in HER, are discussed below.

6.2 Why 2D Materials for Electrochemical CO_2 Reduction and H_2 Production?

Two-dimensional materials are the unique class of materials exhibiting fewer layers than that of the bulk material and having at least one constraint dimension compared to three-dimensional (3D) materials, for example, graphene, graphite, etc. Graphene was the first 2D material discovered, although it was theoretically established in 1940 [30]. Graphene exhibits a very high specific area, layered structure, high conductivity, and very exceptional stability, which makes it a promising low-cost electrocatalyst candidate for HER and CO_2 reduction reactions [31]. The 2D nanostructured materials exhibit high surface area (SA), the possibility of tailoring the active sites on the edges, and high tolerance to impurities in electrolytes. Numerous 2D nanomaterials have received researcher's attention for energy conversion and storage since the discovery of 2D graphene [32, 33]. The structural tuning and well-controlled exposure of specific crystal facets of these 2D nanostructured materials not only improves their activity but also enhances their stability [4, 12, 34]. Among the different 2D heterogeneous transition metal and non-metal species developed as electrocatalysts for CO_2 reduction, nanostructured 2D sulfide and nitrogen-doped graphene (NG) have shown encouraging results [1, 5, 35]. Efforts are being made to further improve their activity and stability by tailoring their surface features, which depend on the synthesis methods adopted for their preparation. Among all 2D materials, the transition metal dichalcogenides (TMDs) and transition metal carbides (TMCs) exhibiting outstanding and unique characteristics have the potential to replace noble metals in electrocatalytic applications [6, 12, 36]. Tungsten diselenide (WSe_2), molybdenum disulfide (MoS_2), molybdenum diselenide ($MoSe_2$), tungsten disulfide (WS_2), tungsten diselenide (WSe_2), hexagonal boron nitride, 2D boron, 2D silicon, and 2D germanium are all examples of TMDs reported for electrocatalytic CO_2 reduction reactions and H_2 production [4, 16, 37, 38]. The molybdenum carbide (Mo_2C) and tungsten carbide (WC) among the TMCs have shown promising results for the replacement of Pt in electrocatalysis of water, especially in acidic mediums due to their high resemblance to Pt [39–41]. The recent advancements in 2D-based electrocatalysts for CO_2 reduction and HER electrocatalysts are now discussed here. Further, the important parameters that influence the performance of 2D materials for CO_2 reduction and HER activity are also discussed in detail.

6.3 2D Materials for CO_2 Reduction

6.3.1 Transition Metal Dichalcogenides (TMDs) for CO_2 Reduction

Transition metal dichalcogenides (TMDs), a unique class of layered (2D) materials, exhibit a sandwich-like structure composed of one layer of metal cations in between two layers of chalcogen anions, as shown in Figure 6.1a [42, 43]. Molybdenum disulfide (MoS_2) represents a class of TMDs used as a catalyst for CO_2 reduction due to high HER activity with greater stability [44]. Asadi et al. [12, 44] studied the role of metal atoms at edge sites on the high intrinsic activity and stability of nanostructured TMDs as the exceptional class of materials for CO_2 reduction. Theoretically, MoS_2 does not serve as an active catalyst for CO_2 reduction, but with the assistance of ionic liquid (IL) as a co-catalyst has shown surprising results. The terminated Mo-edge sites of bulk MoS_2 have shown high efficiency toward CO_2 reduction attributed to its relatively low work function (3.9 eV), more metallic nature, and d-electron density, respectively [44, 45]. It was also revealed that the IL co-catalyst is necessarily required to inhibit undesirable HER to favor the reduction of CO_2 into valuable chemicals and fuel. Abbasi et al. [46] reported that the CO_2 reduction rate of 2D-MoS_2 can be enhanced by modifying the surface features and electronic structure by doping with niobium (Nb) and tantalum (Ta). Both Nb and Ta form NbS_2 and TaS_2 covalent bonds with MoS_2 without altering the lattice parameters of the original host. The electronic structure is completely modified due to the different valence electrons of Mo, Nb, and Ta. The host MoS_2 exhibits semiconducting behavior due to more electrons in the valence state compared to NbS_2 and TaS_2, which show a metallic character due to fewer electrons in the valence states of Nb and Ta, respectively. Increasing the doping percentage of Nb enhances the current density compared to pristine MoS_2. It initially increases the current density to a limit and then decreases, if the doping percentage is increased beyond 5%, as shown in Figure 6.1b.

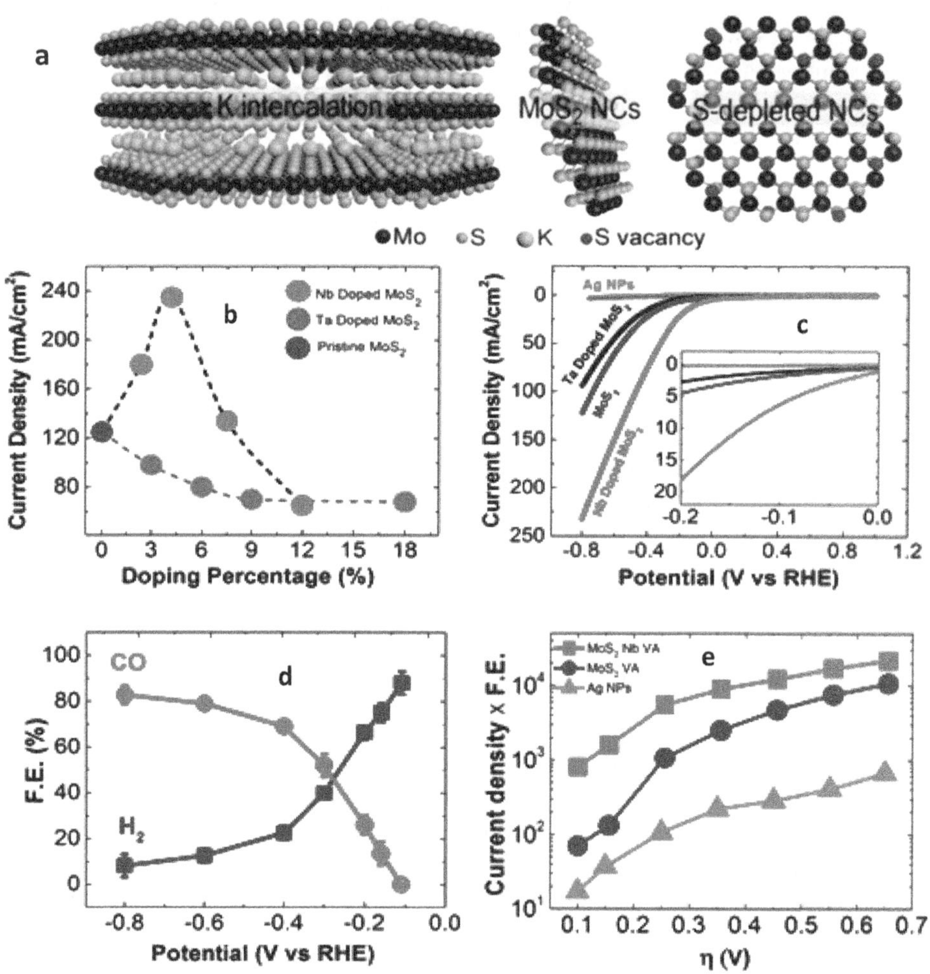

Figure 6.1 (a) Schematic representation of the preparation of monolayered MoS_2 NCs. Reproduced with permission from [45]. Copyright 2017 Elsevier Ltd, CO_2 reduction activity of pristine and doped MoS_2; (b) effect of doping concentration on current density; (c) polarization plots; (d) FE of Nb-doped MoS_2 at various potentials; and (e) CO formation partial current density of Ag NPs, pristine and doped MoS_2. Reproduced from L. Lin et al., 2017 / with permission of Elsevier.

In the same way, Ta was also doped in MoS_2 to promote CO_2 reduction activity, but it exhibited lower performance than pristine MoS_2. The deviation from the optimal electronic properties of Ta-doped MoS_2 lowers CO_2 reduction performance [46]. The polarization plots (Figure 6.1c) also revealed the higher activity of Nb-doped MoS_2 compared to silver nanoparticles (Ag NPs), Ta-doped MoS_2, and pristine MoS_2. Nb-doped MoS_2 shows a Faradic efficiency (FE) range of 12–82% in converting CO_2 into a selective CO and H_2 mixture in the voltage window of −0.16 to −0.8 V, as shown in Figure 6.1d. For CO formation, the partial current density of Nb-doped MoS_2 is higher than pristine MoS_2 and Ag NPs (Figure 6.1e). The current density of Nb-doped MoS_2 is only a magnitude higher than pristine MoS_2 at lower potential and approaching a similar nature at higher potential. In contrast, the current density is two magnitudes higher than the Ag NPs over the potential range [47]. These results indicate the synergistic effects of MoS_2 efficiency in converting CO_2 to CO. Also, the other 2D TMDs, for example WSe_2 nanoflakes (NFs), exhibited better CO_2 reduction activity than bulk MoS_2 structures and other species [12]. This higher activity is attributed to the exposure of more active edge sites in 2D nanoarchitecture. The theoretical studies with different characteristic features determine that the superior activity of WSe_2 NFs is attributed to lower charge transfer (R_{ct}) with considerably lower work function (3.52 eV) [4]. The presence of W atoms on edge sites in WSe_2 NFs provided extended stability up to 27 h, as revealed by chronoamperometry. The other species of TMDs, $MoSe_2$ (molybdenum phosphide (MoP), and the TMDs supported catalyst have shown promising results for CO_2 reduction, but the superior performance is exhibited by MoS_2 and WSe_2 nanostructures [46, 48]. Along with metal species, the 2D non-metal species, especially graphene, have attained considerable research interest in this field due to their unique physical and chemical characteristics [49, 50].

6.3.2 Metal-Free Carbon Material for CO_2 Reduction

The 2D wonder material, graphene, has been considered as a cost-effective and efficient catalyst for electrochemical CO_2 reduction due to its high specific surface area (HSSA) and very high electrical conductivity [49, 51]. However, pure graphene is chemically less reactive toward CO_2 reduction reaction. The network formed due to delocalized π-bonding makes the adsorption of reaction intermediates feasible on the surface of the graphene and lowers the CO_2 reduction performance [5]. Chai and Guo [52] confirmed the adsorption of a metastable CO_2 species on the surface of undoped graphene. To boost the electrocatalytic CO_2 reduction reaction, there is an essential need to tailor and modify the electronic structure of graphene. The incorporation of heteroatoms in graphene is a promising approach for altering the electronic structure and enhancing the electrochemical CO_2 reduction reaction activity. The incorporation of boron (B) in the graphene matrix resulted in an asymmetric charge and spin density throughout the 2D structure [53]. The B and carbon (C) atoms exhibiting positive spin density are the highly active sites for the chemisorption of CO_2 reduction reaction intermediates. More recently, NG has shown efficient results in the reduction of CO_2 to CO in aqueous electrolytes due to synergistic N-doping effects [54]. The defect-induced NG selectively converts CO_2 into CO without forming any hydrocarbons due to the weak interaction between N–C and CO. This weak interaction favors the desorption of CO from the N–C surface [54]. The incorporation of N in the carbon matrix forms three different kinds of dopant species such as graphitic N, pyridinic N, and pyrrolic N. Wu et al. [54] also revealed the role of N content in graphene on CO_2 reduction activity. The N content in graphene was varied by heating the prepared samples at different temperatures. The results revealed that the N content decreases with an increase in synthesis temperature as it removes some of the unstable N phases [55]. The comparison of CO_2 reduction activity of various NG heated at various temperatures is shown in Figure 6.2. The sample heat-treated at 800°C (NG-800) exhibits a higher Faradaic efficiency (FE) for CO and $HCOO^-$ compared to other NG catalysts (Figure 6.2a–b). The Faradic efficiency (FE) for CO with the corresponding potential and temperature-dependent N content in the NG catalyst is shown in Figure 6.2c.

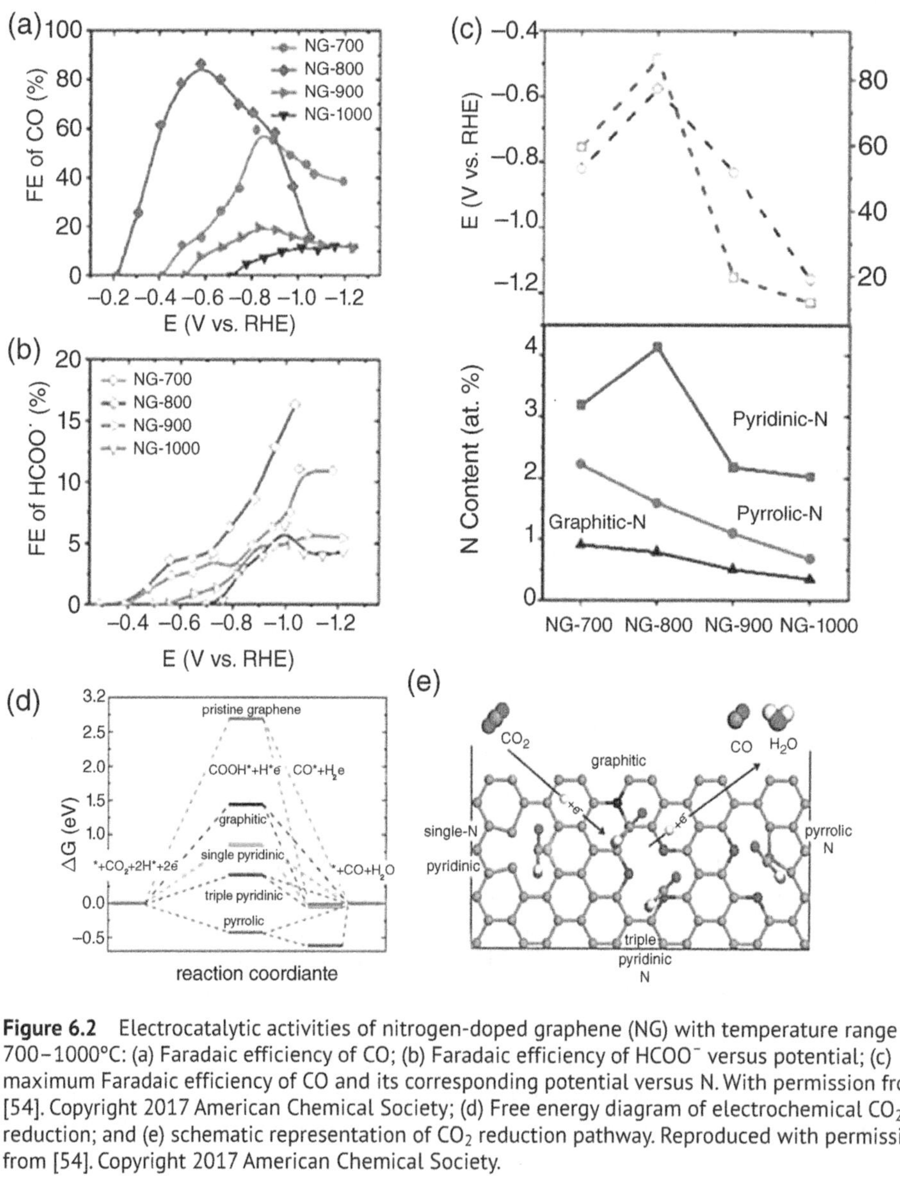

Figure 6.2 Electrocatalytic activities of nitrogen-doped graphene (NG) with temperature range 700–1000°C: (a) Faradaic efficiency of CO; (b) Faradaic efficiency of $HCOO^-$ versus potential; (c) maximum Faradaic efficiency of CO and its corresponding potential versus N. With permission from [54]. Copyright 2017 American Chemical Society; (d) Free energy diagram of electrochemical CO_2 reduction; and (e) schematic representation of CO_2 reduction pathway. Reproduced with permission from [54]. Copyright 2017 American Chemical Society.

The sample NG-800 having more N content, exhibited higher FE and revealed that the NG tailors the electronic structure in a favorable way for enhanced CO_2 reduction. Moreover, the selective product formation due to synergistic effects of N-doping is very useful for converting CO_2 into CO without forming unuseful intermediates. The FE and anodic potentials shift with the increase in pyridinic-N content. No such effect was observed for pyrrolic and graphitic-N. The results revealed that pyridinic-N is the most active site for CO_2 reduction. It is important to note that the contribution from other N-type defects cannot be ruled out. The energy diagram (Figure 6.2d) shows that all the three dopants reduce the free energy barrier for adsorption of CO_2 reduction species as compared to the pristine graphene. The defected N sites are the reaction sites in pyridinic and pyrrolic N, whereas the positively charged C atom adjacent to N in graphitic N is the more active site. The $COOOH^*$ gets attached to the defected N sites in pyrrolic and pyridinic N, respectively. The triple pyridinic N dopant lowers the energy barrier of $COOH^*$ adsorption more than the single pyrrolic N and graphitic N, and the corresponding reaction pathway for the reduction of CO_2 into CO and H_2O is shown in Figure 6.2e. The results confirm that the pyridinic N is the most active species compared to the other two. For long-term testing and for commercial applicability of the product, the sample with higher N content (NG-800) was further tested for CO_2 reduction activity and stability at various potentials.

The results revealed that CO_2 reduction started at very low potentials in the presence of the electrocatalyst (NG-800) and showed a higher FE of 85% at higher overpotentials. The stability of the tests revealed the potential of the electrocatalyst for longer durations and is superior to metal catalysts [56]. However, the reaction mechanism of NG materials for CO_2 reduction reaction is not clear and is still a subject of debate. Certain reports suggest that the graphitic N site in the NG possesses the lower energy barrier for CO_2 reduction reaction and is the most active site [52]. In the same way, Liu et al. [57] highlighted the role of pyrrolic N in the reduction of CO_2 by reducing CO to HCOOH with a very low overpotential (0.24 V). The generation of the defects in the graphene is unavoidable, and defect control is an effective method to alter the electronic structure and improve the catalytic properties of graphene [7, 58]. Graphitic carbon nitride (g-C_3N_4) has received considerable attention as an electrocatalyst in CO_2 reduction due to synergistic effects of C–N bonding features, induced N defects, and layered (2D) structures resembling that of graphene. The enhanced reductive conduction bands in ultra-thin (~1 nm thick) and surface polarized g-C_3N_4 have shown higher CO_2 reduction activity [59]. However, certain metal-free carbon structures (although doped with N) endure poor CO_2 reduction performance when operated at low overpotentials. Pan et al. [60] demonstrated that sulfur (S) addition in N-doped carbon improves the CO_2 reduction performance and enhances the selective formation of CO on the N–C surface. FE of 92% for CO was obtained for N, S co-doped carbon species at a very small overpotential of 490 mV. The synthesized electrocatalyst is a promising candidate for improving the electrocatalytic CO_2 reduction performance as compared to its S-free N–C counterparts. The incorporation of S in N–C enhances the higher density of pyridinic N as a highly active and efficient site for CO_2 reduction. The experimental and theoretical results revealed that the doping of N and S together in a carbon matrix paves the way to synthesize high-performance carbon catalysts for CO_2 reduction.

Single-atom catalysts (SACs) with metal (M) atoms distributed on a solid support, especially graphene, emerged as another important class of materials in catalysis science, especially in CO_2 reduction reactions [61]. The unique electronic properties and uniform geometric structure of SACs endows higher selectivity and efficiency for electrochemical CO_2 reduction [62]. The two adjacent armchair graphene layers bridging the edge-hosted moieties are the real active site for CO_2 reduction [63]. In the case of metal incorporated NG/carbon species, the M-N_{2+2}-C_8 site favors the dissociation of the reaction intermediate COOOH* and the adsorption of CO* at the M site during the reduction of CO_2. Moreover, the hydroxyl (OH*) bonds to the C atom having a dangling bond with the adjacent N atom. Certain metal species like Fe-N-C and Co-N-C exhibited more positive potential for CO_2 reduction than that of NG. After incorporation of metal species in the 2D carbon matrix, their superior performance toward CO_2 reduction reaction is observed. This has encouraged scientists to study the metal and metal oxide, especially the multivalent and stable transition metal (TM) oxide materials.

6.3.3 Transition Metal Oxides (TMOs) for CO_2 Reduction

Transition metal/metal oxide and their composites have attained special attention as active electrocatalysts for CO_2 reduction due to their high abundance in nature, non-toxicity, and lower cost [64, 65]. Certain TM oxide (TMO) species have shown promising results for CO_2 reduction, but the presence of lower active sites and the poor electrical conductivity in the TMOs is still prevailing. Efforts are being made to improve their practical applicability for altering their structural and surface features for improved CO_2 reduction performance. One of the efficient ways to improve the properties and features of TMOs is to decrease their thickness to lower atomic levels. It not only improves the density of states (DOS) but also increases the charge density near the conduction band edges (CBE) [43]. The ultra-thin oxide layers in the 2D form not only improve the electronic conductivity for rapid transport of charge carriers but also expose a lot of co-coordinatively unsaturated surface atoms as active centers for CO_2 reduction reaction [66]. The layer thickness plays an important role in the CO_2 reduction of the synthesized ultrathin Co_3O_4 layers. The ultrathin Co_3O_4 layers with a thickness of 1.72 nm exhibited a CO_2 reduction current of 0.68 mA cm^{-2} at 0.88 V (vs. SCE). The CO_2 reduction reaction performance of the ultrathin layer of 1.72 nm thick is 1.5 and 20 times higher than those of 3.51-nm-thick Co_3O_4 layers and bulk Co_3O_4, respectively [66, 67]. Moreover, the ultrathin Co_3O_4 layers revealed a higher Faradaic efficiency (FE) of over 60% and extraordinary stability of 20 h toward CO_2 reduction reaction. Partially oxidized Co-based sheets (4 atoms thick 0.84 nm) revealed promising applications toward CO_2 reduction due to the higher number of active surface sites and superior electrical conductivity resulting from the metallic nature of Co [43, 64]. The partially oxidized Co 4-atom layers showed a reduction peak at −0.85 V (vs. SCE) with a current density of 10.59 mA cm^{-2} in Na_2SO_4 aqueous solution [43]. The results were 10, 40, and 260 times higher than pure Co 4-atom-thick layers, partially oxidized bulk Co, and bulk Co layers, respectively [43]. This higher activity is attributed to the higher electrochemical surface area (ECSA) of partially oxidized Co 4-atom layer electrocatalyst material. The partially oxidized layers Co 4-atom-thick possessed almost the same ECSA as that of pure Co_4-atomic-thick layers, but the former exhibited 10 times higher activity than the latter [43]. The results demonstrated more enhanced intrinsic activity of Co_3O_4 than atomic Co for the CO_2 electroreduction. The defect induced oxides have shown higher activity than the pure species, especially the oxygen (II) vacant species, which decreases the activation barrier by stabilizing the $HCOO^-$ intermediate species and lowers the onset potential from 0.81 to 0.78 V. The results indicate that the TMs/TMOs nanostructured 2D atomic level species with controlled manipulation of oxidation states can be a new route to develop highly efficient electrocatalyst for CO_2 reduction [43].

6.3.4 2D Metal Carbides/Nitrides and MXene for CO_2 Reduction

At present, a new category of 2D transition metal carbide/nitride has emerged as the very good catalyst species known as MXenes [7]. MXenes are an efficient class of electrocatalyst species for CO_2 reduction. MXenes, a secondary class of TMCs/TMNs, precisely exhibit a layered structure that can be successfully formed by etching the MAX phases, where M is the transition metal element, A is the main group sp element, and X represents C or N [68, 69]. Although MXenes have been successfully explored for other reactions like hydro processing and water splitting, their affinity for CO_2 reduction is also reported. This is because of the 2D structure of MXenes, where an easy electron transfer mode exists. MXenes are synthesized from MAX phases [40]. By adopting a suitable etching procedure to remove the "A" layer from the MAX phase, MXenes can be attributed, as shown in Figure 6.3a, and the corresponding changes in the microstructures of the Ti-based MAX phases to MXene are shown in Figure 6.3b.

The layered structure of MXene phases is viable in microstructures (Figure 6.3c–h), and the corresponding transmission electron microscope (TEM) images of MXenes are shown in Figure 6.3i–l. The numeric in Figure 6.3i shows the increase in the number of layers from monolayer to 2, 3, and more. The thinnest layers are most likely transparent and facilitate the charge transfer to improve the electroactivity of MXenes. Li et al. [70] synthesized single-layered Ti_3C_2 MXene (a TMC species) by etching Ti_3AlC_2 precursor with aqueous potassium hydroxide (KOH). The higher metallic conductivity and hydrophilicity of $Ti_3C(OH)_2$ is induced to replace aluminum (Al) with OH in the parent precursor during etching. This variation in structural feature widens the scope of these MXenes for energy conversion applications. Zhang et al. [71] theoretically investigated the electrochemical CO_2 reduction activity for various oxygen-terminated MXenes (Ti_2CO_2, V_2CO_2, and $Ti_3C_2O_2$). The 2D Ti_2CO_2 exhibited superb catalytic performance where the surface oxygen vacancy (Ov) is the active center for lowering the energy barrier for CO_2 reduction reactions. Ma et al. [11] studied the role of N-doping on $Ti_3C_2T_x$ MXene on CO_2 reduction. The doping of N in the Ti-based MXenes favors the formation of carbon nitride and (CH_4) as a reaction product, whereas Co nanoparticle-supported Ti_3C_2Tx (Co–Ti_3C_2Tx) has shown the activity for the formation of CO during CO_2 hydrogenation. The results also revealed the formation of TiO_2 nanoparticles on the surface of Ti_3C_2Tx during N-doping. The formed phase becomes strongly attached to Co nanoparticles and improves the reaction activity to a greater extent. The reducibility of the Co and TiO_2 nanoparticles shifts the product selectivity from CO to CH_4. This also revealed the surface modification of the 2D MXenes for regulating the product distribution of CO_2 reduction. Later, different MXenes were fabricated and tested as CO_2 reduction catalysts, especially those having the formula M_3C_2. Transition metal carbides (TMCs) and nitrides (TMNs) or MXenes from Group IV to VI, mostly Cr_3C_2 and Mo_3C_2, revealed highly selective efficiency for CO_2 reduction to CH_4 [72]. The surface termination of Cr and Mo carbides with O and OH groups significantly reduces the energy barrier for the transformation of CO_2 to CH_4 via reduction reaction. This provides a different perspective for the use of these 2D layered metal-carbon-containing species as efficient catalysts for electrochemical reduction of CO_2, which have been highly explored for HER and ORR.

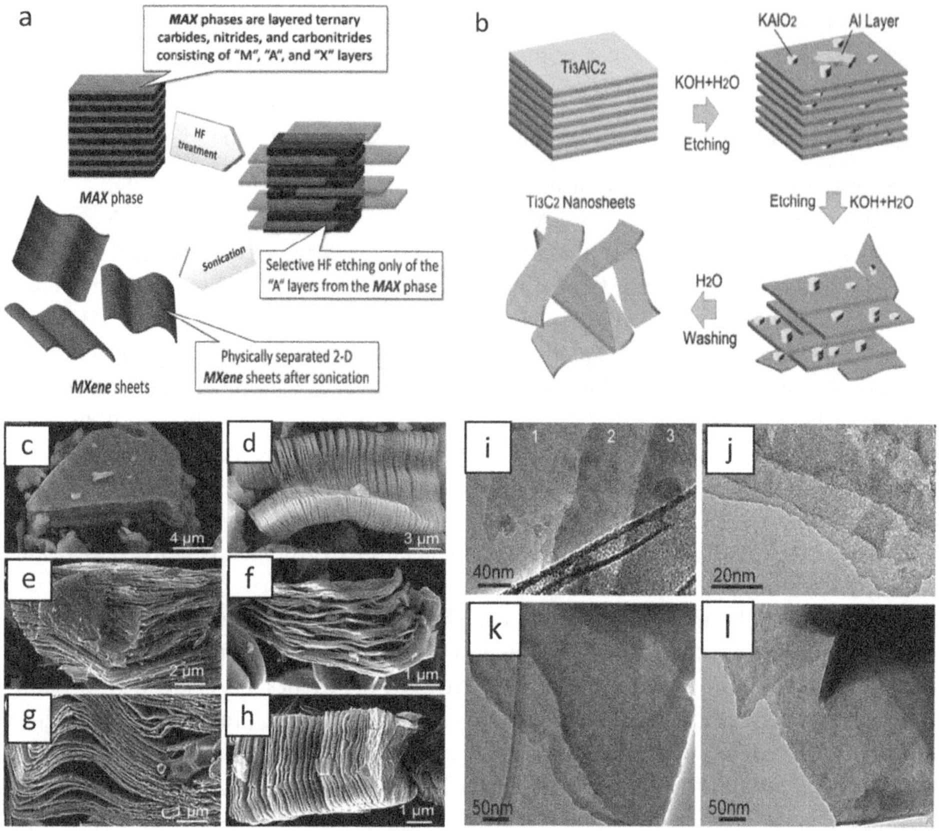

Figure 6.3 (a) Schematic representation of Mxene from MAX phase. Reproduced with permission from [69]. Copyright 2012 American Chemical Society; and (b) synthesis and structural features of Ti_3C_2 MXene. Reproduced with permission from [70]. Copyright 2017 American Chemical Society. Scanning electron microscope (SEM) microstructures of MAX phase. (c) Ti_3AlC_2 layered MAX phases [71], Naguib et al. (2012), Reproduced with permission from American Chemical Society; (d) hydrofluoric acid (HF) treated Ti_3AlC_2; (e) after HF treatment; (f) Ta_4AlC_3 after HF treatment; (g) TiNbAlC after HF treatment; (h) Ti_3AlCN after HF treatment and transmission electron microscope (TEM) micrographs; (i) Ti_3C_2 layers formed after HF treatment of Ti_3AlC_2 at room temperature for 22 h assuming a single sheet (monolayer) is the most transparent part of the sample; (j) similar layers of Ti3CNx exfoliated from Ti_3AlCN by HF treatment; (k) TiNbAlC after HF treatment; and (l) Ta_4AlC_3 after HF treatment. Reproduced with permission from Ref. [69]. Copyright 2012 American Chemical Society.

6.4 2D Materials for H_2 Production

6.4.1 Graphene-Based Materials as Electrocatalysts for the HER

Pristine graphene is catalytically inert for HER due to its flat structure [73]. As a result, graphene is subjected to various doping, functionalization, and strain operations in order to fine tune its chemical characteristics. Graphene is a non-bandgap semiconductor material with excellent mobility and conductivity. It is made up of a single sheet of graphite.

Graphene derivatives like reduced graphene oxide (rGO) can be made using the Hummers process by making graphene oxide (GO), then reduced and utilized as a precursor for a variety of graphene-based products. In addition to the huge research efforts in studying the traditional carbon-based electrocatalysts, new forms of graphene-based electrocatalysts have been created and shown to boost catalytic performance. Huang et al. [74] studied the effect of graphene on the electrochemical HER performance of Ni-graphene (Ni-G) composites in an alkaline medium. The addition of graphene improved the surface area and electrical conductivity of the sample and hence showed improved HER activity. The onset overpotential and current exchange density of Ni-G/304 SS for the HER were 75 mV and 1.43 mA cm^{-2}, respectively. Akyüz et al. [75] synthesized GO-N_3 and RGO-N_3 decorated with cobalt (Co) phthalocyanine complex bearing terminal alkyne moieties (TA-CoPc). GCE/GO-N_3/TA-CoPc and GCE/RGO-N_3/TA-CoPc both functioning as active HER electrocatalysts. On the GCE/RGO-N_3/TA-CoPc electrode, the overpotential of HER decreased to 340 mV at pH 3.40. Yang et al. [76] reported molybdenum carbide nanoparticles decorated with graphene shells having nitrogen dopants (MoC_x@C^{-1}) synthesized through annealing and etching processes. The graphene shells entirely cover the MoC_x core. The synergistic effects of extremely diffusive MoC_x cores, graphene structure, and the N dopants are responsible for the highly effective electrocatalytic activity and cyclic stability of MoC_x@C^{-1} for water splitting. Wang et al. [77] synthesized the Ni-rGO composite and investigated its HER activity. It shows outstanding HER activity with a low Tafel slope of 77 mV dec^{-1} and small overpotentials of only 36 mV to reach 10 mA cm^{-2}. Figure 6.4a shows the Ni nanoparticles covered with rGO sheets.

The high electrical conductivity, high specific surface area, and synergistic harmonic impact between Ni particles and rGO sheets were responsible for exceptional HER performance. Sun et al. [78] reported graphene/graphene nanoribbon (GGNR) hybrid aerogels, which were used as a framework for few-layered MoS_2. The schematic representation of the synthesis process is shown in Figure 6.4b. The prepared sample shows a Tafel slope of 49 mV dec^{-1} as well as a low overpotential of 183 mV at 10 mA cm^{-2} in acidic media. Cardoso et al. [80] successfully synthesized AuPd and FePd; two Pd-based alloy nanoparticles were assembled on reduced graphene oxide (rGO) and their HER activity was investigated. At 25°C, the onset overpotentials of the rGO-$Au_{48}Pd_{52}$ and rGO-$Fe_{48}Pd_{52}$ composites were 80 and 140 mV, respectively. Qu et al. [81] investigated the influence of curvature of graphene on HER performance. They found that the curvature has a significant impact on their HER activity and doping. Also, N and Mo-doped WGs are catalytically effective against HER at specific compressions. Further, high Bader charge transfer, which is produced synergistically by the curvature, is directly associated with high HER performance at various compression ratios. Chen et al. [82] synthesized Co@N-CNTs@rGO composite in which Co nanoparticles are embedded within the N-doped carbon nanotubes grafted on rGO nanosheets. The linked N-CNTs@rGO can control the H adsorption on Co to an appropriate degree while maintaining structural stability to enhance HER activity, according to the density functional theory (DFT) calculations. The prepared sample showed an overpotential of 108 mV and 87 mV (at a current density of 10 mA cm^{-2}) in 1 M KOH and 0.5 M H_2SO_4 mediums, respectively. Wang et al. [79] grew $Co_{0.85}Se$ hollow nanospheres on NG nanosheets through the hydrothermal process. $Co_{0.85}Se$@NG_3 exhibited very low overpotentials of −227 mV (at 10 mV cm^{-2}) and a Tafel slope of 36.1 mV dec^{-1},

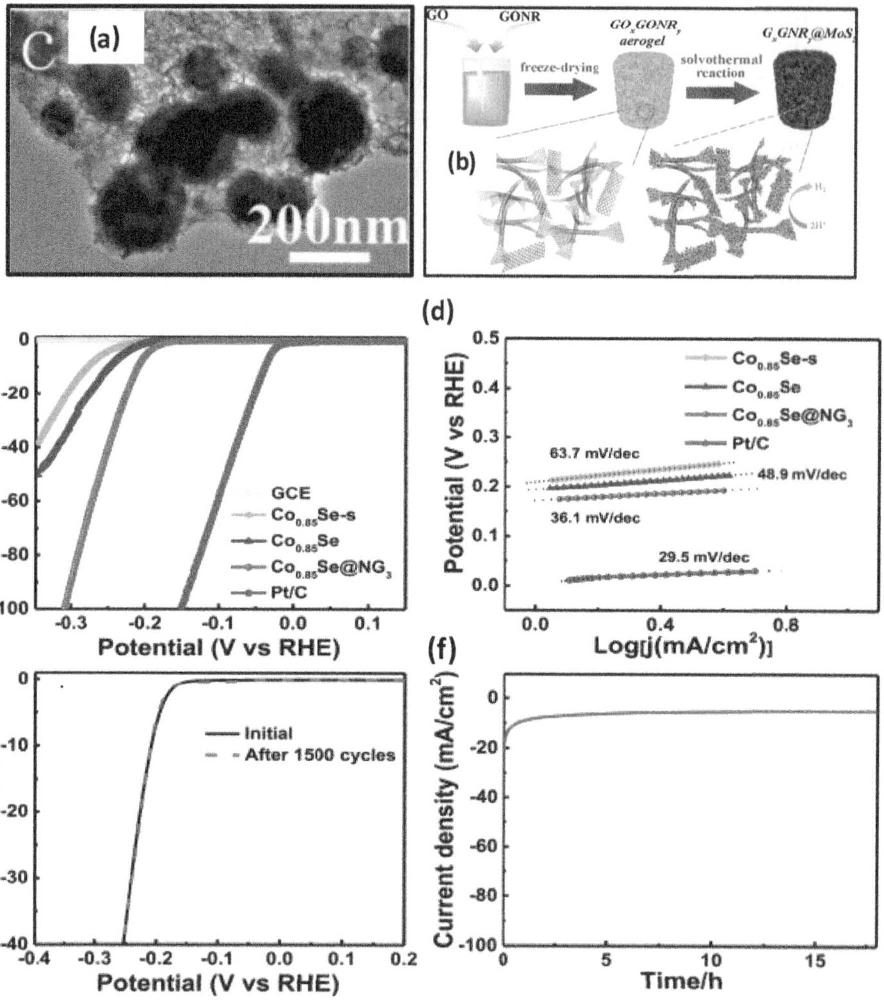

Figure 6.4 (a) TEM image of Ni-rGO composite [79], Wang et al. (2017), Reproduced with permission from ELSEVIER; (b) schematic diagram of the synthesis process of GxGNRy@MoS_2 hybrids. Reproduced with permission from [78]. Copyright 2017 Elsevier Ltd; (c) linear sweep voltammetry (LSV) curves; (d) Tafel curves of glassy carbon electrode (GCE), $Co_{0.85}$Se-s, $Co_{0.85}$Se, $Co_{0.85}$Se@NG_3, and Pt/C samples; (e) LSV curves of $Co_{0.85}$Se@NG_3 before and after 1500 cycles; and (f) time dependence of current density of $Co_{0.85}$Se@NG3. Reproduced with permission from [79]. Copyright 2018 Elsevier Ltd.

respectively. Also, the prepared sample showed excellent cyclic stability up to 1500 cyclic voltammetry (CV) cycles, as shown in Figure 6.4c–f. The novel hybrid design made up of hollow nano spherical $Co_{0.85}$Se evenly anchored on an NG network is responsible for better HER performance. Ye et al. [83] synthesized single Pt atoms anchored on aniline-stacked graphene (Pt SASs/AG) via the microwave reduction method. The prepared sample (44% Pt) showed a very low onset overpotential of 12 mV at 10 mA cm^{-2} in an acidic medium.

Such excellent HER activity can be attributed to the high electrical conductivity of graphene. Further, Zhu et al. [84] synthesized $Ni_{0.85}Se$ nanospheres uniformly decorated on reduction of graphene oxide ($Ni_{0.85}Se$/rGO) hybrid using a facile hydrothermal process. The FESEM and TEM micrographs revealed that $Ni_{0.85}Se$ nanospheres consist of several nanosheets. In addition, the $Ni_{0.85}Se$ nanospheres were uniformly decorated on rGO sheets that resulted in a high specific surface area (Figure 6.5a–d). The prepared catalyst showed an overpotential of 128 mV at a current density of 10 mA cm^{-2} and a small Tafel slope of 91 mV dec^{-1}. Also, it showed high current retention of about 97.7% even after 1500 cycles. Teich et al. [85] reported 3D MoS_2/graphene heterostructures synthesized via the atomic layer deposition (ALD) method. An ideal annealing temperature of 500°C is found to increase electrocatalytic performance with an overpotential of 180 mV and a low Tafel slope of 47 mV dec^{-1}. They found that the density of defects/catalytic sites and the crystallinity of the sample are mutually exclusive.

Khiarak et al. [86] grew CoNi metal-organic framework (MOF) nanosheets on high-quality porous conductive graphene (PCG) through template-assisted chemical vapor deposition followed by the solvothermal method. PCG's highly porous 3D structure with open channels ranging in size from 200 to 500 nm offered a large active surface area and made gas evolution easier. The prepared sample showed an overpotential of 265 mV at 10 mA cm^{-2} and a low Tafel slope of 44.5 mV dec^{-1}. According to the findings, the bi-metal MOF particles boosted ECSA by around 12 times and supplied several active catalytic sites for HER. Sung Chul Yi et al. [87] prepared $Co_{0.5}MoS_2$ and nitrogen sulfur-doped graphene oxide composite through a simple hydrothermal method. It was found that $Co_{0.5}MoS_2$ possesses a 2D nanosheets structure. The $Co_{0.5}MoS_2$/N, S-rGO composite showed a low overpotential of 178 mV at a current density of 10 mA cm^{-2} and a Tafel slope of 63.8 mV dec^{-1}. Jung et al. [88] reported graphene oxide (GO) for the development of nitrogen-coordinated atomic metal sites (M—N—C) synthesized via the Hummers method. The synthesis of Co-N moieties might be substantially facilitated by a high number of basal oxygen functional groups near anchoring cobalt sites; however, the production of Co-N moieties was hampered by a small number of basal plane oxygen species. The degree of Co-N saturation was highly associated with the HER activity trend of the prepared Co/GO_x-T catalysts. The prepared sample Co/$GO_{35.7}$¯750 showed an overpotential of about 210 mV at a current density of 10 mA cm^{-2} (Figure 6.5e). The overpotentials of Co/$GO_{35.7}$-T are lower than those of Co/$GO_{27.8}$¯T, showing that the large proportion of Co-N moieties in Co/$GO_{35.7}$¯T resulted an efficient hydrogen production (Figure 6.5f). Figure 6.5g shows that HER performance is directly proportional to the quantity of Co-N sites and exhibited excellent cyclic stability up to 5000 CV cycles (Figure 6.5h).

6.4.2 Transition Metal Dichalcogenides (TMDs) as Electrocatalysts for the HER

Two-dimensional TMDs have recently received interest in photocatalytic water splitting applications due to their outstanding chemical, optical, electrical, and mechanical characteristics [6]. The layers that are piled together have a Van der Waals force between them. TMDs as HER catalysts have made remarkable developments in recent years, owing to increased basic knowledge of the catalytic process and advances in chemical, electrical,

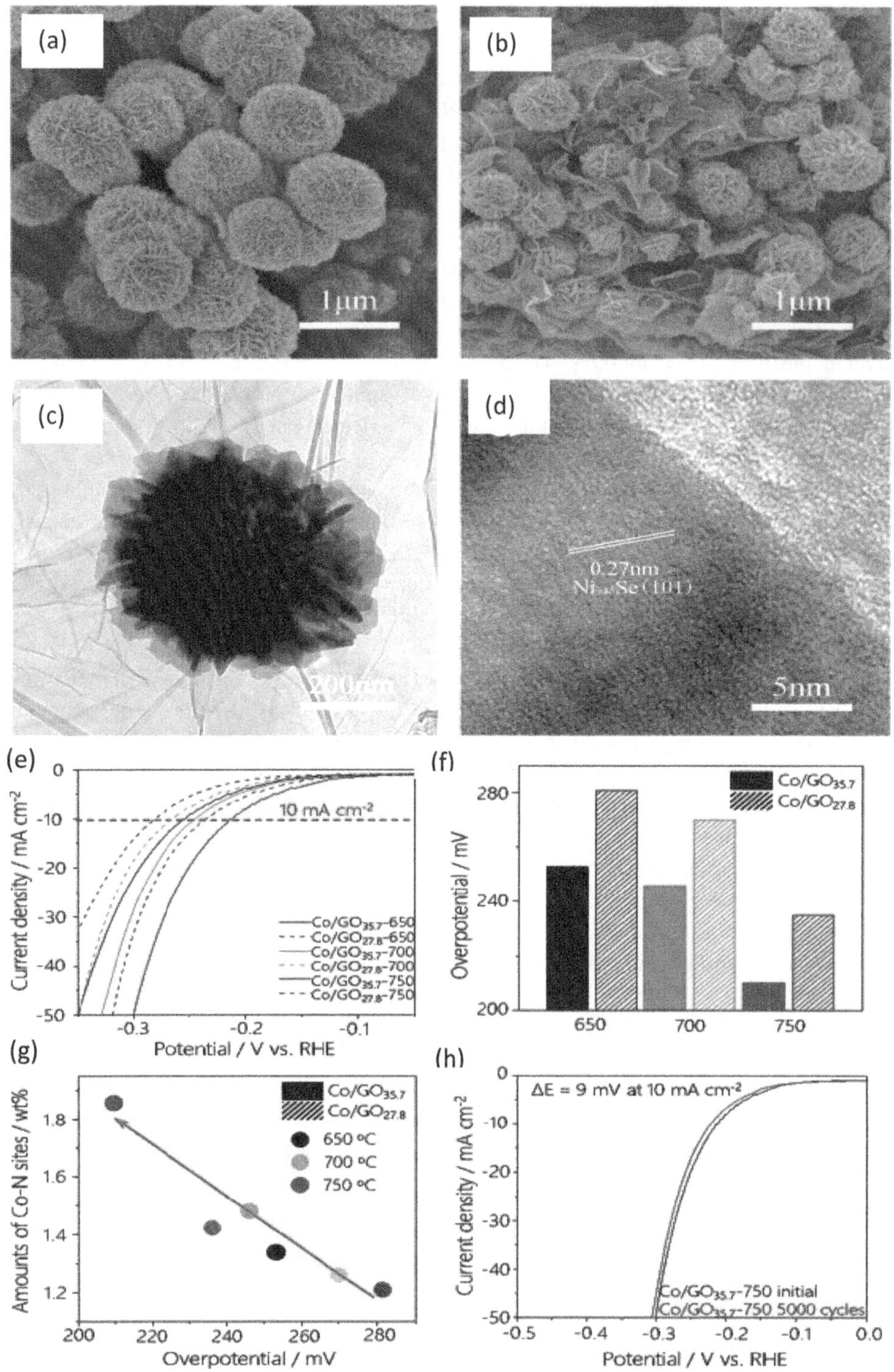

Figure 6.5 (a) SEM micrographs of (a) $Ni_{0.85}Se$ and (b) $Ni_{0.85}Se$/rGO hybrids; (c) TEM micrograph of $Ni_{0.85}Se$/rGO hybrid; (d) High resolution transmission electronic microscope (HRTEM) micrograph of $Ni_{0.85}Se$/rGO hybrid [86], Zhu et al. (2020), Reproduced with permission from ELSEVIER; (e) Polarization curves; (f) HER overpotentials; (g) graph between amounts of Co-N sites and the HER overpotentials; and (h) stability polarization curves of Co/GO samples at different annealing temperatures.

and structural tailoring of TMDs. TMDs were first used for electrochemical water splitting in the late 1970s [89]. Further, a-few-layer TMDs have exceptional characteristics due to surface effects and quantum-confinement states, whereas bulk TMDs lack these features [90]. The bulk semiconducting TMDs exhibited indirect bandgaps, while the monolayer TMDs possessed direct bandgaps. It indicates that they might be employed as effective photocatalysts with higher photoluminescence (PL). The two basic phases of WS_2 and MoS_2 are 1T and 2H, which have differing characteristics. Different methods exist to achieve these phases based on their structure and design [91]. Generally, there are two methods for the synthesis of TMDs: bottom-up and top-down [92, 93]. However, we can tailor the edges, crystallinity, and layer numbers of TMDs to use them as extremely effective catalysts utilizing a combination of these procedures.

Zhang et al. [94] reported interface engineering of heterostructures of MoS_2 and Ni_3S_2 on nickel foam. The prepared heterostructure showed excellent HER and OER activity in the 1 M KOH electrolyte. It exhibited a low Tafel slope of 83.1 mV dec^{-1} with long-term cyclic stability up to 10 h. Huang et al. [95] synthesized a CNT@$MoSe_2$ nanohybrid through a single-step solvothermal reaction. It showed outstanding HER performance with a low onset potential of −0.07 V vs. reversible hydrogen electrode (RHE) and a very small Tafel slope of 58 mV dec^{-1}. Due to its well-designed hierarchical architecture, it allows for not only homogeneous dispersal of $MoSe_2$ nanosheets on the large surface area of the highly conductive carbon nanotubes (CNTs) but also close interaction between $MoSe_2$ and CNTs. Sun et al. [96] prepared 1 T-phase enriched WS_2|P catalyst via a combination of solvothermal and chemical vapor deposition (CVD) methods. The sample exhibited stacking of surface nanosheets, which provides greater edge exposure to achieve a successful hydrogen evolution.

The prepared catalyst showed excellent electrochemical HER performance in acidic and basic mediums. It exhibited a low onset overpotential of 125 mV under 10 mA cm^{-2} with a low Tafel slope of 73.73 mV dec^{-1}. The development of the 1 T-phase WS_2|P catalyst, which provides higher HER active sites and improves the intrinsic electronic conductivity, is credited with the high HER activity. Wang et al. [97] successfully synthesized graphene-like WSe_2 nanosheets via a one-pot solvothermal method. It was observed that the WSe_2 nanosheets are highly porous and have a graphene-like structure. The prepared WSe_2 nanosheets exhibited an outstanding HER performance with a low Tafel slope of 78 mV dec^{-1}, a small onset overpotential of ~150 mV, and a very high cyclic stability. Yanrong Li et al. [98] reported a WSe_2/CNTs hybrid network (WCHN) consisting of a few layers of WSe_2 nanosheets interlaced with CNTs through a one-pot solvothermal approach. It has a small Tafel slope of 59.7 mV dec^{-1} and a low onset overpotential of 120 mV. The unique network topology created by 2D WSe_2 nanosheets interlaced with one-dimensional (1D) highly conductive CNTs is responsible for the considerable improvement in the HER performance of WCHN. Pramoda et al. [99] synthesized MoS_2–Borocarbonitride (BC_7N) Nanocomposites and studied their photochemical HER activity. The electrochemical and photochemical HER performance of the nanocomposite was examined using a variety of spectroscopic techniques, including fluorescence labeling. The G/BCN–MoS_2 (1:2) exhibited a low onset overpotential of 30 mV and an overpotential of 35 mV at a current density of 10 mV cm^{-2}. It showed a small Tafel slope of 33 mV dec^{-1} with long-term cyclic stability.

Shi et al. [100] reported thickness-tunable tantalum disulfide (TaS_2) flakes through a simple chemical vapor deposition route. The 2H-TaS_2 triangle having a thickness of~20 nm was observed in the microstructure analysis. The 2H-TaS_2 exhibited a low overpotential of 65 mV to achieve a current density of 10 mA cm^{-2} and a small Tafel slope of 33 mV dec^{-1}, which is very close to Pt (~ 31 mV dec^{-1}). Seok et al. [101] reported 2D layered semi-metallic $MoTe_2$ through the flux method. It showed a low overpotential of −0.356 V at 10 mA cm^{-2} and a small Tafel slope of 127 mV dec^{-1}. Zhao et al. [102] synthesized MoS_2 nanoflowers doped with FeS_2 via a hydrothermal process. MoS_2 crystals have a regular shape that consists of a nanosheet with a large number of MoS_2 layers (Figure 6.6A–C).

The nanoflower is made up of tens to hundreds of gently curved petals that develop perpendicular to the center, increasing the number of edges and corner atoms (Figure 6.6D–F). This electrode requires just a 136 mV overpotential to provide a current density of 10 mA cm^{-2} and keeps the catalytic performance stable for at least 24 h. Toh et al. [103] prepared MoS_2

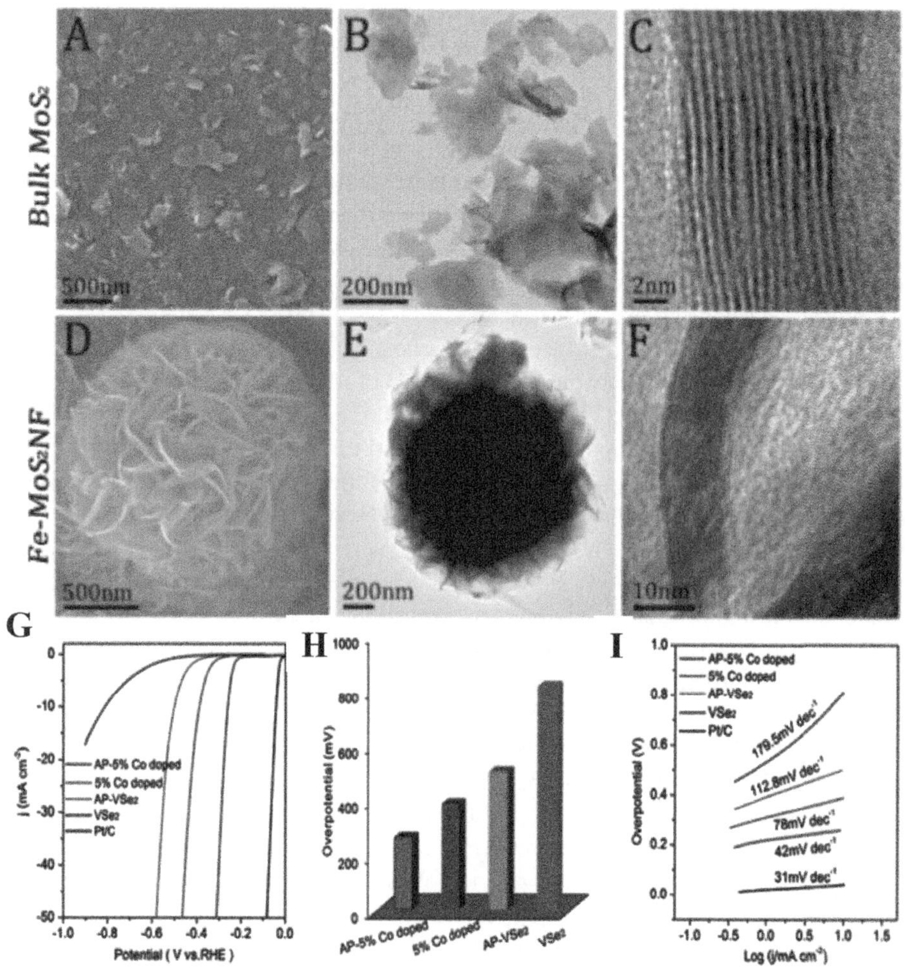

Figure 6.6 SEM, TEM, and HRTEM micrographs of (A–C) [104], Zhao et al. (2017), Reproduced with permission from ELSEVIER pure MoS_2 and (D–F) $FeMoS_2NF$; (G) LSV curves; (H) overpotential comparison of different electrocatalysts; and (I) Tafel plots of prepared VSe_2-based electrodes. Adapted from [104]. Reproduced with permission from [104]. Copyright 2017 Elsevier Ltd.

and WS_2 in their 3R phase and investigated their electrochemical HER activity. The 3R-MoS_2 and 3R-WS_2 showed a Tafel slope of 113 and 111 mV dec^{-1}, respectively. Fang et al. [104] synthesized hexagonal $CoSe_2$ nanosheets (NSs) on N-doped reduced graphene oxides (rGO) through the hydrothermal method. The hexagonal shape of $CoSe_2$ provides a chemically favorable basal plane and additional active sites for HER. It requires a low overpotential of 172 mV to achieve a current density of 10 mA cm^{-2}, and the very small Tafel slope is 35.2 mV dec^{-1}. Qianfan Zhang et al. [105] developed a simple solvothermal ion exchange process to develop S-doped $CoSe_2$ nanosheets. The prepared sample exhibited a low overpotential of 88 mV at a current density of 10 mA cm^{-2}. It exhibited a Tafel slope of 50 mV dec^{-1} and long-term cyclic stability. The resulting S-$CoSe_2$ has a porous lamellar structure, and the S-to-Se atomic ratio in the finished product can be adjusted by changing the preparation conditions. Further, Zhou et al. [106] prepared an S-$CoSe_2$/$CoMoO_4$ hybrid via the hydrothermal method and studied the HER activity. It showed an overpotential of 177 mV to afford a current density of 10 mA cm^{-2} and a Tafel slope of 54 mV dec^{-1}. The results demonstrate that the electrocatalyst has a high electrocatalytic activity for the HER because of the closely linked nature of $CoMoO_4$ and $CoSe_2$/DETA, the aid of an S heteroatom, and the unique nanosheet shape. Zhu et al. [107] successfully synthesized Pt-decorated VS_2 electrocatalyst via hydrothermal treatment. It needed only an overpotential of 77 mV to achieve a current density of 10 mA cm^{-2} with a Tafel slope of 40.13 mV dec^{-1}. Further, theoretical studies show that Pt decorating can improve the absorption behavior of H atoms and charge-transfer kinetics, hence improving HER performance. Kumar et al. [108] developed an efficient and cost-effective hydrothermal method to synthesize nanodiscs (NDs)-like VS_2. The electrochemical characteristics of NDs-like structures consisting of thin nanoplates containing active sites were discovered to be highly influenced by them. It showed highly efficient HER activity with a Tafel slope of 79 mV dec^{-1}. However, it exhibited a large overpotential of 440 mV to achieve a current density of 10 mA cm^{-2}. Wang et al. [109] synthesized TM@VS_2s (TM = Ti–Ni) atoms and investigated their HER catalytic performance. Because of the charge modulation in S sites, the HER activity of TM@VS_2s is substantially superior to those of pure VS_2. Further, Bohua Dong et al. [110] reported 2D colloidal VSe_2 nanosheets synthesized via a one-pot colloidal method. The derived AP-5 percent Co-doped VSe_2 has shown improved electrocatalytic activity and good HER endurance. This is due to the synergistic impact of clean surfaces, defects or cracks, and heteroatom dopants on the basal plane. It showed an onset overpotential of 160 mV and a Tafel slope of 42 mV dec^{-1} (Figure 6.6G–I). Ling et al. [111] successfully synthesized layered WTe_2 (semi-metallic) for active HER electrocatalysis. According to the XPS analysis, the edge structure enhances the formation of W–O and Te–O bonds. The prepared electrocatalyst exhibited excellent electrochemical activity with an overpotential of 350 mV at 10 mA cm^{-2} and a Tafel slope of 96 mV dec^{-1}. Yan et al. [112] prepared perpendicularly anchored $ReSe_2$ nanoflakes on rGO nanosheets through the single-step hydrothermal method. The rGO support offers a growth substrate for few-layered $ReSe_2$ nanoflakes while allowing for abundant exposure of $ReSe_2$ edge/corner sites, which is ideal for increasing catalytic activity of $ReSe_2$@rGO. The synthesized material ($ReSe_2$@rGO) showed excellent electrochemical performance with a low overpotential of 145.3 mV at 10 mA cm^{-2} and a small Tafel slope of 40.7 mV dec^{-1}. Li et al. [113] synthesized a 1 T-MoS_2/N-doped $NiSe_2$ composite through a solvothermal process. Here, $NiSe_2$ nanoparticles provided co-catalytic support to MoS_2 nanosheets. It needed only an overpotential of 94 mV to obtain a current density of 10 mA cm^{-2}. Also, it exhibited a Tafel slope of 58 mV dec^{-1}. The abundance of exposed active edge sites resulting from the ultra-small size and widely scattered shape of 1T-MoS_2 nanosheets is responsible for such high HER results. Li et al. [114] studied the influence of S vacancies and structure on the HER performance of MoS_2. They observed that two distinct regions of surface S vacancies correspond to two distinct HER regimes. Also, it was observed that S-stripping big vacancies improve HER performance by forming domains of undercoordinated molybdenum (Mo), but at a slower pace. The prepared sample MoS_2-6H showed a small Tafel slope of 85 mV dec^{-1}. Gopalakrishnan et al. [115] reported ultrathin NbS_2 nanoflakes synthesized via a facile ambient pressure annealing technique. The optical bandgap was found to be 1.49 eV, indicating that 3R-NbS_2 is semiconducting. The NbS_2/rGO hybrid exhibited an excellent catalytic activity with an onset potential of 0.15 eV and a small Tafel slope of 84 mV dec^{-1}.

6.5 Challenges in 2D Materials for CO_2 Reduction and H_2 Production

Although many 2D nanostructured materials have made profound progress in electrocatalytic reduction of CO_2, certain challenges are prevailing, which limits their practical applications. First, commercial production of these 2D nanomaterials with controlled morphology and low agglomeration is still prevailing due to strong in-plane bonds, electrostatic interaction, and π–π stacking. Second, the specific component modification and structural tailoring at the atomic level for improving the electrocatalyst's selectivity, activity, and stability for CO_2 reduction and H_2 production for industrial applications is a challenging job. Third, the non-universal catalyst loading, substrates, and electrolyte play a key role in enhancing CO_2 reduction performance and H_2 production ability. The aqueous electrolytes are not feasible for CO_2 reduction as they facilitate the competing HER, and also, the membrane resistance restricts the rapid charge and mass transferability [116]. Various efforts are being made to overcome these issues, like the *in-situ* characterization of materials at the time of processing to get the actual facts that deteriorate the electrocatalyst's performance and lead to the formation of H_2 during the CO_2 reduction process. It will also provide information about the real reaction pathways followed for CO_2 reduction and H_2 production. The uniform structural tailoring of 2D materials at the nano range is essential for improving CO_2 reduction and H_2 production, and their performance at the same time is still prevailing. Theoretical studies suggested that the doping of another metal in the 2D materials has shown promising results for improving the electrocatalytic performance [117]. However, the doping of certain metals and the formation of their oxide species restrict their applicability. Efforts are being made for technological breakthroughs for the commercial application of 2D materials as low cost and efficient electrocatalysts for CO_2 reduction and H_2 production.

6.6 Conclusion and Outlook

The role of structural, morphological, and surface features of 2D materials for electrocatalytic CO_2 reduction and H_2 production has been detailed. The nanostructuring, tailoring of surface features, and incorporation of metal/non-metal species enhance the electrocatalytic activity. The emerging nanostructured 2D electrocatalysts, a special class of materials MXenes, atomically thin TMs/TMOs, nanostructured TMDs, and metal-free 2D materials, have shown extraordinary electrolytic activity toward CO_2 reduction. The role of edge sites and the influence of the layered 2D materials with improved activity and stability has been highlighted. The oxide species have shown less efficiency toward H_2 production due to low conductivity, whereas metal carbide and sulfides have shown promising results to replace Pt due to their similar electronic features. The exceptional results and characteristics of the wonder material graphene, having a high surface area toward CO_2 reduction and H_2 production as catalyst or support, have been elaborated on in detail. Finally, challenges prevailing for the large production of nanostructured 2D materials with uniform features have been predicted. Surface engineering of these 2D materials at the nano scale with new approaches will be explored to enhance their activity for longer durations. Further, composite structures with tailored surfaces can be synthesized and predicted on the basis of experimental and theoretical studies. An investigation based on theoretical and experimental results together will significantly advance this research area and pave the way to address the environmental issues caused by CO_2 emissions. Moreover, a new energy source (H_2) produced from renewable resource water can be commercialized.

References

1 Kumar, N., Ray, S.S., Kumar, S. et al. (2020). Polypyrrole-promoted RGO–MoS_2 nanocomposites for enhanced photocatalytic conversion of CO_2 and H_2O to CO, CH_4, and H_2 products. *ACS Appl. Energy Mater.* 3: 9897–9909. https://doi.org/10.1021/acsaem.0c01602.

2 Hernandez, S., Farkhondehfal, M.A., Sastre, F. et al. (2017). Syngas production from electrochemical reduction of CO_2: Current status and prospective implementation. *Green Chem.* 19: 2326–2346. https://doi.org/10.1039/C7GC00398F.

3 Mahvelati-Shamsabadi, T. and Lee, B.K. (2020). Photocatalytic H_2 evolution and CO_2 reduction over phosphorus-doped g-C_3N_4 nanostructures: Electronic, optical, and surface properties. *Renew. Sustain. Energy Rev.* 130: 109957. https://doi.org/10.1016/j.rser.2020.109957.

4 Liu, J., Guo, C., Vasileff, A. et al. (2017). Nanostructured 2D materials: Prospective catalysts for electrochemical CO_2 reduction. *Small Methods* 1: 1600006–1600014. https://doi.org/10.1002/smtd.201600006.

5 Sun, Z., Ma, T., Tao, H. et al. (2017). Fundamentals and challenges of electrochemical CO_2 reduction using two-dimensional materials. *Chem.* 3: 560–587. https://doi.org/10.1016/j.chempr.2017.09.009.

6 Manzeli, S., Ovchinnikov, D., Pasquier, D. et al. (2017). 2D transition metal dichalcogenides. *Nat. Rev. Mater.* 28.2: 1–15. https://doi.org/10.1038/natrevmats.2017.33.

7 Zhu, X., and Li, Y. (2019). Review of two-dimensional materials for electrochemical CO_2 reduction from a theoretical perspective. *Wiley Interdiscip. Rev. Comput. Mol. Sci.* 9: 1–13. https://doi.org/10.1002/wcms.1416.

8 Stux, A.M., Laberty-Robert, C., and Swider-Lyons, K.E. (2008). Pechini synthesis and characterization of molybdenum carbide and nickel molybdenum carbide. *J. Solid State Chem.* 181: 2741–2747. https://doi.org/10.1016/j.jssc.2008.06.050.

9 Liu, J.H., Yang, L.M., Ganz, E. et al. (2019). Electrocatalytic reduction of CO_2 by two-dimensional transition metal porphyrin sheets. *J. Mater. Chem. A.* 7: 11944–11952. https://doi.org/10.1039/c9ta01188a.

10 Choi, S., Kim, Y.J., Jeon, J. et al. (2019). Scalable two-dimensional lateral metal/semiconductor junction fabricated with selective synthetic integration of transition-metal-carbide (Mo_2C)/-Dichalcogenide (MoS_2). *ACS Appl. Mater. Interfaces.* 11: 47190–47196. https://doi.org/10.1021/acsami.9b13660.

11 Ma, J., Jiang, Q., Zhou, Y. et al. (2021). Tuning the chemical properties of Co–$Ti_3C_2T_x$ MXene materials for catalytic CO_2 reduction. *Small* 17: 1–11. https://doi.org/10.1002/smll.202007509.

12 Mohammad, A., Kibum, K., Cong, L. et al. (2016). Nanostructured transition metal dichalcogenide electrocatalysts for CO_2 reduction in ionic liquid. *Science (80)* 353: 464–467.

13 Chen, W.-F., Iyer, S., Sasaki, K. et al. (2013). Biomass-derived electrocatalytic composites for hydrogen evolution. *Energy Environ. Sci.* 6: 1818. https://doi.org/10.1039/c3ee40596f.

14 Gupta, U. and Rao, C.N.R. (2017). Hydrogen generation by water splitting using MoS_2 and other transition metal dichalcogenides. *Nano Energy* 41: 49–65. https://doi.org/10.1016/j.nanoen.2017.08.021.

15 Mir, R.A. and Pandey, O.P. (2020). An ecofriendly route to synthesize C-Mo_2C and C/N-Mo_2C utilizing waste polyethene for efficient hydrogen evolution reaction (HER) activity and high performance capacitors. *Sustain. Energy Fuels.* 4: 655–669. https://doi.org/10.1039/c9se00516a.

16 Tong, R., Ng, K.W., Wang, X. et al. (2020). Two-dimensional materials as novel co-catalysts for efficient solar-driven hydrogen production. *J. Mater. Chem. A* 8: 23202–23230. https://doi.org/10.1039/d0ta08045d.

17 Han, Q., Liu, K., Chen, J. et al. (2003). Hydrogen evolution reaction on amorphous Ni-S-Co alloy in alkaline medium. *Int. J. Hydrogen Energy.* 28: 1345–1352. https://doi.org/10.1016/S0360-3199(03)00043-0.

18 Liu, T., Liu, R., Li, Q. et al. (2020). Theoretical insight into the role of defects and facets in the selectivity of products in water oxidation over Bismuth Vanadate ($BiVO_4$). *ACS Sustain. Chem. Eng.* 8: 1980–1988. https://doi.org/10.1021/acssuschemeng.9b06404.

19 Elbaz, L., Kreller, C.R., Henson, N.J. et al. (2014). Electrocatalysis of oxygen reduction with platinum supported on molybdenum carbide-carbon composite. *J. Electroanal. Chem.* 720–721: 34–40. https://doi.org/10.1016/j.jelechem.2014.02.023.

20 Ghobrial, S., Kirk, D.W., and Thorpe, S.J. et al. (2019). Amorphous Ni-Nb-Y alloys as hydrogen evolution electrocatalysts. *Electrocatalysis* 10: 243–252. https://doi.org/10.1007/s12678-019-00519-4.

21 Greeley, J., Jaramillo, T.F., Bonde, J. et al. (2006). Computational high-throughput screening of electrocatalytic materials for hydrogen evolution. *Nat. Mater.* 5: 909–913. https://doi.org/10.1038/nmat1752.

22 Nørskov, J.K., Bligaard, T., Rossmeisl, J. et al. (2009). Towards the computational design of solid catalysts. *Nat. Chem.* 1: 37–46. https://doi.org/10.1038/nchem.121.

23 Alhajri, N.S., Yoshida, H., Anjum, D.H. et al. (2013). Synthesis of tantalum carbide and nitride nanoparticles using a reactive mesoporous template for electrochemical hydrogen evolution, *J. Mater. Chem. A* 1: 12606. https://doi.org/10.1039/c3ta12984e.

24 Gonçalves, J.M., Ireno Da Silva, M., Angnes, L. et al. (2020). Vanadium-containing electro and photocatalysts for the oxygen evolution reaction: A review. *J. Mater. Chem. A* 8: 2171–2206. https://doi.org/10.1039/c9ta10857b.

25 Mahmood, N., Yao, Y., Zhang, J.W. et al. (2018). Electrocatalysts for hydrogen evolution in alkaline electrolytes: Mechanisms, challenges, and prospective solutions. *Adv. Sci.* 5: 1700464. https://doi.org/10.1002/advs.201700464.

26 Hasani, A., Tekalgne, M., Van Le, Q. et al. (2019). Two-dimensional materials as catalysts for solar fuels: Hydrogen evolution reaction and CO_2 reduction. *J. Mater. Chem. A* 7: 430–454. https://doi.org/10.1039/C8TA09496A.

27 Yang, J. and Shin, H.S. (2014). Recent advances in layered transition metal dichalcogenides for hydrogen evolution reaction. *J. Mater. Chem. A* 2: 5979–5985. https://doi.org/10.1039/C3TA14151A.

28 Zhang, X. and Xie, Y. (2013). Recent advances in free-standing two-dimensional crystals with atomic thickness: Design, assembly and transfer strategies. *Chem. Soc. Rev.* 42: 8187–8199. https://doi.org/10.1039/C3CS60138B.

29 Ganguly, P., Harb, M., Cao, Z. et al. (2019). 2D nanomaterials for photocatalytic hydrogen production. *ACS Energy Lett.* 4: 1687–1709. https://doi.org/10.1021/acsenergylett.9b00940.

30 Burgers, J.M. (1940). Geometrical considerations concerning the structural irregularities to be assumed in a crystal. *Proc. Phys. Soc.* 52: 23. https://doi.org/10.1088/0959-5309/52/1/304.

31 Li, J., Zhao, Z., Ma, Y. et al. (2017). Graphene and their hybrid electrocatalysts for water splitting. *ChemCatChem.* 9: 1554–1568. https://doi.org/10.1002/CCTC.201700175.

32 Zhang, H. (2015). Ultrathin two-dimensional nanomaterials. *ACS Nano.* 9: 9451–9469. https://doi.org/10.1021/ACSNANO.5B05040.

33 Yang, Z., Ren, J., Zhang, Z. et al. (2015). Recent advancement of nanostructured carbon for energy applications. *Chem. Rev.* 115: 5159–5223. https://doi.org/10.1021/CR5006217.

34 Chen, C., Hu, J., Yang, X. et al. (2012). Ambient-stable black phosphorus-based 2D/2D S-scheme heterojunction for efficient photocatalytic CO_2 reduction to syngas. *ACS Appl. Mater. Interfaces* 13: 20162–20173. https://doi.org/10.1021/acsami.1c03482.

35 Wang, H., Chen, Y., Hou, X. et al. (2016), Nitrogen-doped graphenes as efficient electrocatalysts for the selective reduction of carbon dioxide to formate in aqueous solution. *Green Chem.* 18: 3250–3256. https://doi.org/10.1039/c6gc00410e.

36 Powar, N.S., Hiragond, C.B., Bae, D. et al. (2022). Two-dimensional metal carbides for electro- and photocatalytic CO_2 reduction: Review. *J. CO_2 Util.* 55: 101814–101834. https://doi.org/10.1016/j.jcou.2021.101814.

37 Zaman, S. and Smith, K.J. (2012). A review of molybdenum catalysts for synthesis gas conversion to alcohols: Catalysts, mechanisms and kinetics. *Catal. Rev. – Sci. Eng.* 54: 41–132. https://doi.org/10.1080/01614940.2012.627224.

38 Shanker, G.S., Biswas, A., and Ogale, S. (2021). 2D materials and their heterostructures for photocatalytic water splitting and conversion of CO_2 to value chemicals and fuels. *J. Phys. Energy* 3: 022003–022037. https://doi.org/10.1088/2515-7655/abdcab.

39 Gao, W., Shi,Y., Zhang, Y. et al. (2016). Molybdenum carbide anchored on graphene nanoribbons as highly efficient all-pH hydrogen evolution reaction electrocatalyst. *ACS Sustain. Chem. Eng.* 4: 6313–6321. https://doi.org/10.1021/acssuschemeng.6b00859.

40 Liu, Y., Kelly, T.G., Chen, J.G. et al. (2013). Metal carbides as alternative electrocatalyst supports. *ACS Catal.* 3: 1184–1194. https://doi.org/10.1021/cs4001249.

41 Regmi, Y.N., Waetzig, G.R., Duffee, K.D. et al. (2015). Carbides of group IVA, VA and VIA transition metals as alternative HER and ORR catalysts and support materials. *J. Mater. Chem. A* 3: 10085–10091. https://doi.org/10.1039/c5ta01296a.

42 Jaramillo, T.F., Jørgensen, K.P., Bonde, J. et al. (2007). Identification of active edge sites for electrochemical H_2 evolution from MoS_2 nanocatalysts. *Science (80)* 317: 100–102. https://doi.org/10.1126/science.1141483.

43 Lin, L., Miao, N., Huang, J. et al. (2017). Nanostructured 2D materials: Prospective catalysts for electrochemical CO reduction *Nano Energy* 38: 544–552. https://doi.org/10.1016/J.NANOEN.2017.06.008.

44 Asadi, M., Kumar, B., Behranginia, A. et al. (2014). Robust carbon dioxide reduction on molybdenum disulphide edges. *Nat. Commun.* 5: 1–8. https://doi.org/10.1038/ncomms5470.

45 Liu, J., Guo, C., Vasileff, A. et al. (2017). Nanostructured 2D materials: Prospective catalysts for electrochemical CO_2 reduction. *Small Methods* 1: 1–7. https://doi.org/10.1002/smtd.201600006.

46 Abbasi, P., Asadi, M., Liu, C. et al. (2017). Tailoring the edge structure of molybdenum disulfide toward electrocatalytic reduction of carbon dioxide. *ACS Nano* 11: 453–460. https://doi.org/10.1021/acsnano.6b06392.

47 Salehi-Khojin, A., Jhong, H.R.M., Rosen, B.A. et al. (2013). Nanoparticle silver catalysts that show enhanced activity for carbon dioxide electrolysis. *J. Phys. Chem. C* 117: 1627–1632. https://doi.org/10.1021/jp310509z.

48 Lv, R., Robinson, J.A., Schaak, R.E. et al. (2015). Transition metal dichalcogenides and beyond: Synthesis, properties, and applications of single- and few-layer nanosheets. *Acc. Chem. Res.* 48: 56–64. https://doi.org/10.1021/ar5002846.

49 Sun, Z. and Hu, Y.H. (2021). 3D graphene materials from the reduction of CO_2. *Accounts Mater. Res.* 2: 48–58. https://doi.org/10.1021/accountsmr.0c00069.

50 Hasani, A., Teklagne, M.A., Do, H.H. et al. (2020). Graphene-based catalysts for electrochemical carbon dioxide reduction. *Carbon Energy* 2: 158–175. https://doi.org/10.1002/cey2.41.

51 Zhang, X., Wang, W., and Yang, Z. (2020). CO_2 reduction on metal- and nitrogen-codoped graphene: Balancing activity and selectivity via coordination engineering. *ACS Sustain. Chem. Eng.* 8: 6134–6141. https://doi.org/10.1021/acssuschemeng.0c01923.

52 Chai, G.L. and Guo, Z.X. (2016). Highly effective sites and selectivity of nitrogen-doped graphene/CNT catalysts for CO_2 electrochemical reduction. *Chem. Sci.* 7: 1268–1275. https://doi.org/10.1039/c5sc03695j.

53 Sreekanth, N., Nazrulla, M.S., Vineesh, T.V. et al. (2015). Metal-free boron-doped graphene for selective electroreduction of carbon dioxide to formic acid/formate. *Chem. Commun.* 51: 16061–16064. https://doi.org/10.1039/c5cc06051f.

54 Wu, J., Liu, M., Sharma, P.P. et al. (2016). Incorporation of nitrogen defects for efficient reduction of CO_2 via two-electron pathway on three-dimensional graphene foam. *Nano Lett.* 16: 466–470. https://doi.org/10.1021/acs.nanolett.5b04123.

55 Schlo, R., Arrigo, R., and Ha, M. (2008). *Dynamic surface rearrangement and thermal stability of nitrogen functional groups on carbon nanotubes.* 40: 4891–4893. https://doi.org/10.1039/b812769g.

56 Chen, Y., Li, C.W., and Kanan, M.W. (2012). Aqueous CO 2 reduction at very low overpotential on oxide-derived Au nanoparticles. 0–3. *J. Am. Chem. Soc.* 134: 19969–19972.

57 Liu, Y., Zhao, J., and Cai, Q. (2016). Pyrrolic-nitrogen doped graphene: A metal-free electrocatalyst with high efficiency and selectivity for the reduction of carbon dioxide to formic acid: A computational study. *Phys. Chem. Chem. Phys.* 18: 5491–5498. https://doi.org/10.1039/c5cp07458d.

58 Terrones, H., Lv, R., and Terrones, M. (2012). The role of defects and doping in 2D graphene sheets and 1D nanoribbons. *Rep. Prog. Phys.* 75: 062501. https://doi.org/10.1088/0034-4885/75/6/062501.

59 Zhang, B., Zhao, T.J., Feng, W.J. et al. (2018). Polarized few-layer g-C_3N_4 as metal-free electrocatalyst for highly efficient reduction of CO_2. *Nano Res.* 11: 2450–2459. https://doi.org/10.1007/s12274-017-1866-y.

60 Pan, F., Li, B., Deng, W. et al. (2019). Promoting electrocatalytic CO_2 reduction on nitrogen-doped carbon with sulfur addition. *Appl. Catal. B Environ.* 252: 240–249. https://doi.org/10.1016/j.apcatb.2019.04.025.

61 Qiao, B., Wang, A.L., Yang, X. et al. (2011). Single-atom catalysis of CO oxidation using Pt 1/FeO_x. *Nat. Chem.* 3: https://doi.org/10.1038/nchem.1095.

62 Zhao, C., Wang, Y., Zhao, C. et al. (n.d.). Solid-diffusion synthesis of single-atom catalysts directly from bulk metal for efficient CO_2 reduction. *Joule* 3: 584–594. https://doi.org/10.1016/j.joule.2018.11.008.

63 Feng, Z., Wang, G., Wu, G. et al. (2018). Unveiling active sites of CO_2 reduction on nitrogen-coordinated and atomically dispersed iron and cobalt catalysts, *ACS Catal.* 8: 3116–3122. https://doi.org/10.1021/acscatal.8b00398.

64 Gao, S., Lin, Y., Jiao, X. et al. (2016). Partially oxidized atomic cobalt layers for carbon dioxide electroreduction to liquid fuel. *Nature* 529: 68–71. https://doi.org/10.1038/nature16455.

65 Li, F., Chen, L., Knowles, G.P. et al. (2017). Hierarchical mesoporous SnO_2 nanosheets on carbon cloth: A robust and flexible electrocatalyst for CO_2 reduction with high efficiency and selectivity. *Angew. Chem. Int. Ed.* 56: 505–509. https://doi.org/10.1002/anie.201608279.

66 Gao, S., Jiao, X., Sun, Z. et al. (2016). Ultrathin $Co3O_4$ layers realizing optimized CO_2 electroreduction to formate Angewandte. *Angew. Chem. Int. Ed.* 55: 698–702. https://doi.org/10.1002/anie.201509800.

67 Jin, H., Guo, C., Liu, X. et al. (2018). Emerging two-dimensional nanomaterials for electrocatalysis. *Chem. Rev.* 118: 6337–6408. https://doi.org/10.1021/acs.chemrev.7b00689.

68 Naguib, M., Mochalin, V.N., Barsoum, M.W. et al. (2014). 25th anniversary article: MXenes: A new family of two-dimensional materials. *Adv. Mater.* 26: 992–1005. https://doi.org/10.1002/adma.201304138.

69 Naguib, M., Mashtalir, O., Carle, J. et al. (2012). Two-dimensional transition metal carbides. *ACS Nano.* 6: 1322–1331. https://doi.org/10.1021/nn204153h.

70 Li, G., Tan, L., Wu, B. et al. (2017). Highly efficiently delaminated single-layered MXene nanosheets with large lateral size. *Langmuir* 33: 9000–9006. https://doi.org/10.1021/acs.langmuir.7b01339.

71 Zhang, X., Zhang, Z., Li, J. et al. (2017). Ti_2CO_2 MXene: A highly active and selective photocatalyst for CO_2 reduction. *J. Mater. Chem. A* 5: 12899–12903. https://doi.org/10.1039/c7ta03557h.

72 Li, N., Chen, X., Ong, W.J. et al. (2017). Understanding of electrochemical mechanisms for CO_2 capture and conversion into hydrocarbon fuels in transition-metal carbides (MXenes). *ACS Nano.* 11: 10825–10833. https://doi.org/10.1021/acsnano.7b03738.

73 Wang, H., Maiyalagan, T., and Wang, X. (2012). Review on recent progress in nitrogen-doped graphene: Synthesis, characterization, and its potential applications. *ACS Catal.* 2: 781–794. https://doi.org/10.1021/CS200652Y.

74 Guo Huang, Y., Lin Fan, H., Kai Chen, Z. et al. (2016). The effect of graphene for the hydrogen evolution reaction in alkaline medium. *Int. J. Hydrogen Energy* 41: 3786–3793. https://doi.org/10.1016/J.IJHYDENE.2015.12.113.

75 Akyüz, D., Keskin, B., Şahintürk, U. et al. (2016). Electrocatalytic hydrogen evolution reaction on reduced graphene oxide electrode decorated with cobaltphthalocyanine. *Appl. Catal. B Environ.* 188: 217–226. https://doi.org/10.1016/J.APCATB.2016.02.003.

76 Yang, X., Feng, X., Tan, H. et al. (2016). N-Doped graphene-coated molybdenum carbide nanoparticles as highly efficient electrocatalysts for the hydrogen evolution reaction. *J. Mater. Chem. A* 4: 3947–3954. https://doi.org/10.1039/C5TA09507G.

77 Wang, L., Li, Y., Xia, M. et al. (2017). Ni nanoparticles supported on graphene layers: An excellent 3D electrode for hydrogen evolution reaction in alkaline solution. *J. Power Sources* 347: 220–228. https://doi.org/10.1016/J.JPOWSOUR.2017.02.017.

78 Sun, Z., Fan, W., and Liu, T. (2017). Graphene/graphene nanoribbon aerogels as tunable three-dimensional framework for efficient hydrogen evolution reaction. *Electrochim. Acta.* 250: 91–98. https://doi.org/10.1016/J.ELECTACTA.2017.08.009.

79 Wang, H., Wang, X., Yang, D. et al. (2018). Co0.85Se hollow nanospheres anchored on N-doped graphene nanosheets as highly efficient, nonprecious electrocatalyst for hydrogen evolution reaction in both acid and alkaline media. *J. Power Sources* 400: 232–241. https://doi.org/10.1016/J.JPOWSOUR.2018.08.027.

80 Cardoso, J.A.S.B., Amaral, L., Metin, D.S.P. et al. (2017). Reduced graphene oxide assembled Pd-based nanoalloys for hydrogen evolution reaction. *Int. J. Hydrogen Energy* 42: 3916–3925. https://doi.org/10.1016/J.IJHYDENE.2016.11.100.

81 Qu, Y., Ke, Y., Shao, Y. et al. (2018). Effect of curvature on the hydrogen evolution reaction of graphene. *J. Phys. Chem. C* 122: 25331–25338. https://doi.org/10.1021/ACS.JPCC.8B06750/SUPPL_FILE/JP8B06750_SI_001.PDF.

82 Chen, Z., Wu, R., Liu, Y. et al. (2018). Ultrafine Co nanoparticles encapsulated in carbon-nanotubes-grafted graphene sheets as advanced electrocatalysts for the hydrogen evolution reaction. *Adv. Mater.* 30: 1802011. https://doi.org/10.1002/ADMA.201802011.

83 Ye, S., Luo, F., Zhang, Q. et al. (2019). Highly stable single Pt atomic sites anchored on aniline-stacked graphene for hydrogen evolution reaction. *Energy Environ. Sci.* 12: 1000–1007. https://doi.org/10.1039/C8EE02888E.

84 Zhu, M., Yan, Y., Yan, Q. et al. (2020). *In situ* growth of Ni_0.85Se on graphene as a robust electrocatalyst for hydrogen evolution reaction. *Int. J. Hydrogen Energy* 45: 10486–10493. https://doi.org/10.1016/J.IJHYDENE.2019.03.092.

85 Teich, J., Dvir, R., Henning, A. et al. (2020). Light and complex 3D MoS_2/graphene heterostructures as efficient catalysts for the hydrogen evolution reaction. *Nanoscale* 12: 2715–2725. https://doi.org/10.1039/C9NR09564K.

86 Khiarak, B.N., Hasanzadeh, M., and Simchi, A. (2021). Electrocatalytic hydrogen evolution reaction on graphene supported transition metal-organic frameworks. *Inorg. Chem. Commun.* 127: 108525. https://doi.org/10.1016/J.INOCHE.2021.108525.

87 Kumaran, Y., Maiyalagan, T., and Yi, S.C. (2021). An efficient $CoMoS_2$ nanosheets on nitrogen, sulfur dual doped reduced graphene oxide as an electrocatalyst for the hydrogen evolution reaction. *Int. J. Energy Res.* 45: 17397–17407. https://doi.org/10.1002/ER.5394.

88 Jung, J.Y., Hong, Y.L., Kim, J.G. et al. (2021). New insight of tailor-made graphene oxide for the formation of atomic Co-N sites toward hydrogen evolution reaction. *Appl. Surf. Sci.* 563: 150254. https://doi.org/10.1016/J.APSUSC.2021.150254.

89 Lin, L., Sherrell, P., Liu, Y. et al. (2020). Engineered 2D transition metal dichalcogenides: A vision of viable hydrogen evolution reaction catalysis. *Adv. Energy Mater.* 10: 1903870. https://doi.org/10.1002/AENM.201903870.

90 Kumar, A. and Xu, Q. (2018). Two-dimensional layered materials as catalyst supports. *ChemNanoMat.* 4: 28–40. https://doi.org/10.1002/CNMA.201700139.

91 Splendiani, A., Sun, L., Zhang, Y. et al. (2010). Emerging photoluminescence in monolayer MoS_2. *Nano Lett.* 10: 1271–1275. https://doi.org/10.1021/NL903868W.

92 Chou, S.S., Kaehr, B., Kim, J. et al. (2013). Chemically exfoliated MoS_2 as Near-infrared photothermal agents. *Angew. Chemie Int. Ed.* 52: 4160–4164. https://doi.org/10.1002/ANIE.201209229.

93 Zhang, W., Huang, J.K., Chen, C.H. et al. (2013). High-gain phototransistors based on a CVD MoS_2 monolayer. *Adv. Mater.* 25: 3456–3461. https://doi.org/10.1002/ADMA.201301244.

94 Zhang, J., Wang, T., Pohl, D. et al. (2016). Interface engineering of MoS_2/Ni_3S_2 heterostructures for highly enhanced electrochemical overall-water-splitting activity. *Angew. Chemie – Int. Ed.* 55: 6702–6707. https://doi.org/10.1002/ANIE.201602237.

95 Huang, Y., Lu, H., Gu, H. et al. (2015). CNT@$MoSe_2$ hybrid catalyst for efficient and stable hydrogen evolution. *Nanoscale* 7: 18595–18602. https://doi.org/10.1039/C5NR05739F.

96 Sun, L., Gao, M., Jing, Z. et al. (2022). 1 T-phase enriched P-doped WS_2 nanosphere for highly efficient electrochemical hydrogen evolution reaction. *Chem. Eng. J.* 429: 132187. https://doi.org/10.1016/J.CEJ.2021.132187.

97 Wang, X., Chen, Y., Zheng, B. et al. (2017). Graphene-like WSe_2 nanosheets for efficient and stable hydrogen evolution. *J. Alloys Compd.* 691: 698–704. https://doi.org/10.1016/J.JALLCOM.2016.08.305.

98 Wang, X., Chen, Y., Qi, F. et al. (2016). Interwoven WSe_2/CNTs hybrid network: A highly efficient and stable electrocatalyst for hydrogen evolution. *Electrochem. Commun.* 72: 74–78. https://doi.org/10.1016/J.ELECOM.2016.09.007.

99 Pramoda, K., Ayyub, M.M., Singh, N.K. et al. (2018). Covalently bonded MoS_2-borocarbonitride nanocomposites generated by using surface functionalities on the nanosheets and their remarkable HER activity. *J. Phys. Chem. C* 122: 13376–13384. https://doi.org/10.1021/acs.jpcc.7b10782.

100 Shi, J., Wang, X., Zhang, S. et al. (2017). Two-dimensional metallic tantalum disulfide as a hydrogen evolution catalyst. *Nat. Commun.* 81.8: 1–89. https://doi.org/10.1038/s41467-017-01089-z.

101 Seok, J., Lee, J.H., Cho, S. et al. (2017). Active hydrogen evolution through lattice distortion in metallic $MoTe_2$. *2D Mater.* 4: 025061. https://doi.org/10.1088/2053-1583/AA659D.

102 Zhao, X., Ma, X., Lu, Q. et al. (2017). FeS_2-doped MoS_2 nanoflower with the dominant 1T-MoS_2 phase as an excellent electrocatalyst for high-performance hydrogen evolution. *Electrochim. Acta.* 249: 72–78. https://doi.org/10.1016/J.ELECTACTA.2017.08.004.

103 Toh, R.J., Sofer, Z., Luxa, J. et al. (2017). 3R phase of MoS_2 and WS_2 outperforms the corresponding 2H phase for hydrogen evolution. *Chem. Commun.* 53: 3054–3057. https://doi.org/10.1039/C6CC09952A.

104 Fang, X., Wang, Z., Kang, S.L. et al. (2020). Hexagonal $CoSe_2$ nanosheets stabilized by nitrogen-doped reduced graphene oxide for efficient hydrogen evolution reaction. *Int. J. Hydrogen Energy* 45: 1738–1747. https://doi.org/10.1016/J.IJHYDENE.2019.11.059.

105 Xue, N., Lin, Z., Li, P. et al. (2020). Sulfur-doped $CoSe_2$ porous nanosheets as efficient electrocatalysts for the hydrogen evolution reaction. *ACS Appl. Mater. Interfaces* 12: 28288–28297. https://doi.org/10.1021/acsami.0c07088.

106 Zhou, Q., Peng, X., Zhong, L. et al. (2020). $CoSe_2$ nanobelt coupled with $CoMoO_4$ nanosheet as efficient electrocatalysts for hydrogen and oxygen evolution reaction. *Environ. Sci. Ecotechnology* 1: 100004. https://doi.org/10.1016/J.ESE.2019.100004.

107 Zhu, J., Cai, L., Yin, X. et al. (2020). Enhanced electrocatalytic hydrogen evolution activity in Single-Atom Pt-decorated VS_2 nanosheets. *ACS Nano* 14: 5600–5608. https://doi.org/10.1021/acsnano.9b10048.

108 Kumar, G.M., Ilanchezhiyan, P., Cho, H.D. et al. (2020). Ultrathin VS_2 nanodiscs for highly stable electro catalytic hydrogen evolution reaction. *Int. J. Energy Res.* 44: 811–820. https://doi.org/10.1002/ER.4892.

109 Wang, L., Gao, W., Chen, X. et al. Charge-modulated VS_2 monolayer for effective hydrogen evolution reaction. *J. Phys. Chem. C* 125: 12011–12041. https://doi.org/10.1021/ACS.JPCC.1C02352/SUPPL_FILE/JP1C02352_SI_001.PDF.

110 Wang, R., Guo, Z., Tan, X. et al. (2021). Atmosphere plasma treatment and Co heteroatoms doping on basal plane of colloidal 2D VSe_2 nanosheets for enhanced hydrogen evolution. *Int. J. Hydrogen Energy* 46: 32425–32434. https://doi.org/10.1016/J.IJHYDENE.2021.07.087.

111 Ling, N., Zheng, S., Lee, Y. et al. (2021). Active hydrogen evolution on the plasma-treated edges of WTe_2. *APL Mater.* 9: 061108. https://doi.org/10.1063/5.0048525.

112 Yan, Y., Xu, S., Li, H. et al. (2021). Perpendicularly anchored $ReSe_2$ nanoflakes on reduced graphene oxide support for highly efficient hydrogen evolution reactions. *Chem. Eng. J.* 405: 126728. https://doi.org/10.1016/J.CEJ.2020.126728.

113 Li, Z., Ma, X., Wu, L. et al. (2021). Synergistic effect of cocatalytic $NiSe_2$ on stable 1T-MoS_2 for hydrogen evolution. *RSC Adv.* 11: 6842–6849. https://doi.org/10.1039/D1RA00506E.

114 Li, L., Qin, Z., Ries, L. et al. (2019). Role of sulfur vacancies and undercoordinated Mo regions in MoS_2 nanosheets toward the evolution of hydrogen. *ACS Nano* 13: 6824–6834. https://doi.org/10.1021/ACSNANO.9B01583/SUPPL_FILE/NN9B01583_SI_001.PDF.

115 Gopalakrishnan, D., Lee, A., Thangavel, N.K. et al. (2017). Facile synthesis of electrocatalytically active NbS_2 nanoflakes for an enhanced hydrogen evolution reaction (HER). *Sustain. Energy Fuels* 2: 96–102. https://doi.org/10.1039/C7SE00376E.

116 Rosen, B.A., Salehi-Khojin, A., Thorson, M.R. et al. (2011). Ionic liquid-mediated selective conversion of CO_2 to CO at low overpotentials. *Science (80)* 334: 643–644. https://doi.org/10.1126/science.1209786.

117 Hong, X., Chan, K., Tsai, C. et al. (2016). How doped MoS_2 breaks transition-metal scaling relations for CO_2 electrochemical reduction. *ACS Catal.* 6: 4428–4437. https://doi.org/10.1021/acscatal.6b00619.

7

Hybrid Materials for CO_2 Reduction and H_2 Generation

Anupma Thakur[1] *and Pooja Devi*[2]

[1] *Discipline of Chemical Engineering, Indian Institute of Technology Gandhinagar, Gujarat 382355, India*
[2] *Materials Science and Sensor Application, Central Scientific Instruments Organisation, Chandigarh 160030, India*

7.1 Introduction

The fast-expanding global energy consumption based on traditional polluting fuels has put immense strain on our environment's health and, as a result, the long-term viability of our planet's animal and plant life [1–5]. The globe must transition to alternate sources of environmentally beneficial, carbon-neutral, clean, and renewable energy as soon as possible [4]. To address this, one of the most promising technologies now being investigated is to produce hydrogen energy from water using solar/electrical energy, which has a limitless supply of both water and sunlight/electricity. Since the 1970s, hydrogen fuel as an ideal energy source via the carbon-neutral water splitting process has been a hot focus of research [2]. Indeed, the photo/electrocatalytic process for creating hydrogen from water is rapidly gaining traction as a potentially significant solution for meeting our future energy demands [3]. In practice, electrocatalytic water has an efficiency of 70–80% (with 20–30% conversion loss), and it is expected to grow to 82–86% by 2030 [6]. However, the efficiency of solar energy conversion to electricity (e.g., with solar cells) required for electrolysis is still far lower, particularly in the case of low-cost solar cell modules, signaling that much more research is needed. At the same time, CO_2 produced into the atmosphere by various polluting energy sources might be reduced and/or turned into value-added compounds, improving atmospheric air quality even more [7, 8]. There is an emerging demand to reduce CO_2 emissions to valuable fuels or chemicals using sunlight/electricity in this direction. Photo/electrocatalysts that are efficient, strong, and inexpensive are required for both of these research pathways [8]. Noble metals (e.g., Pt, Pd, Au, and Ir), which are exceedingly expensive, are being used as high-cost state-of-the-art co-catalysts for photo/electrocatalytic hydrogen (H_2) evolution reaction and carbon dioxide (CO_2) reduction. One possible answer to these problems is to develop unique and designed materials based on a deep understanding of and insights into materials and surface chemistry, and then employ them as photo/electrocatalysts for H_2 production or CO_2 reduction [9]. In order to find suitable photo/electrocatalysts, researchers have focused their efforts in recent years on the synthesis of a variety of low-dimensional materials (0D, 1D, 2D, or their hybrids) and the creation of diverse architectures incorporating heterostructures or hetero-interfaces [10, 11]. Due to their unique dimensionality-dependent properties, two-dimensional (2D) materials and their hybrids, have evolved as

Green Energy Harvesting: Materials for Hydrogen Generation and Carbon Dioxide Reduction,
First Edition. Edited by Pooja Devi.

a special class that demonstrates pretty impressive conversion efficiency on this journey of photo/electrocatalytic hydrogen (H_2) production and carbon dioxide (CO_2) reduction [12–14]. The typical plasmonic metals (Au, Ag, Cu, etc.) have a low catalytic activity for green fuel applications. In order to create superior hybrid plasmonic nanocatalysts, plasmonic components have been interfaced with other nanomaterials in recent years. By harnessing solar energy, plasmonic materials can improve the performance of electrocatalytic systems [15]. This can lead to lower electrical energy consumption and increased product production, as seen by a lower overpotential or higher current density. The introduction of a plasmonic material can also change the chemical landscape, affecting selectivity by initiating or boosting reactions that would otherwise be thermodynamically and/or kinetically blocked. Several configurations of hybrid plasmonic bimetallic nanoparticles have been used in recent years to improve H_2 production and CO_2 reduction by plasmon excitation, ranging from alloys, core–shell, partly coated structures, and self-assembled catalysts. Also the research into molecular electrocatalysts for CO_2 reduction and hydrogen production has accelerated in recent years. The immobilization of molecular catalysts on the electrode surface reduces the amount of catalyst required, and has been shown in some studies to improve the rate of interfacial electron transfer from the semiconductor to the catalyst. Electropolymerization is one of the most efficient ways of adhering molecular catalysts to semiconductor surfaces for CO_2 reduction and hydrogen production reaction pathways. Further, based on the field of dye-sensitized solar cells, an innovative way to manufacturing cathodes for CO_2 reduction and hydrogen production has evolved, which uses supramolecular structures to achieve long-range charge separated states [16]. Instead of employing a small bandgap semiconductor as the light absorber, a broad bandgap material is utilized as the hole transporting substrate, and a molecular photosensitizer is employed as the light collecting material in this type of hybrid cathode [17]. Recent research trends addressing the utilization of hybrid materials for water splitting and CO_2 reduction applications are addressed and analyzed in this chapter. Here, we will center on the considerable advances in the development of hybrid nanomaterials toward the production of green fuels. The research progress (in terms of scientific publications and citations) related toward water splitting and CO_2 reduction applications is increasing continuously, as evident from Figure 7.1.

7.2 Fundamental Theory

7.2.1 Principles of CO_2 Reduction

The chemical conversion of CO_2 into fuels is a technique of recycling carbon dioxide, which could help to mitigate climate change while also extending the life of oil reservoirs for other uses [18]. Photo/electrocatalytic CO_2 reduction has a number of advantages, including the ability to operate at room temperature under ambient circumstances [19]. As a result, there has been a lot of research interest in finding effective photo/electrocatalysts. Despite significant scientific advances, further research is needed to develop photo/electrocatalysts with low overpotential, high efficiency, selectivity, and stability [19]. Different electrochemical pathways corresponding to

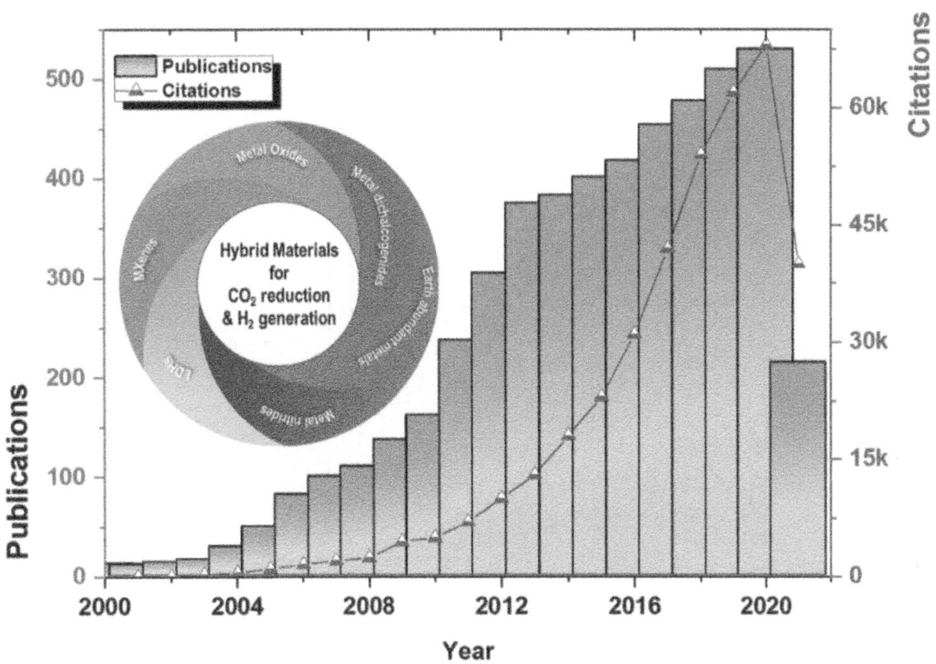

Figure 7.1 Impression of the research contributions related to energy harvesting and storage applications [Web of Science].

standard electrode potentials are used in the electrochemical CO_2 reduction process (vs. SHE) are as presented in Figure 7.2 [20]:

a) 2-electron reduction process (conversion of CO_2 into HCOOH, CO, or $H_2C_2O_4$)
$CO_2 + 2H^+ + 2e^- \rightarrow HCOOH \quad [E° = -0.250\ V]$
$CO_2 + 2H^+ + 2e^- \rightarrow CO + H_2O \quad [E° = -0.106\ V]$
$2CO_2 + 2H^+ + 2e^- \rightarrow H_2C_2O_4 \quad [E° = -0.500\ V]$

b) 4-electron reduction process (conversion of CO_2 into HCHO)
$CO_2 + 4H^+ + 4e^- \rightarrow HCHO + H_2O \quad [E° = -0.070\ V]$

c) 6-electron reduction process (conversion of CO_2 into CH_3OH)
$CO_2 + 6H^+ + 6e^- \rightarrow CH_3OH + 2H_2O \quad [E° = 0.016\ V]$

d) 8-electron reduction process (conversion of CO_2 into CH_4)
$CO_2 + 8H^+ + 8e^- \rightarrow CH_4 + 2H_2O \quad [E° = 0.169\ V]$

e) 12-electron reduction process (conversion of CO_2 into CH_2CH_2 or C_2H_5OH)
$2CO_2 + 12H^+ + 12e^- \rightarrow CH_2CH_2 + 4\ H_2O \quad [E° = 0.064\ V]$
$2CO_2 + 12H^+ + 12e^- \rightarrow C_2H_4OH + 3H_2O \quad [E° = 0.084\ V]$

Overpotential, photo/current density, aqeous stability, Faradaic efficiency, and Tafel slope are used to evaluate the effectiveness of photo/electrocatalysts that drive the photo/electrochemical CO_2 reduction process [19]. Despite recent breakthroughs in electrocatalysts, a number of obstacles remain, necessitating additional research into the fundamental physics of electrochemical CO_2 reduction. Metal-free electrocatalysts are less expensive than metal

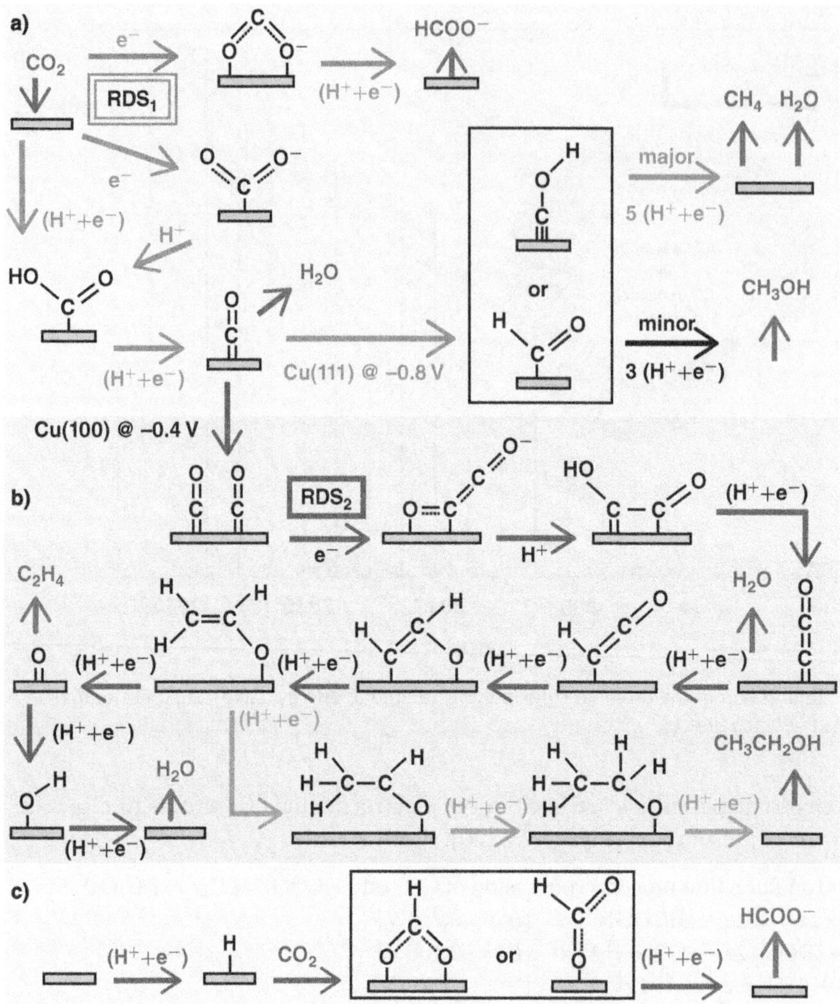

Figure 7.2 Schematic representation of electrochemical CO_2 reduction: (a) pathways from CO_2 to CO, CH_4, CH_3OH, and $HCOO^-$; (b) pathways from CO_2 to ethylene, and (c) pathways from CO_2 to formate. (with permission from [3]. American Chemical Society).

catalysts; however, they are less stable [19]. The investigation of efficient, low-cost, and large-surface-area hybrid materials continues for effective photo/electrochemical CO_2 reduction.

7.2.2 Principles of H_2 Generation

Natural photosynthesis is imitated by the photo-electrocatalytic water-splitting reaction process. It has the ability to transform photon energy into chemical fuel. As shown in Figure 7.3, a photocatalyst absorbs photons with an energy equal to or greater than that of its bandgap and transfers that energy to the band electrons. The negative-electron (e^-) and

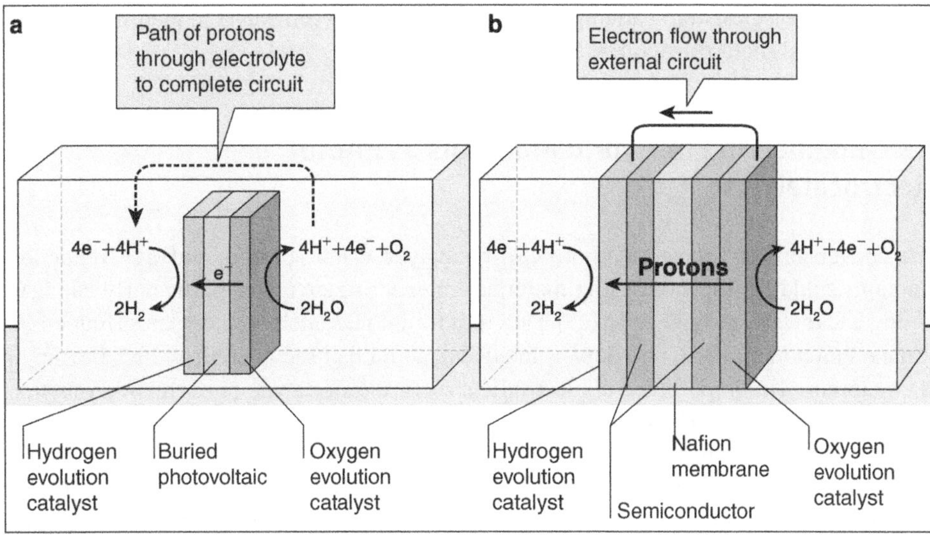

Figure 7.3 Schematic representation of artificial photosynthesis systems for solar-to-hydrogen production. (with permission from [24]. Springer, Nature).

positive-hole (h^+) pairs are formed when electrons in the photocatalyst's valence band excite to the conduction band, accompanying the holes left in the valence band. Additionally, these photo-excited electrons and holes are transmitted to the photocatalyst's surface for reduction and oxidation reactions, respectively [21]. Water redox's conduction and valence bands are vital in overall water splitting. According to theoretical calculations, the minimum photon energy thermodynamically required to drive the reaction is 1.23 eV [21]. "Artificial photosynthesis" is another term for the photo-electrochemical (PEC) water-splitting process. A PEC cell is made up of two or three electrodes immersed in an aqueous electrolyte, having at least one photoactive anode/cathode or both when exposed to sunlight [22]. The photo-electrochemical water-splitting half-reactions are as follows [23]:

$$2H_2O + 4h\nu \rightarrow O_2 + 4H^+ + 4e^- \; [\Delta G = 237 \text{ kJ/mol}, \Delta E^\circ = 1.23 \text{ V/e}] \text{ (OER)}$$
$$4H^+ + 4e^- \rightarrow 2H_2 \text{ (HER)}$$
$$2H_2O \rightarrow 2H_2 + O_2 \; [\Delta G = 237 \text{ kJ/mol}, \Delta E^\circ = 1.23 \text{ V/e}] \text{ (Overall)}$$

where HER and OER correspond to the hydrogen evolution reaction and oxygen evolution reaction, respectively. Under standard-state circumstances, water oxidation is an endothermic reaction that requires $\Delta G^\circ = 237$ kJ mol of H_2 or $\Delta E^\circ = -1.23$ V energy.

The reaction pathways of photo-electrochemical (PEC) water splitting are: i) solar light is absorbed by the photoanode (cathode), resulting in the creation of electron and hole pairs; ii) electrons and holes are separated; iii) electrons are transported to the cathode via an external circuit and holes to the photoanode surface; and iv) OER and HER occur at the photoanode and cathode, respectively. The absence of suitably economical, efficient, and stable materials as photoelectrodes is the fundamental impediment to commercialization of these technologies. Many attempts have been made in recent years to identify the right mix of materials heterostructures, but with limited success.

7.3 Engineering of Hybrid Materials as Photo/Electrocatalysts

The conversion efficiencies of photo/electrocatalytic water splitting and CO_2 reduction reactions could be improved using materials engineering methods, particularly bandgap tuning and interface engineering [25]. Different techniques, including thickness reduction, surface functionalization, and dopant inclusion, could be used to control the bandgap of 2D materials. The importance of electronic structure on catalytic performance was first discovered for transition metals, where the adsorption energy is highly linked to the density of d-states at the Fermi level [26]. The higher the center of the d-band about the Fermi level's energy, the stronger the binding. As a result, modifying the band structure of a material allows for the creation of more active catalysts. DFT simulations make it possible to screen a large number of candidate catalysts quickly to find the optimal catalyst composition, structure, and morphology for achieving the desired band structure [27]. Simultaneously, significant experimental efforts are focused on engineering band structure by controlled synthesis of functional materials. Quantum confinement and plasmon resonance effects improve photon absorption by shifting the valence and conduction bands, which improves the photocatalytic performance of materials. Further, the activity, selectivity, and Faradaic efficiency of the photo/electroctalytic reactions can be improved by stimulating the surface by preferentially selecting crystal facets and phase engineering. According to DFT simulations, the electronic structure of functional materials, which is profoundly influenced by the type of edges, phase, and composition of functional materials, has a significant impact on catalytic activity for H_2 generation and CO_2 reduction [28]. Dangling bonds, crystal lattice mismatches, and low-quality interfaces are all significant difficulties in the fields of electrocatalysis and photocatalysis that must be solved in order to engineer materials for photo/electrocatalytic applications. The promise of producing high-quality interfaces between materials with ideally aligned band offsets and efficient light absorption motivates the study of van der Waals (vdW) heterostructures of semiconducting low-dimensional materials [29]. This recommends that photoelectrodes that absorb the bulk of the solar spectrum could use vdW heterostructures to efficiently break up excitons and transport electrons or holes to catalyst active sites.

7.3.1 Structure and Phase Engineering

Fundamentally, because photo/electrocatalysis is mostly a surface-related phenomenon, material surface, structure, and phase engineering significantly impact catalytic performance. Many ultrathin 2D semiconductors have a large surface area with different metal or nonmetal functional groups that are coordinated. As a result, simple chemical treatment methods can easily bind diverse functional groups on these nanostructured surfaces [29]. These approaches allow multiple functional groups to be attached to the surface, allowing surface structures and surface charges to be modulated to the required specifications. This flexibility also enables more control over the surface composition as well as the creation and adjustment of unique interface features. Further, as compared to their bulk counterparts, these ultrathin-layered structures show superior charge separation and charge transfer processes, as well as a longer photogenerated carrier lifetime [29, 30].

7.3.2 Doping

As demonstrated in the case of photo/electrocatalytic water splitting and CO_2 reduction processes, flaws play an intriguing role in 2D materials. The intrinsic properties of 2D materials can be considerably affected by adding several sorts of defects, such as structural stability, electronic DOS alteration, electrical resistivity, nearest neighboring atoms at the metal center (co-ordination number), and carrier density/mobility [30]. As a result, the presence of a variety of surface-related defects in 2D photocatalysts, such as anions (mainly), cation vacancies, and their benefits have been studied explicitly [30]. According to the literature, defect-engineered 2D materials outperform conventional 2D materials for hydrogen synthesis from water and solar fuels/chemicals from CO_2 reduction via photo/electrocatalysis.

7.3.3 Co-Catalyst Modification

Another well-studied technique for improving catalytic performance is to combine a co-catalyst with a photocatalyst [31]. Thin sheets of 2D materials also feature a significant number of many active sites on their surface due to their 2D structure, which is the result of a large number of under-coordinated surface atoms and a large specific surface area [32]. These improve the photo-overall catalyst's catalytic activity and efficiency. In light illumination and electrolyte, these 2D co-catalyst-based composites or heterojunctions considerably improve photocatalytic performance and photocatalyst durability [33]. The electronic band structures of the catalyst and co-catalyst, as well as a variety of materials issues such as the synthetic conditions for the two, their morphological differences, and the materials' and interfaces' accessibility to illumination, are all taken into account when designing a co-catalyst. In light illumination, these 2D co-catalyst-based composites or heterojunctions considerably improve photocatalytic performance and photocatalyst durability [32]. The nature of the electronic states at the interfaces has an impact on charge transmission across interfaces. The light scattering effects that must be managed are also controlled by the final morphological features. The enhancements are stated in various research publications, although it is not always obvious which feature contributed specifically to a given enhancement.

7.4 Hybrid Materials for Hydrogen Production via Water Splitting

Broadband light absorption, better charge carrier generation and separation, and effective production of H_2 and O_2 from H_2O in separated compartments are all required for efficient photo/electrocatalytic systems to be realized (HER, OER) [34]. The progress in photocatalysts/photoelectrodes is discussed in this section in the light of their use in solar-powered water-splitting devices, as presented in Figure 7.4. Water splitting has received a lot of

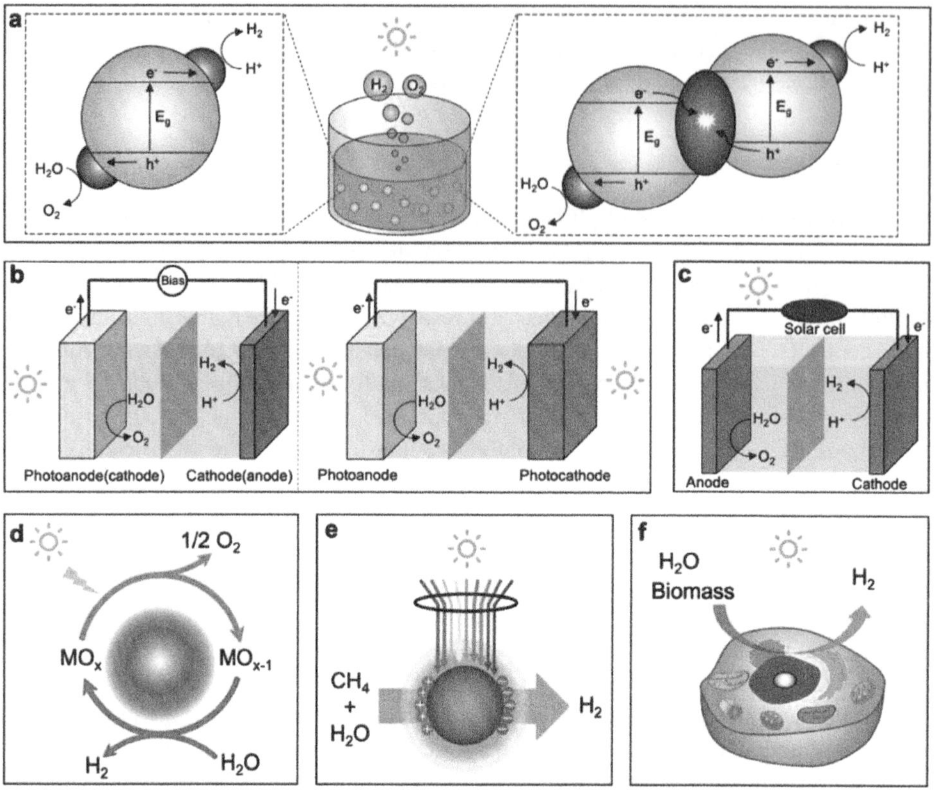

Figure 7.4 Representation of various solar-driven H_2 production approaches [1] / American Chemical Society / CC BY-4.0.

attention recently, and it has been dubbed one of the most essential answers to energy concerns in the next decades. Since Honda and Fujishima's pioneering work, the key processes in water splitting, water reduction, and oxidation, have been thoroughly explored [35]. Over the last three decades, many semiconductors with heterogeneous and homogeneous structures have been described and extensively explored for water splitting. Various research organizations have produced numerous notable breakthroughs and research efforts in engineering semiconductor-based photo/electrodes to date, both internationally and nationally. Transition metal oxides with metal cations are the most common water-splitting photocatalysts. The metal oxides of Ti^{4+}, Zr^{4+}, Nb^{5+}, Ta^{5+}, Mo^{6+}, W^{6+}, Ga^{3+}, In^{3+}, Ge^{4+}, Sn^{4+}, and Sb^{5+} have been investigated for solar-water splitting [36–39]. Metal oxides have been extensively used as photoelectrodes and co-catalysts for photo-electrochemical water splitting, according to the reviewed literature [40]. The performance, limitations, and potentials of metal oxides-based photoelectrodes and co-catalysts are also highlighted. Metal chalcogenide photocatalysts are mostly made up of metal cations such as Cu, Ag, Zn, Cd, Ga, In, Ge, and Sn [41, 42].

Multinary metal chalcogenides have lately been studied in addition to binary metal chalcogenides. The literature summarizes the most recent reports on new photoactive materials [43–70]. Noble metals are widely used as water-splitting electrocatalysts today. At the anode, Ir and Ru are used to boost the OER for the water oxidation reaction [71–73]. Pt is a well-known hydrogen evolution electrocatalyst that is utilized at the cathode to improve the HER for the water reduction reaction [6]. Further, Ru, Rh, Pd, Os, Ir, Pt, Ag, and Au nanoparticles are excellent candidates for H_2- or O_2-evolution co-catalysts [74]. Primarily, Ag and Au nanoparticles also play a role in the enhancement of the H_2-production rate under visible light irradiation due to (SPR) effects [75]. Cr, Co, Ni, Mn, Fe, and Cu are examples of first-row transition metals and their oxides that have been identified as suitable earth-abundant co-catalysts for hydrogen generation [76]. A tandem-junction GaAs–InGaP light harvester in combination with an NiMo as HER co-catalyst and nickel as OER co-catalyst has been acknowledged as extremely efficient with 8.6% solar-to-hydrogen conversion efficiency under one-sun illumination [5]. Other successful photosystems have been documented as extremely efficient, namely, a buried triple-junction silicon light harvester with cobalt oxy-hydroxide as a co-catalyst for OER and ternary NiMoZn HER co-catalyst with an energy efficiency of 2.5% under one-sun illumination [5]. The recent discovery of low-cost p-type semiconductors (e.g., $Cu_2BaSn(S,Se)_4$ and Sb_2Se_3) has aroused interest in the concept of solar-driven hydrogen synthesis via PEC water splitting with earth-abundant semiconductors. They can be viewed as better contenders for competing with present conventional steam reforming techniques for hydrogen production in terms of cost and practicality. A cutting-edge photovoltaic electrocatalyst (PVEC) combination, which has the largest economic potential, is the most fundamental configuration. In keeping with this premise, perovskite solar cells are also pioneering technologies due to their ease of manufacture and high open-circuit voltage, which is favorable for driving photo-assisted water splitting with a single photo-harvester. They can split water without an external bias when connected in series to the electrodes of monolithic PEC devices. In the burgeoning field of organic photoelectrochemical (OPEC) devices, recent rapid developments in organic materials have resulted in a wide range of high-performance donor polymers (PEDOT: PSS, BPhen, BDT-TT) and non-fullerene-based acceptors. Similar reports on hybrid materials have been published in the recent years for hydrogen production, as tabulated in Table 7.1 [45, 77–90]. The use of expensive, rare metals in hybrid plasmonic (Au, Ag) catalyst systems is a bottleneck in the development of long-term large-scale photo/electrode systems. In this direction, hybrid plasmonic catalyst systems have ascended with new plasmonic materials such as Al, Cu, and Mg, which have been unexplored in this domain. Further, hybrid photo/electrode configurations of these materials are fascinating and rapidly rising, and should be investigated further. Direction of research could combine plasmonic-material assisted catalysis with long-term, large-scale, low overpotential, high energy conversion efficiency, revealing their potential for industrial use. Along with developments in photovoltaic and photo-electrochemical devices, the present evolution in tandem and multijunction solar cell designs can aid the development of high-performance OPEC devices in the future as a means of producing higher photovoltage for effective solar harvesting systems. PEC systems coupled to PV, which have recently appeared, constitute a large step forward for water-splitting technology on its way to device realization and commercial viability in the near future, especially in terms of scale-up and long-term use. In this scenario, engineers and material scientists are encouraged to collaborate to overcome the current problems of developing highly effective solar water-splitting systems.

Table 7.1 Hybrid materials reported for hydrogen production.

Photo/electrocatalyst	Electrolyte	Performance	Reference
p-$GaInP_2$ /GaAs	3 M H_2SO_4	CD: 120 mA cm^{-2} STH: 12.4% S: 3 h	[77]
ZnS/$CuGaSe_2$	0.1 M Na_2SO_4	CD: 4.35 mA cm^{-2} STH: 0.25%	[78]
g-C_3N_4/WO_3	0.5 M Na_2SO_4	ER: 3120 μmol g h^{-1} CD: 7 mA cm^{-2}	[79]
Bifunctional catalyst electrode (Fe- and O-doped Co_2P–CoFePO) metal hydroxide grown on nickel foam (NF)	1 M KOH	CD: 10 mA cm^{-2}	[45]
g-C_3N_4/CdS/r-GO	Water + 10% (v/v) TEOA	ER: 4800 μmol g h^{-1} CD: 8.2 mA cm^{-2}	[80]
Cu_2O/Al_2O_3/TiO_2	0.1 M $NaClO_4$	CD: 5.7 mA cm^{-2} STH: 7% S: 5 h	[84]
CdSe-pillared ZnCr-LDH nanohybrid	0.1 M NaOH	ER: 374 mmol gh^{-1} CD: 10 mA cm^{-2}	[81]
CdS/ $CuGaSe_2$	0.1 M Na_2SO_4	CD: 8.1 m cm^{-2} STH: 1.2% S: 8 h	[85]
CdS-NSs/RGO and WO_3 NSs	0.5 M Na_2SO_4	ER: 119.4 ± 4.0 μmol h^{-1}	[82]
g-C_3N_4/ZnO	0.5 M Na_2SO_4	CD: 0.2 mA cm^{-2} STH: 0.07% S: 5 h	[86]
$BiVO_4$@$ZnInS_4$/Ti_3C_2	0.1 M NaOH	ER: 102 μmol h^{-1}	[83]
ZnO/$ZnFe_2O_4$	0.1 M Na_2SO_4	CD: 2 mA cm^{-2} STH: 0.1% S: 2 h	[87]
Ti_3C_2/ZnS or $Zn_xCd_{1-x}S$	0.5 M Na_2SO_4	ER: 14342 μmol h	[89]
$AgSbS_2$-ZnO	1 M Na_2S	CD: 5.86 mA cm^{-2} STH: 0.7% S: 2 h	[88]
Pt/$CuGaS_2$ and RGOCoO_x/$BiVO_4$ (Z-scheme)	0.5 M Na_2SO_4	ER: 3.45 μmol g h^{-1}	[90]

CD: Current Density; ER: Evolution Rate; STH: Solar-to-Hydrogen Conversion Efficiency; S: Stability

Further, the design and manufacture of new PEC devices could lead to strategies such as unsupported cell design and tandem absorber structure that could boost total water-splitting efficiency by converting solar energy to hydrogen. It seems evident that practical solar water splitting and hydrogen for real-world applications will be attainable in the not-too-distant future, thanks to the progress of PEC devices.

7.5 Hybrid Materials for CO_2 Reduction

As photo/electrocatalysts for CO_2 reduction, metal oxides and nitrides, carbon-based materials (graphene, graphene oxide), h-BNs, transition metal chalcogenides, carbon nitrides (C_3N_4), oxynitrides, oxyhalides, metal carbides, and layered double hydroxides (LDHs) have all been used. Metal oxides have been extensively studied for photo/electrocatalytic CO_2 reduction in recent decades [91–95]. In this context, TiO_2 has been the subject of much investigation since it is simple to synthesis, has suitable functions, is durable, and is both nontoxic and affordable [91, 92]. Also, the large surface area of single-layer Bi_2WO_6 permits stronger CO_2 adsorption and enhanced light absorption due to reduced thickness, resulting in superior activity [95]. Additionally, the ultra-thin nature enables for rapid charge transport from the interior to the surface, which extends the lifetime of photoexcited charge carriers and prevents charge recombination. Also, niobates ($SnNb_2O_6$) and metal oxide gallates ($ZnGa_2O_4$) have been investigated as electrocatalysts for CO_2 reduction [93, 94]. Two-dimensional metal chalcogenides have recently received a lot of attention in the context of photocatalytic CO_2 reduction due to their excellent optical and electrical capabilities. Because of its unique electrical and optical properties, MoS_2 is useful as a non-noble metal catalyst or as a co-catalyst (when combined with other semiconductors) for greater CO_2 reduction [96]. As shown in Figure 7.5, the hybrid structure of TiO_2-graphene and few-layers MoS_2 has good CO_2 reduction activity, with a superior CO yield of 14.5 times that of pure TiO_2 and superior selectivity of 97% for CO_2 reduction [4]. Other chalcogenides, layered SnS_2 and CuS, exhibited efficient CO_2 reduction activity and produced CO under visible light illumination [97]. The photoactivity of g-C_3N_4 has also been explored as a CO_2 reduction co-catalyst [98].

Figure 7.6 shows that a hybrid structure made up of AgBr nanoparticles, g-C_3N_4, and N-doped graphene had better activity for CO_2 reduction into CH_3OH and HCHO than pure g-C_3N_4 [2]. Graphene and its derivatives, including GO and r-GO, are also well explored in the context of CO_2 reduction activity due to their high surface area, electrical conductivity, optical transparency, and flexibility. Graphene derivatives have also been used as co-catalysts for CO_2 reduction by combining them with semiconductors. The photocatalytic CO_2 reduction into methanol of r-GO/CdS (2D/2D) nanocomposite was examined in a study [99]. Similarly, under visible-light illumination, r-GO-coated gold (Au) nanoparticles showed good catalytic conversion of CO_2 into formic acid (HCOOH) [100]. Further, single-unit-cell ultrathin BiOBr films show remarkable photocatalytic conversion of CO_2 into CO [101].

In another study, crystalline ultrathin SiC nanosheets offer minimum charge carrier recombinations, resulting in efficient conversion of CO_2 into CH_4 with good selectivity of >90% [102]. According to DFT studies, oxidized Ti_2CO_2 MXene has a 0.91 eV bandgap,

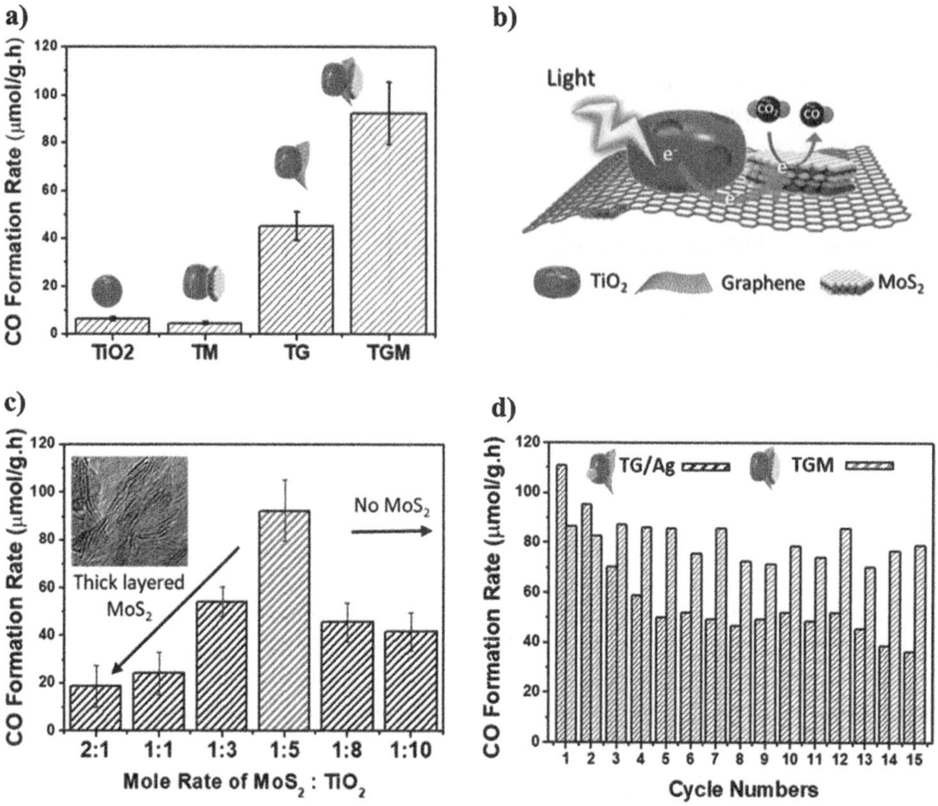

Figure 7.5 Research findings (a and c) CO formation rate Jung et al. (2018), Adapted with permission from American Chemical Society (b) Possible CO_2 conversion mechanism and (d) Stability tests. (with permission from [4]. Americal Chemical Society).

making it excellent for photocatalytic CO_2 reduction to HCOOH [103]. Further, surface-alkalization on the surface of Ti_3C_2 MXene as a co-catalyst boosts CO_2 electrocatalytic reduction due to hydroxyl-rich functional groups that provide abundant reaction sites for CO_2 adsorption [104]. The recent findings suggest that MXenes could be used as both a photocatalyst and a co-catalyst in CO_2 reduction. In this direction, engineering of these materials will be required for a better understanding and enhanced efficiency for CO_2 reduction. Other studies using molecular catalysts-based hybrid photo/electrodes began to appear lately, after the initial reports of photo/electrocatalytic CO_2 reduction [17, 105]. In one case, a 100% Faradaic efficiency (FE) for CO generation utilizing a CdTe absorber with a cobalt phthalocyanine complex was demonstrated by a group of researchers [106]. Similarly, another complex, namely $Ni(cyclam)^{2+}$ and their derivatives, has been investigated with a range of p-type semiconductors to be extremely selective for CO_2 reduction in water to CO [105]. In another study, Cr_2O_3, N,Zn-Fe_2O_3, TiO_2, Ta_2O_5, GaP, and InP with Ru catalysts have been demonstrated as hybrid photo/electrodes for selective CO_2 reduction in water to HCOOH [107, 108]. The similar reports on hybrid materials have been published in the recent years for carbon dioxide reduction, as tabulated in Table 7.2 [104, 109–120].

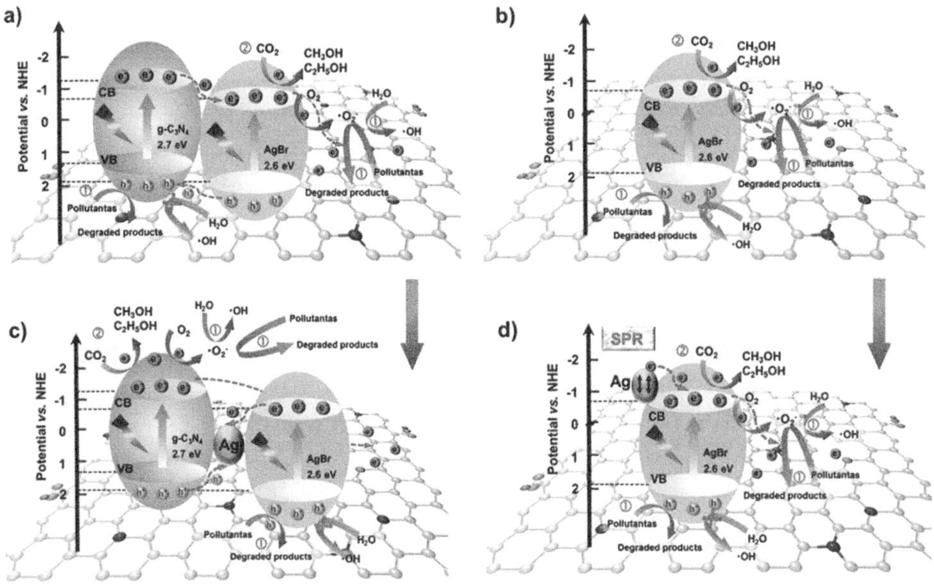

Figure 7.6 Schematic representation of photodegradation of organic pollutants and photoreduction of CO_2 over ACNNG-50 nanocomposite. (with permission from [2]. John Wiley & Sons.)

The plasmonic metal-organic frameworks (MOFs) hybrid system, producing methanol and formic acid as major products by photo-electrocatalytic carbon dioxide reduction with near-100% selectivity, are worth discussing. Under irradiation from a solar simulator, AuNPs integrated in a zeolitic imidazolate framework (ZIF-67) produced methanol [120, 121]. In another case, hybrids of porphyrin frameworks loaded with AuNPs produced formic acid by photo-electrocatalytic carbon dioxide reduction [122]. These two literature reports demonstrate hybrids of plasmonic–MOFs that can be used to photo-electrocatalytically reduce CO_2 into green fuels. The plasmonic materials-based hybrid systems have also been demonstrated for the CO_2 photoelectro reduction. Because the photothermal effect could allow CO_2 activation by lowering the reaction energy barrier, thus the plasmonic excitation is believed to be advantageous for CO_2 reduction at the semiconductor–catalyst interface. In a study, a metal–metal Ag/Cu hybrid plasmonic photocatalyst system is reported to increase the CO_2 photo/electroreduction to CO by 90% under illumination [123]. Similarly, the partial current density values of the CO_2 photo-electro reduction to CO and HCOOH increased by 260% and 100%, respectively, utilizing a Cu/p-NiO metal–semiconductor hybrid system [124]. Owing to the published literature on CO_2 reduction, the most important goal is to significantly enhance photo/electrode stability from the existing state-of-the-art (~20 h) to thousands of hours of operation for their commercialization. This will necessitate improvements in the photo/electrocatalyst, as well as the hybrid interface design. Despite being used in some of the very first studies recently in the literature, polymerized and molecular catalysts have consistently resulted in the highest aqeous stabilities, but this comes at the expense of high orientation control, making it difficult to

Table 7.2 Hybrid materials reported for CO_2 reduction.

Photo/electrocatalyst	Electrolyte	Product Quantification	Reference
Bi_2WO_6/RGO/g-C_3N_4	0.1 M potassium carbonate	~15.96 μmo h g^{-1} (CO_2 to CO)	[109]
Pt NPs/g-C_3N_4	0.1 M NaCl	13.03 mmol g (CO_2 to CH_4)	[120]
g-C_3N_4/NiAl-LDH heterojunction	Water and high pure CO_2	~11.2 μmol h g^{-1} (CO_2 to CO)	[110]
AlGaN/GaN	0.1 M NaCl	~70% (CO_2 to HCOOH)	[114]
MnO_2/g-C_3N_4	0.1 M potassium carbonate	~8.2 μmol h g^{-1} (CO_2 to CO)	[111]
Graphene oxide (rGO)/ protonated g-C_3N_4 (pCN)	0.1 M potassium carbonate	~80% (CO_2 to CH_4)	[115]
Co_4@g-C_3N_4 hybrid	0.1 M sodium carbonate	~107 μmol h g^{-1} (CO_2 to CO)	[113]
TiO_2/C_3N_4	0.5 M $KHCO_3$	~80% (CO_2 to CH_3OH)	[116]
Ti_3C_2-OH/P25 hybrids	0.1 M potassium carbonate	~11.7 μmol h g^{-1} (CO_2 to CO)	[104]
CoAg/CN	0.5 M $KHCO_3$	~85% (CO_2 to CH_4)	[117]
g-C_3N_4/$FeWO_4$	0.5 M Na_2SO_3 and high pure CO_2	~27.7 μmol h g^{-1} (CO_2 to CO)	[112]
InGaN/two Si p-n junctions	0.5 M $KHCO_3$	~83% (CO_2 to CH_3OH)	[118]
Mg-Al-LDH/C_3N_4	0.1 M sodium carbonate	~85% (CO_2 to CH_4)	[119]

design effective charge transfer interfaces. Also, great advances have been made to coat photoabsorbers with thick metal oxide layers (such as TiO_2, ZnO) that enable stable water splitting for ~90 h, and it will be exciting to see if a combination of these approaches can achieve both maximum light absorption and stabilization of a semiconductor–catalyst interface for photo-electrocatalytic CO_2 reduction.

7.6 Summary and Outlook

In summary, due to dimensionality and a specific unique set of physical properties, 2D materials-based hybrids have attracted significant attention among the different materials systems explored for photo/electrocatalysis. Due to their huge specific surface area, which results in a large number of exposed active sites, outstanding durability, high absorption capability, and cost-effectiveness, 2D materials are strong competitors for practical and scalable H_2 generation and CO_2 reduction (earth-abundant materials, simple synthetic methods for fabrications). Nano-hybrids of layered metal oxides, graphene, carbon nitride, chalcogenides, phosphorene, LDHs, and MXenes are examples of material systems that are moving in this direction. Because of the large interfacial contact area, which effectively reduces the recombination process and charge transfer resistance, stable engineered heterostructures based on 2D earth-abundant materials as photocatalysts and co-catalysts result in exceptional energy conversion efficiencies compared to bulk materials. Additionally, a significant feature in achieving a high current enhancement is the interface between components in hybrid nanostructures that can improve the hot electron–hole separation efficiency. Further research efforts should be made to investigate materials with high electron mobility or long electron diffusion length as potential plasmonic metal coupling candidates. In the future, researchers can attempt to sketch some broad concepts, newer experimental situations to test emerging 2D materials for scalable H_2 production and CO_2 reduction. As a result, predictive modeling and integrated theoretical and practical studies are highly sought to narrow the outlines and develop an effective photo/electrocatalytic system for efficient H_2 production and CO_2 reduction.

References

1 Song, H., Luo, S., Huang, H. et al. (2022). Solar-driven hydrogen production: Recent advances, challenges, and future perspectives. *ACS Energy Letters* 7(3): 1043–1065.

2 Li, H., Gan, S., Wang, H. et al. (2015). Intercorrelated superhybrid of AgBr supported on graphitic-C_3N_4-decorated nitrogen-doped graphene: High engineering photocatalytic activities for water purification and CO_2 reduction. *Advanced Materials* 27(43): 6906–6913.

3 Kortlever, R., Shen, J., Schouten, K.J.P. et al. (2015). Catalysts and reaction pathways for the electrochemical reduction of carbon dioxide. *The Journal of Physical Chemistry Letters* 6(20): 4073–4082.

4 Jung, H., Cho, K.M., Kim, K.H. et al. (2018). Highly efficient and stable CO_2 reduction photocatalyst with a hierarchical structure of mesoporous TiO_2 on 3D graphene with few-layered MoS_2. *ACS Sustainable Chemistry & Engineering* 6(5): 5718–5724.

5 Reece, S.Y., Hamel, J.A., Sung, K. et al. (2011). Wireless solar water splitting using silicon-based semiconductors and earth-abundant catalysts. *Science* 334(6056): 645–648.

6 Carmo, M., Frit, D.L., Mergel, J. et al. (2013). A comprehensive review on PEM water electrolysis. *International Journal of Hydrogen Energy* 38(12): 4901–4934.

7 Rafiee, A., Khalilpour, K.R., Milani, D. et al. (2018). Trends in CO_2 conversion and utilization: A review from process systems perspective. *Journal of Environmental Chemical Engineering* 6(5): 5771–5794.

8 Artz, J., Müller, T.E., Thenert, K. et al. (2018). Sustainable conversion of carbon dioxide: An integrated review of catalysis and life cycle assessment. *Chemical Reviews* 118(2): 434–504.

9 Jiang, C., Moniz, S.J., Wang, A. et al. (2017). Photoelectrochemical devices for solar water splitting: Materials and challenges. *Chemical Society Reviews* 46(15): 4645–4660.

10 Liu, J., Wei, Z., and Shangguan, W. (2019). Defects engineering in photocatalytic water splitting materials. *ChemCatChem* 11(24): 6177–6189.

11 Saraswat, S.K., Rodene, D.D., and Gupta, R.B. (2018). Recent advancements in semiconductor materials for photo-electrochemical water splitting for hydrogen production using visible light. *Renewable and Sustainable Energy Reviews* 89: 228–248.

12 Kilic, M.E. and Lee, K-R. (2021). Tetrahex carbides: Two-dimensional group-IV materials for nanoelectronics and photocatalytic water splitting. *Carbon* 174: 368–381.

13 Marschall, R. (2021). 50 years of materials research for photocatalytic water splitting. *European Journal of Inorganic Chemistry* 2021(25): 2435–2441.

14 Du, J., Ma, Y-Y., Tan, H. et al. (2021). Progress of electrochemical CO_2 reduction reactions over polyoxometalate-based materials. *Chinese Journal of Catalysis* 42(6): 920–937.

15 Ezendam, S., Herran, M., Nan, L. et al. (2022). Hybrid plasmonic nanomaterials for hydrogen generation and carbon dioxide reduction. *ACS Energy Letters* 7(2): 778–815.

16 Tamaki, Y., Tokuda, K., Yamazaki, Y. et al. (2019). Ruthenium picolinate complex as a redox photosensitizer with wide-band absorption. *Frontiers in Chemistry* 7: 327.

17 Garcia Osorio, D.A., Neri, G., and Cowan, A.J. et al. (2021). Hybrid photocathodes for carbon dioxide reduction: Interfaces for charge separation and selective catalysis. *ChemPhotoChem* 5(7): 595–610.

18 Zhang, Y., Guo, S-X., Zhang, X. et al. (2020). Mechanistic understanding of the electrocatalytic CO_2 reduction reaction: New developments based on advanced instrumental techniques. *Nano Today* 31: 100835.

19 Yaashikaa, P., Kumar, P.S., Varjani, S.J. et al. (2019). A review on photochemical, biochemical and electrochemical transformation of CO_2 into value-added products. *Journal of CO_2 Utilization* 33: 131–147.

20 Yuan, Y., Lu, J., Duan, Z., et al. eds. (2019). Hand-held water quality monitoring system based on laser-induced fluorescence. *Asia Communications and Photonics Conference 2019*. Optical Society of America.

21 Wang, Q., Hisatomi, T., Jia, Q. et al. (2016). Scalable water splitting on particulate photocatalyst sheets with a solar-to-hydrogen energy conversion efficiency exceeding 1%. *Nature Materials* 15(6): 611–615.

22 Ager, J.W., Shaner, M.R., Walczak, K.A. et al. (2015). Experimental demonstrations of spontaneous, solar-driven photoelectrochemical water splitting. *Energy & Environmental Science* 8(10): 2811–2824.

23 Ismail, A.A. and Bahnemann, D.W. (2014). Photochemical splitting of water for hydrogen production by photocatalysis: A review. *Solar Energy Materials and Solar Cells* 128: 85–101.

24 Roger, I., Shipman, M.A., and Symes, M.D. (2017). Earth-abundant catalysts for electrochemical and photoelectrochemical water splitting. *Nature Reviews Chemistry* 1(1): 1–13.

25 Hu, J., Guo, Z., Mcwilliams, P.E. et al. (2016). Bandgap engineering in a 2D material for solar-to-chemical energy conversion. *Nano Letters* 16(1): 74–79.

26 Bhosale, R., Tonda, S., Kumar, S., and Ogale S.B. (2020). Two-dimensional materials for photocatalytic water splitting and CO_2 reduction. In *Micro and Nano Technologies, Nanostructured Photocatalysts*, pp. 173–227. Elsevier. ISBN 9780128178362.

27 Faraji, M., Yousefi, M., Yousefzadeh, S. et al. (2019). Two-dimensional materials in semiconductor photoelectrocatalytic systems for water splitting. *Energy & Environmental Science* 12(1): 59–95.

28 Kumar, P., Boukherroub, R., and Shankar K. (2018). Sunlight-driven water-splitting using two-dimensional carbon based semiconductors. *Journal of Materials Chemistry A* 6(27): 12876–12931.

29 Chu, S., Wang, Y., Guo, Y. et al. (2013). Band structure engineering of carbon nitride: In search of a polymer photocatalyst with high photooxidation property. *Acs Catalysis* 3(5): 912–919.

30 Cao, S., Low, J., Yu, J. et al. (2015). Polymeric photocatalysts based on graphitic carbon nitride. *Advanced Materials* 27(13): 2150–2176.

31 Ran, J., Gao, G., Li, F-T. et al. (2017). Ti_3C_2 MXene co-catalyst on metal sulfide photo-absorbers for enhanced visible-light photocatalytic hydrogen production. *Nature Communications* 8(1): 1–10.

32 Chhowalla, M., Shin, H.S., Eda, G. et al. (2013). The chemistry of two-dimensional layered transition metal dichalcogenide nanosheets. *Nature Chemistry* 5(4): 263–275.

33 Pan, Y., Qian, Y., Zheng, X. et al. (2021). Precise fabrication of single-atom alloy co-catalyst with optimal charge state for enhanced photocatalysis. *National Science Review* 8(1): 224.

34 Eriksson, E. and Gray, E.M. (2017). Optimization and integration of hybrid renewable energy hydrogen fuel cell energy systems: A critical review. *Applied Energy* 202: 348–364.

35 Fujishima, A. and Honda, K. (1972). Electrochemical photolysis of water at a semiconductor electrode. *Nature* 238(5358): 37–38.

36 Sayama, K and Arakawa, H. (1993). Photocatalytic decomposition of water and photocatalytic reduction of carbon dioxide over zirconia catalyst. *The Journal of Physical Chemistry* 97(3): 531–533.

37 Kudo, A and Hijii, S. (1999). H_2 or O_2 evolution from aqueous solutions on layered oxide photocatalysts consisting of Bi^{3+} with 6s2 configuration and d0 transition metal ions. *Chemistry Letters* 28(10): 1103–1104.

38 Liao, L., Zhang, Q., Su, Z. et al. (2014). Efficient solar water-splitting using a nanocrystalline CoO photocatalyst. *Nature Nanotechnology* 9(1): 69.

39 Paracchino, A., Laporte, V., Sivula, K. et al. (2011). Highly active oxide photocathode for photoelectrochemical water reduction. *Nature Materials* 10(6): 456.

40 Yang, Y., Niu, S., Han, D. et al. (2017). Progress in developing metal oxide nanomaterials for photoelectrochemical water splitting. *Advanced Energy Materials* 7(19): 1700555.

41 Holmes, M.A., Townsend, T.K., and Osterloh, F.E. (2012). Quantum confinement controlled photocatalytic water splitting by suspended CdSe nanocrystals. *Chemical Communications* 48(3): 371–373.

42 Kalyanasundaram, K., Borgarello, E., Duonghong, D. et al. (1981). Cleavage of water by visible-light irradiation of colloidal CdS solutions: Inhibition of photocorrosion by RuO_2. *Angewandte Chemie International Edition in English* 20(11): 987–988.

43 Zhang, Z., Gao, C., Wu, Z. et al. (2016). Toward efficient photoelectrochemical water-splitting by using screw-like SnO_2 nanostructures as photoanode after being decorated with CdS quantum dots. *Nano Energy* 19: 318–327.

44 Dias, P., Vilanova, A., Lopes, T. et al. (2016). Extremely stable bare hematite photoanode for solar water splitting. *Nano Energy* 23: 70–79.

45 Duan, J., Chen, S., Vasileff, A. et al. (2016). Anion and cation modulation in metal compounds for bifunctional overall water splitting. *ACS Nano* 10(9): 8738–8745.

46 Zhang, W., Eperon, G.E., and Snaith, H.J. (2016). Metal halide perovskites for energy applications. *Nature Energy* 1: 16048.

47 Kim, J.K., Shi, X., Jeong, M.J. et al. (2018). Enhancing Mo:$BiVO_4$ solar water splitting with patterned Au nanospheres by Plasmon-induced energy transfer. *Advanced Energy Materials* 8(5): 1701765.

48 Li, M., Yang, Y., Ling, Y. et al. (2017). Morphology and doping engineering of Sn-doped hematite nanowire photoanodes. *Nano Letters* 17(4): 2490–2495.

49 Luo, J., Steier, L., Son, M-K. et al. (2016). Cu_2O nanowire photocathodes for efficient and durable solar water splitting. *Nano Letters* 16(3): 1848–1857.

50 Varadhan, P., Fu, H-C., Priante, D. et al. (2017). Surface passivation of GaN nanowires for enhanced photoelectrochemical water-splitting. *Nano Letters* 17(3): 1520–1528.

51 Pu, Z., Luo, Y., Asiri, A.M. et al. (2016). Efficient electrochemical water splitting catalyzed by electrodeposited nickel diselenide nanoparticles-based film. *ACS Applied Materials & Interfaces* 8(7): 4718–4723.

52 Kieslich, G., Cerretti, G., Veremchuk, I. et al. (2016). A chemist's view: Metal oxides with adaptive structures for thermoelectric applications. *Physica Status Solidi (a)* 213(3): 808–823.

53 Wang, L., Xia, L., Wu, Y. et al. (2016). Zr-doped β-In_2S_3 ultrathin nanoflakes as photoanodes: Enhanced visible-light-driven photo-electrochemical water splitting. *ACS Sustainable Chemistry & Engineering* 4(5): 2606–2614.

54 Urbain, F., Smirnov, V., Becker, J-P. et al. (2016). Multijunction Si photocathodes with tunable photovoltages from 2.0 V to 2.8 V for light induced water splitting. *Energy & Environmental Science* 9(1): 145–154.

55 Wang, R., Luo, Y., Chen, Z. et al. (2016). The effect of loading density of nickel-cobalt sulfide arrays on their cyclic stability and rate performance for supercapacitors. *Science China Materials* 59(8): 629–638.

56 Kawasaki, S., Takahashi, R., Yamamoto, T. et al. (2016). Photo-electrochemical water splitting enhanced by self-assembled metal nanopillars embedded in an oxide semiconductor photoelectrode. *Nature Communications* 7(1): 1–6.

57 Yang, J., Cooper, J.K., Toma, F.M. et al. (2017). A multifunctional biphasic water splitting catalyst tailored for integration with high-performance semiconductor photoanodes. *Nature Materials* 16(3): 335–341.

58 Choi, M., Lee, J.H., Jang, Y.J. et al. (2016). Hydrogen-doped Brookite TiO_2 nanobullets array as a novel photoanode for efficient solar water splitting. *Scientific Reports* 6: 36099.

59 Pei, Y., Pei, R., Liang, X. et al. (2016). CdS-nanowires flexible photo-detector with Ag-nanowires electrode based on non-transfer process. *Scientific Reports* 6: 21551.

60 Qi, H., Wolfe, J., Fichou, D. et al. (2016). Cu_2O photocathode for low bias photoelectrochemical water splitting enabled by NiFe-layered double hydroxide Co-catalyst. *Scientific Reports* 6: 30882.

61 Dos Santos, W.S., Rodriguez, M., Afonso, A.S. et al. (2016). A hole inversion layer at the $BiVO_4/Bi_4V_2O_{11}$ interface produces a high tunable photovoltage for water splitting. *Scientific Reports* 6(1): 1–9.

62 Sun, C., Dong, Q., Yang, J. et al. (2016). Metal–organic framework derived $CoSe_2$ nanoparticles anchored on carbon fibers as bifunctional electrocatalysts for efficient overall water splitting. *Nano Research* 9(8): 2234–2243.

63 De, S., Zhang, J., Luque, R. et al. (2016). Ni-based bimetallic heterogeneous catalysts for energy and environmental applications. *Energy & Environmental Science* 9(11): 3314–3347.

64 Xu, X., Bao, Z., Tang, W., Wu, H., Pan, J., Hu, J. and Zeng, H. (2017). Surface states engineering carbon dots as multi-band light active sensitizers for ZnO nanowire array photoanode to boost solar water splitting. *Carbon* 121: 201–208.

65 Mehta, A., Pooja, D., Thakur, A. et al. (2017). Enhanced photocatalytic water splitting by gold carbon dot core shell nanocatalyst under visible/sunlight. *New Journal of Chemistry* 41(11): 4573–4581.

66 Sang, L., Lin, J., and Zhao, Y. (2017). Preparation of carbon dots/TiO_2 electrodes and their photo-electrochemical activities for water splitting. *International Journal of Hydrogen Energy* 42(17): 12122–12132.

67 Shinde, S., Sami, A., and Lee, J-H. (2016). Sulfur mediated graphitic carbon nitride/S-Se-graphene as a metal-free hybrid photocatalyst for pollutant degradation and water splitting. *Carbon* 96: 929–936.

68 Pan, Z., Zheng, Y., Guo, F. et al. (2017). Decorating CoP and Pt nanoparticles on graphitic carbon nitride nanosheets to promote overall water splitting by conjugated polymers. *ChemSusChem* 10(1): 87–90.

69 Su, J., Geng, P., Li, X. et al. (2016). Graphene-linked graphitic carbon nitride/TiO_2. *Journal of Applied Electrochemistry* 46(8): 807–817.

70 Fan, X., Wang, T., Gao, B. et al. (2016). Preparation of the TiO_2/Graphic Carbon Nitride Core–Shell array as a photoanode for efficient photoelectrochemical water splitting. *Langmuir* 32(50): 13322–13332.

71 Schlögl, R. ed. (2015). Systemic aspects of the transition to sustainable energy. *EPJ Web of Conferences 2015*, vol. 98, p. 04001. EDP Sciences.

72 Tee, S.Y., Lee, C.J.J., Dinachali, S.S. et al. (2015). Amorphous ruthenium nanoparticles for enhanced electrochemical water splitting. *Nanotechnology* 26(41): 415401.

73 Lee, Y., Suntivich, J., May, K.J. et al. (2012). Synthesis and activities of rutile IrO_2 and RuO_2 nanoparticles for oxygen evolution in acid and alkaline solutions. *The Journal of Physical Chemistry Letters* 3(3): 399–404.

74 Li, X., Yu, J., Low, J. et al. (2015). Engineering heterogeneous semiconductors for solar water splitting. *Journal of Materials Chemistry A* 3(6): 2485–2534.

75 Linic, S., Christopher, P., and Ingram, D.B. (2011). Plasmonic-metal nanostructures for efficient conversion of solar to chemical energy. *Nature Materials* 10(12): 911.

76 Fukuzumi, S., Hong, D., and Yamada, Y. (2013). Bioinspired photocatalytic water reduction and oxidation with earth-abundant metal catalysts. *The Journal of Physical Chemistry Letters* 4(20): 3458–3467.

77 Khaselev, O. and Turner, J.A. (1998). A monolithic photovoltaic-photoelectrochemical device for hydrogen production via water splitting. *Science* 280(5362): 425–427.

78 Kim, J., Minegishi, T., Kobota, J. et al. (2012). Enhanced photoelectrochemical properties of $CuGa_3$ Se_5 thin films for water splitting by the hydrogen mediated co-evaporation method. *Energy & Environmental Science* 5(4): 6368–6374.

79 Yu, W., Chen, J., Shang, T. et al. (2017). Direct Z-scheme g-C_3N_4/WO_3 photocatalyst with atomically defined junction for H_2 production. *Applied Catalysis B: Environmental* 219: 693–704.

80 Tonda, S., Kumar, S., Gawli, Y. et al. (2017). g-C_3N_4 (2D)/CdS (1D)/rGO (2D) dual-interface nano-composite for excellent and stable visible light photocatalytic hydrogen generation. *International Journal of Hydrogen Energy* 42(9): 5971–5984.

81 Zhang, G., Lin, B., Yang, W. et al. (2015). Highly efficient photocatalytic hydrogen generation by incorporating CdS into ZnCr-layered double hydroxide interlayer. *RSC Advances* 5(8): 5823–5829.

82 Huang, Y., Liu, Y., Zhu, D. et al. (2016). Mediator-free Z-scheme photocatalytic system based on ultrathin CdS nanosheets for efficient hydrogen evolution. *Journal of Materials Chemistry A* 4(35): 13626–13635.

83 Du, X., Zhao, T., Xiu, Z. et al. (2020). $BiVO_4$@$ZnIn_2S_4$/Ti_3C_2 MXene quantum dots assembly all-solid-state direct Z-Scheme photocatalysts for efficient visible-light-driven overall water splitting. *Applied Materials Today* 20: 100719.

84 Morales-Guio, C.G., Tilley, S.D., Vrubel, H. et al. (2014). Hydrogen evolution from a copper (I) oxide photocathode coated with an amorphous molybdenum sulphide catalyst. *Nature Communications* 5: 3059.

85 Zhang, L., Minegishi, T., Kubota, J. et al. (2014). Hydrogen evolution from water using Ag x Cu_{1-x} $GaSe_2$ photocathodes under visible light. *Physical Chemistry Chemical Physics* 16(13): 6167–6174.

86 Xiao, J., Zhang, X., and Li, Y. (2015). A ternary g-C_3N_4/Pt/ZnO photoanode for efficient photo-electrochemical water splitting. *International Journal of Hydrogen Energy* 40(30): 9080–9087.

87 Yang, T., Xue, J., Tan, H. (2018). Highly ordered ZnO/$ZnFe_2O_4$ inverse opals with binder-free heterojunction interfaces for high-performance photoelectrochemical water splitting. *Journal of Materials Chemistry A*6(3): 1210–1218.

88 Han, J., Liu, Z., Guo, K. et al. (2015). $AgSbS_2$ modified ZnO nanotube arrays for photoelectrochemical water splitting. *Applied Catalysis B: Environmental* 179: 61–68.

89 Ran, J., Gao, G., Li, F.T. et al. (2017). Ti_3C_2 MXene co-catalyst on metal sulfide photo-absorbers for enhanced visible-light photocatalytic hydrogen production. *Nature Communications* 8(1): 1–10.

90 Iwase, A., Yoshino, S., Takayama, T. et al. (2016). Water splitting and CO_2 reduction under visible light irradiation using Z-scheme systems consisting of metal sulfides, CoO_x-loaded $BiVO_4$, and a reduced graphene oxide electron mediator. *Journal of the American Chemical Society* 138(32): 10260–10264.

91 Habisreutinger, S.N., Schmidt-Mende, L., and Stolarczyk, J.K. (2013). Photocatalytic reduction of CO_2 on TiO_2 and other semiconductors. *Angewandte Chemie International Edition* 52(29): 7372–7408.

92 Ma, Y., Wang, X., Jia, Y. et al. (2014). Titanium dioxide-based nanomaterials for photocatalytic fuel generations. *Chemical Reviews* 114(19): 9987–10043.

93 Zhu, S., Liang, S., Bi, J. et al. (2016). Photocatalytic reduction of CO_2 with H_2O to CH_4 over ultrathin SnNb $2O_6$ 2D nanosheets under visible light irradiation. *Green Chemistry* 18(5): 1355–1363.

94 Liu, Q., Wu, D., Zhou, Y. et al. (2014). Single-crystalline, ultrathin $ZnGa_2O_4$ nanosheet scaffolds to promote photocatalytic activity in CO_2 reduction into methane. *ACS Applied Materials & Interfaces* 6(4): 2356–2361.

95 Liang, L., Lei, F., Gao, S. et al. (2015). Single unit cell bismuth tungstate layers realizing robust solar CO_2 reduction to methanol. *Angewandte Chemie International Edition* 54(47): 13971–13974.

96 Wang, Z. and Mi, B. (2017). Environmental applications of 2D molybdenum disulfide (MoS_2) nanosheets. *Environmental Science & Technology* 51(15): 8229–8244.

97 Jiao, X., Li, X., Jin, X. et al. (2017). Partially oxidized SnS_2 atomic layers achieving efficient visible-light-driven CO_2 reduction. *Journal of the American Chemical Society* 139(49): 18044–18051.

98 Ong, W-J., Tan, -L-L., Ng, Y.H. et al. (2016). Graphitic carbon nitride (g-C_3N_4)-based photocatalysts for artificial photosynthesis and environmental remediation: Are we a step closer to achieving sustainability? *Chemical Reviews* 116(12): 7159–7329.

99 Cho, K.M., Kim, K.H., Park, K. et al. (2017). Amine-functionalized graphene/CdS composite for photocatalytic reduction of CO_2. *ACS Catalysis* 7(10): 7064–7069.

100 Kumar, D., Lee, A., Lee, T. et al. (2016). Ultrafast and efficient transport of hot plasmonic electrons by graphene for Pt free, highly efficient visible-light responsive photocatalyst. *Nano Letters* 16(3): 1760–1767.

101 Ye, L., Jin, X., Liu, C. et al. (2016). Thickness-ultrathin and bismuth-rich strategies for BiOBr to enhance photoreduction of CO_2 into solar fuels. *Applied Catalysis B: Environmental* 187: 281–290.

102 Han, C., Wang, B., Wu, C. et al. (2019). Ultrathin SiC nanosheets with high reduction potential for improved CH_4 generation from photocatalytic reduction of CO_2. *ChemistrySelect* 4(7): 2211–2217.

103 Zhang, X., Zhang, Z., Li, J. et al. (2017). Ti_2CO_2 MXene: A highly active and selective photocatalyst for CO_2 reduction. *Journal of Materials Chemistry A* 5(25): 12899–12903.

104 Ye, M., Wang, X., Liu, E. et al. (2018). Boosting the photocatalytic activity of P25 for carbon dioxide reduction by using a surface-alkalinized titanium carbide MXene as cocatalyst. *ChemSusChem* 11(10): 1606–1611.

105 Bradley, M. and Tysak, T. (1982). p-Type silicon based photoelectrochemical cells for optical energy conversion: Electrochemistry of tetra-azomacrocyclic metal complexes at illuminated. *Journal of Electroanalytical Chemistry & Interfacial Electrochemistry* 135(1): 153–157.

106 Jo'm, B. and Wass, J. (1989). On the photoelectrocatalytic reduction of carbon dioxide. *Materials Chemistry & Physics* 22(3–4): 249–280.

107 Sekizawa, K., Sato, S., Arai, T. et al. (2018). Solar-driven photocatalytic CO_2 reduction in water utilizing a ruthenium complex catalyst on p-type Fe_2O_3 with a multiheterojunction. *ACS Catalysis* 8(2): 1405–1416.

108 Sato, S., Arai, T., Morikawa, T. et al. (2011). Selective CO_2 conversion to formate conjugated with H_2O oxidation utilizing semiconductor/complex hybrid photocatalysts. *Journal of the American Chemical Society* 133(39): 15240–15243.

109 Jo, W.K., Kumar, S., Eslava, S. et al. (2019). Construction of Bi_2WO_6/RGO/g-C_3N_4 2D/2D/2D hybrid Z-scheme heterojunctions with large interfacial contact area for efficient charge separation and high-performance photoreduction of CO_2 and H_2O into solar fuels. *Applied Catalysis B: Environmental* 239: 586–598.

110 Tonda, S., Kumar, S., Bhardwaj, M. et al. (2018). g-C_3N_4/NiAl-LDH 2D/2D hybrid heterojunction for high-performance photocatalytic reduction of CO_2 into renewable fuels. *ACS Applied Materials & Interfaces* 10(3): 2667–2678.

111 Wang, M., Shen, M., Zhang, L. et al. (2019). 2D-2D MnO_2/g-C_3N_4 heterojunction photocatalyst: *In-situ* synthesis and enhanced CO_2 reduction activity. *Carbon* 120: 23–31.

112 Bhosale, R., Jain, S., Vinod, C.P. et al. (2019). Direct Z-scheme g-C_3N_4/$FeWO_4$ nanocomposite for enhanced and selective photocatalytic CO_2 reduction under visible light. *ACS Applied Materials & Interfaces* 11(6): 6174–6183.

113 Zhou, J., Chen, W., Sun, C. et al. (2017). Oxidative polyoxometalates modified graphitic carbon nitride for visible-light CO_2 reduction. *ACS Applied Materials & Interfaces* 9(13): 11689–11695.

114 Lee, M-L., Liao, P-H., Li, G-L. et al. (2019). Enhanced production rates of hydrogen generation and carbon dioxide reduction using aluminum gallium nitride/gallium nitride heteroepitaxial films as photoelectrodes in seawater. *Solar Energy Materials & Solar Cells* 202: 110153.

115 Ong, W-J., Tan, L-L., Chai, S-P. et al. (2015). Surface charge modification via protonation of graphitic carbon nitride (g-C_3N_4) for electrostatic self-assembly construction of 2D/2D reduced graphene oxide (rGO)/g-C_3N_4 nanostructures toward enhanced photocatalytic reduction of carbon dioxide to methane. *Nano Energy* 13: 757–770.

116 Tseng, I., Sung, Y-M., Chang, P-Y. et al. (2019). Anatase TiO_2-decorated graphitic carbon nitride for photocatalytic conversion of carbon dioxide. *Polymers* 11(1): 146.

117 Nazir, R., Kumar, A., Ali, S. et al. (2019). Galvanic exchange as a novel method for carbon nitride supported CoAg catalyst synthesis for oxygen reduction and carbon dioxide conversion. *Catalysts* 9(10): 860.

118 Sekimoto, T., Shinagawa, S., Uetake, Y. et al. (2015). Tandem photo-electrode of InGaN with two Si pn junctions for CO_2 conversion to HCOOH with the efficiency greater than biological photosynthesis. *Applied Physics Letters* 106(7): 073902.
119 Hong, J., Zhang, W., Wang, Y. et al. (2014). Photocatalytic reduction of carbon dioxide over self-assembled carbon nitride and layered double hydroxide: The role of carbon dioxide enrichment. *ChemCatChem* 6(8): 2315–2321.
120 Ong, W-J., Tan, L-L., Chai, S-P. et al. (2015). Heterojunction engineering of graphitic carbon nitride (gC_3N_4) via Pt loading with improved daylight-induced photocatalytic reduction of carbon dioxide to methane. *Dalton Transactions* 44(3): 1249–1257.
121 Becerra, J., Nguyen, D-T., Gopalakrishnan, V-N. et al. (2020). Plasmonic Au nanoparticles incorporated in the zeolitic imidazolate framework (ZIF-67) for the efficient sunlight-driven photoreduction of CO_2. *ACS Applied Energy Materials* 3(8): 7659–7665.
122 Chen, L., Wang, Y., Yu F. et al. (2019). A simple strategy for engineering heterostructures of Au nanoparticle-loaded metal–organic framework nanosheets to achieve plasmon-enhanced photocatalytic CO_2 conversion under visible light. *Journal of Materials Chemistry A* 7(18): 11355–11361.
123 Corson, E.R., Subramani, A., Cooper, J.K. et al. (2020). Reduction of carbon dioxide at a plasmonically active copper–silver cathode. *Chemical Communications* 56(69): 9970–9973.
124 DuChene, J.S., Tagliabue, G., Welch, A.J. et al. (2020). Optical excitation of a nanoparticle Cu/p-NiO photocathode improves reaction selectivity for CO_2 reduction in aqueous electrolytes. *Nano Letters* 20(4): 2348–2358.

8

Possible Ways for CO_2 Reduction into Hydrocarbons

Shelly Singla[1], Pooja Devi[2,], and Soumen Basu[1]*

[1] School of Chemistry and Biochemistry, Thapar Institute of Engineering and Technology, Patiala 147004, India
[2] Materials Science and Sensor Application, Central Scientific Instruments Organization, Chandigarh 160030, India
[*] Corresponding author

8.1 Introduction

Globally, there is a rising need for energy consumption [1] as well as environmental deterioration as a result of rapid industrialization and population increase [2, 3]. At present, nearly 85% of the world's energy demand is fulfilled by the combustion of fossil fuels. By 2050, worldwide energy consumption is expected to reach over 40 Terra Watt (TW), and by 2100, it will have surpassed 60 TW. Therefore, burning of fossil fuels forms a large number of gaseous pollutants, including particulates, greenhouse gases, heavy metals, sulfur oxides, and nitrogen oxides [4]. CH_4, CO_2, and N_2O are released into the atmosphere when fossil fuels are burned, resulting in global warming and climate change with severe implications for the environment and human civilization. When these gases are discharged into the atmosphere, they raise global sea levels and temperatures. CO_2 is predicted to be the main contributor among all greenhouse gases, which is produced by both natural and man-made processes and is responsible for environmental ramifications. However, it is also required for the growth of all plants on the planet, as well as a variety of industrial activities [5]. In a perfect world, the amount of CO_2 generated on the earth would be equal to the consumed amount, ensuring that CO_2 levels stay stable and environmental stability is maintained [6]. Unfortunately, as human industrial activities have become more intense, this equilibrium has been disturbed, resulting in increased CO_2 generation, making global warming of significant concern [7]. According to the United States Environmental Protection Agency (USEPA) reports, CO_2 emissions account for around 80% of all greenhouse gas emissions, with fossil fuel burning and chemical processes accounting for 65% of the total emissions. The Intergovernmental Panel on Climate Change (IPCC) estimate that by 2100, CO_2 levels in the atmosphere would rise to 590 ppm, causing world average temperatures to climb by 1.9^{0}C. To avert the dire effects caused by the emission of CO_2, the IPCC has devised a plan to keep the global warming to 1.5^{0}C and achieve net-zero emissions by 2050 [8]. To reduce the quantity of CO_2 emitted, the first option can be handled by improving energy efficiency or switching to a different main energy source [9]. The substitution of less carbon-rich fossil fuels (oil or natural gas) for a carbon-rich energy carrier (coal) is a very simple alternative for lowering CO_2 emissions [10]. However, moving to non-fossil fuels like hydrogen and renewable energy would result in the greatest CO_2 decrease [11]. The use of CO_2 as a chemical feedstock in various applications is the second option for its reduction. The third option entails the creation of novel CO_2 capture and sequestration technology.

Green Energy Harvesting: Materials for Hydrogen Generation and Carbon Dioxide Reduction,
First Edition. Edited by Pooja Devi.

The many suggested methods either capture and geologically sequester the CO_2 or convert it into low-carbon products [12]. CO_2 conversion and usage appear to be a more appealing and viable answer in today's environment of increased energy needs. CO_2 is currently regarded as a reliable, inexpensive, and plentiful carbon source for use in fuels and chemicals [13]. This technique can reduce the greenhouse gas emissions, help in relief of energy problem, and transform waste into money.

Therefore, a vast range of CO_2 reduction methods has been developed to minimize CO_2 emissions and counteract the greenhouse impact. Chemical, thermal, and photoconversion processes, physical conversion processes, and enzymatic conversion processes (biological and biochemical) are the three primary methods for converting CO_2 into valuable fuels. Many CO_2 conversion methods; e.g., enzymatic, chemical, electrocataytic, photocatalytic, photoelectrocatalytic, etc., are summarized in this chapter, along with their advantages and disadvantages.

8.2 CO_2 Conversion From an Energy Point of View

Evolution of CO_2 is often the end result of fossil hydrocarbon fuel burning. The linear and centro-symmetric (O=C=O) structure of CO_2 molecules is extremely stable [14]. The driving force indicated in Equation 8.1 is the difference in Gibbs's free energy (ΔG) between the product and reactants during a specific process.

$$\Delta G = \Delta H + T\Delta S \tag{8.1}$$

where ΔG, ΔH, ΔS, and T signify the change in energy, enthalpy, entropy, and temperature, respectively.

The CO_2 molecule is extremely thermodynamically stable having $\Delta G^{\circ} = -400\,\text{kJ/mol}$. Also, it is highly oxidizable, exhibiting extremely low reactivity, necessitating the passage of a thermodynamic barrier to activate it [15]. To assure thermodynamically favorable CO_2 dissociation, high pressures and high temperatures (1600–2000 K) have been utilized in the past. However, the use of fossil fuels as an energy source can result in an environmentally harmful method, as well as leading to low energy efficiency [16]. As a result, many people believe that using CO_2 to produce chemicals is not practical or cost-effective. However, a deeper and more detailed thermodynamic free energy study reveals that only the inorganic and organic carbonates containing CO_3 moiety are more stable than CO_2 [17]. This indicates that many CO_2 reactions might proceed without the need for additional energy, since the co-reactant typical of the reaction circumstances, such as at ambient pressure, offers enough energy. The CO_2 conversion reactions are divided into two groups:

a) Carboxylation reactions that do not need a large amount of external energy.
b) Reactions that create CO_2 in a reduced form. A significant quantity of external energy is required in these types of reduction reactions. Heat, electrons, and irradiations/photons can all be used as energy sources, or a combination of them. Formaldehyde, CO, methane, methanol, hydrocarbons, and other gases are the end products of these processes. Thermodynamically, highly stable molecules are created by proton coupled

multi-electron reductions rather than single electron reduction and that is why the former is more preferable to the latter. Due to the occurrence of significant reorganizational energy between the bent radical anion and the linear molecule, the CO_2 to $CO_2^{-\bullet}$ reduction by a single electron system happens at $E^0 = -1.850\,V$, as in Equation 8.2. Reduction of CO_2 into valuable chemicals using proton coupled multi-electron reduction along with standard reduction potentials is summed up in Equations 8.2–8.8 [18]. However, a high amount of external energy is necessary for reduction of CO_2 into valuable products [19]. Therefore, several methods are followed for its conversion.

$$CO_2 + e^- \longrightarrow CO_2^{-\bullet} \qquad -1.850\,V(\text{vs SHE}) \tag{8.2}$$

$$CO_2(g) + H_2O(l) + 2e^- \longrightarrow HCOO^-(aq) + OH^-(aq) \qquad -0.665\,V(\text{vs SHE}) \tag{8.3}$$

$$CO_2(g) + H_2O(l) + 2e^- \longrightarrow CO(g) + 2OH^-(aq) \qquad -0.521\,V(\text{vs SHE}) \tag{8.4}$$

$$CO_2(g) + 3H_2O(l) + 4e^- \longrightarrow HCHO(l) + 4OH^-(aq) \qquad -0.485\,V(\text{vs SHE}) \tag{8.5}$$

$$CO_2(g) + 5H_2O(l) + 6e^- \longrightarrow CH_3OH(l) + 6OH^-(aq) \qquad -0.399\,V(\text{vs SHE}) \tag{8.6}$$

$$CO_2(g) + 6H_2O(l) + 8e^- \longrightarrow CH_4(aq) + 8OH^-(aq) \qquad -0.246\,V(\text{vs SHE}) \tag{8.7}$$

$$2H_2O(l) + 2e^- \longrightarrow H_2(g) + 2OH^-(aq) \qquad -0.414\,V(\text{vs SHE}) \tag{8.8}$$

Apart from these two types of reactions, another conventional method for the conversion of CO_2 into useful hydrocarbons involves non-thermal plasma technology. High-energy electrons with temperatures of 1–10 eV are required in non-thermal plasma technology in order to dissociate the chemical bonds of CO_2 to form hydrocarbons while the overall temperature of the gas is low. Direct breakdown of CO_2 is facilitated due to the non-equilibrium nature of plasma which possesses the ability to bypass thermodynamic barriers. This process is considered advantageous due to its inexpensive nature, ability to conduct physical and chemical reactions at low temperature, and the non-equilibrium nature of plasma. However, requirement of the high energy electrons, requirement of the inert gases; e.g., N_2, He, and Ar, which when reacting with CO_2 generates undesired products, and production of lower yield of hydrocarbons limits the efficacy of this process [20].

Another conventional method for the reduction of CO_2 involves the production of syngas using a dry reforming reaction. This method involves the reaction of CH_4 with CO_2 in stoichiometric amounts and results in the production of syngas. Although this technology has the potential to reduce CO_2 emissions, there are significant obstacles associated with direct fuels and chemicals technologies. Lower yield of the H_2/CO is obtained, therefore it is not a viable H_2 generation alternative, especially when compared to other processes like steam reforming, which consistently yield much higher ratios. Also, a reduced-controllable reaction temperature is required. Moreover, decomposition of methane and CO produces coke which is harmful, so this process is not considered efficacious [21].

8.3 CO_2 Reduction Methods

8.3.1 Enzymatic CO_2 Reduction Method

To combat global warming and climate change, production of valuable chemicals by the conversion of CO_2 is the promising way forward [22]. The beginning of the evolution is marked by the conversion of CO_2 into organic material as it is essential for life [23]. In order to convert CO_2 into organic material, the enzymatic method has gained consideration. For this, the Calvin cycle, 3-hydroxypropionate cycle, reductive acetyl-CoA route, dicarboxylate cycles, 4-hydroxybutyrate cycle, and reductive citric acid cycle are the six major routes that involve the two types of enzymes; i.e., oxidoreductases and lyases for the abatement of CO_2 into valuable products [24]. The specific enzyme determines the fixation of CO_2 in each step of the six cycles, where bicarbonate or CO_2 acts as a carbon source. Hence, CO_2 and bicarbonate should be easily and efficiently convertible to one another [25]. Therefore, the interconversion of CO_2 and bicarbonate in a rapid and precise way is carried out by the enzyme carbonic anhydrase [26].

However, enzymes are currently costly and environmentally sensitive catalysts, therefore reduction of CO_2 by the enzymatic method is limited [27]. Also, cofactor-dependent reactions play a critical role in the conversion of CO_2 in an enzymatic way. However, large-scale uses are severely limited owing to high cost and the poor supply of ion-type cofactors [28]. All of these factors limit the efficacy of the enzymatic method. Therefore, various methods; for example, electrochemical, photocatalytic, and photoelectrocatalytic methods, are explored which possess greater potential for the formation of valuable fuels by the reduction of CO_2 rather than the enzymatic method [29].

8.3.2 Chemical Conversion Process

For the synthesis of the organic compounds, the development of CO_2-fixation processes, which would allow the conversion of waste gas into an economic benefit, is one of the key problems [30]. The lower energy level is the largest barrier to using it as a starting material, as CO_2 is the oxidized state of carbon [31]. To overcome these problems, CO_2 reduction using a chemical conversion process was carried out by the researchers. Therefore, for the CO_2 reduction into valuable products using a chemical conversion process, four major principles are followed, as a high amount of energy is necessary for its conversion.

The materials possessing ring strain, and unsaturated compounds like alkynes, alkenes, and organometallic compounds, etc., are the high-energy materials that can be used as the starting materials for the conversion of CO_2 [32]. Also, carbamates, organic carbonates, esters, lactones, etc., can be used, as they possess the carbon at the higher oxidation state but possess the lower energy [33]. Light in the form of external energy can also be provided for carrying out the CO_2 conversion reaction [34]. Lastly, by the removal of the products, the reactions that are equilibrium controlled can be moved to the product side. For the chemical conversion of CO_2, initially, carbonic or carbamic acids are formed by the reaction of carboxyl or the carboxylate products which are formed by the nucleophilic attack of amines, organometallic compounds, alkoxides, etc., on the CO_2 [35]. Finally, the electrophilic attack on the carbonic or the carbamic acids is converted into organic carbonates and carbamates. Also, the Kolbe–Schmidt reaction and carboxylation of the epoxides can be carried out for formation of salicylic acid and cyclic carbonate esters respectively [36].

Despite a lot of research being done on the chemical reduction of CO_2 into useful fuels, CO_2 utilized as a raw material for chemical synthesis is still in its early stages of development. Also, a lower yield of the products is obtained. Therefore, nowadays, electrocatalytic, photocatalytic, and photoelectrocatalytic methods are greatly explored, which are explained in detail below.

8.3.3 Electrocatalytic CO_2 Reduction

The utilization of renewable electricity for the electrocatalytic CO_2 reduction to CO and other chemical compounds provides a dual function of CO_2 storage and clean energy conversion, making it a popular choice among industry and academia in recent years [37]. Steady chemical energy and useful goods are generated at the same time by the transformation of intermittent electricity [38]. Electrocatalytic CO_2 reduction occurs at the electrolyte–electrode interface and primarily includes the following three stages [39].

1) CO_2 adsorption on an electrocatalyst's surface through chemical means.
2) Through electron and proton transport, rupture, and creation of C = O and C–H bonds respectively.
3) Products desorption from the catalyst's surface.

Electrocatalytic CO_2 reduction produces mixed results, including various species by the reduction of CO_2 in the majority of situations. By using a variety of electrocatalysts, electrocatalytic CO_2 reduction may take place at room temperature and under ambient conditions, to achieve the selective desired products [40]. Due to the inclination of the need for ambient temperature, pressure, ecological friendliness, and a suitable electrocatalyst, the electrocatalytic process is deemed efficacious. Moreover, the total amount of chemicals used can be reduced and the electrolyte is recyclable, owing to which electrocatalytic CO_2 reduction has gained attention. The electrocatalysts used have a large impact on the CO_2 reduction routes [41]. Electrocatalyst performance is strongly influenced by protons. Therefore, a variety of electrocatalysts; e.g., bismuth related complexes, carbonaceous nanomaterials, transition metal dichalcogenides, ZnO, Cu_2O, WO_3, TiO_2, etc., have been used by various researchers. However, WO_3, TiO_2, and ZnO, etc., absorb a small portion of light, along with a small surface area due to which they do not act as efficacious photocatalysts. Hybridizing them with other photocatalysts can lead to increased efficiency and the yield of the products obtained by the reduction of CO_2 has been observed.

Bi nanoparticles tuned over a Bi_2O_3 nanosheet synthesized by the facile hydrothermal method were used to carry out the reduction of CO_2 into formate. High selectivity of ~100%, the current density of 24.4 mA cm^{-2}, and an excellent duration of >24 h were observed when the CO_2 reduction using the electrochemical method was carried out. Good catalytic activity was observed as charge transfer resistance is lowered by the tight contact between Bi nanoparticles. The CO_2 reduction is facilitated by the leftover Bi_2O_3, which presumably enriches CO_2 around the perimeter of Bi electrocatalysts. Moreover, Bi NPs/Bi_2O_3 NSs have a lot of potential for electrocatalytic CO_2 reduction to formate as they are simple to manufacture on a large scale, are inexpensive, and have high performance [42].

It has been observed in the literature that electrocatalytic reduction of CO_2 at very low energy was achieved in aqueous solution by an Mn complex catalyst combined with the multi-walled carbon nanotubes (MWCNTs) with K^+ cations. CO_2 reduction at an overpotential of 100 mV was achieved with a current density of more than 2.0 mA cm^{-2} at −0.39 V vs. RHE at a constant rate for 48 h, and 75.5 μmol of CO was produced with the Faradaic efficacy of 87.3%. The overpotential for abatement of CO_2 greatly reduced in the aqueous solution owing to the electron-accumulating MWCNTs, in combination with surface-adsorbed K^+ ions that created an environment that stabilized CO_2 close to the Mn-complex [43].

ZnO/Cu_2O film electrodes synthesized by the electrodeposition method were reported to generate CH_3OH as the product with a yield of 315.656 μmol cm^{-2}, and Faradaic efficacy of 45% by CO_2 reduction. Fixation of the intermediate products of CO_2 reduction and increment in the sites for CO_2 adsorption on the surface of the electrodes was facilitated by the attachment of ZnO to Cu_2O. The ability for the adsorption of CO_2 on the electrodes became enhanced by the deposition of the appropriate amount of Zn^{2+} ions over the surface of the electrode, which deliberately enhanced its surface area. The reduction of CO_2 was substantially promoted by the synergistic effect of ZnO and Cu_2O [44].

Apart from all the advantages, electrocatalytic CO_2 reduction possesses disadvantages; e.g., the occurrence of hydrogen evolution reaction (HER) occurs as a competitive side reaction in an aqueous solution owing to which the Faradaic efficiency of the desirable products is reduced [45]. The poor solubility of the CO_2 in the water is another obstacle. Moreover, CO_2 reduction reaction's products are constantly influenced by the materials used for the formation of electrodes [46]. Adsorption, desorption, electron loss, and gain are the complicated processes that are entailed in the hydrogenation and activation stage in electrochemical reduction [47]. The entire process is inefficient with significant energy consumption, because the activation process is susceptible to various limitations and generally happens at a high overpotential [48]. Further, the reduction process result is not monolithic, but rather a combination of various compounds, with lower stability of the catalyst, making long-term work impossible [49]. However, photocatalytic and photoelectrocatalytic methods have the tendency to reduce the drawbacks of the electrocatalytic method.

8.3.4 Photocatalytic CO_2 Reduction

Natural photosynthesis, in which plants convert water and CO_2 into oxygen and carbohydrates in the presence of sunlight, is similar to photocatalytic CO_2 reduction with water. Processes viz harvesting of light, formation, and the segregation of the charge carriers, besides catalytic reactions, are involved [50]. Solar energy is transformed into chemical bonds and stored through this process [51]. Reduction of CO_2 into useful chemicals by exploitation of solar light seems appealing, yet difficult due to thermodynamic stability (cleavage of C=O in CO_2 demands high energy of 750 kJ mol^{-1}) and the difficulty to absorb light in the 200–900 nm wavelength range [52]. The use of a suitable photosensitizer is required to address this issue. In a single photocatalyst, the charge carriers are pulled onto the surface of the photocatalyst by Columbic forces owing to which the recombination rate becomes higher [53]. Therefore, the high reintegration of the photogenerated excitons, the narrow range of the absorption of light, and the broad bandgap, together with inadequate redox capacity, lowers the photocatalytic efficacy of the single photocatalyst [54]. Therefore, the two semiconductor photocatalysts are joined together in order to create a heterojunction, as its formation reduces the recombination rate of the charges [55]. Apart from this, heterojunction photocatalysts exhibit adequate redox capability, ability to absorb a broader range of light, have appropriate bandgap, and a high surface area, etc. [56]. Therefore, a wide variaty of heterogeneous catalysts have been explored for the formation of the valuable fuels from the reduction of CO_2.

Photocatalytic abatement of CO_2 using a TiO_2-MnO_x-Pt composite synthesized by the hydrothermal method has been reported in the literature. The CH_4 and CH_3OH were formed as a result of the yield of 104 and 91 μmol m^{-2} respectively when CO_2 reduction was carried out under visible light illumination for 3 h using the ternary photocatalyst. However, rates of 28 and 31 μmol m^{-2} were observed for CH_4 and CH_3OH respectively when CO_2 reduction was carried out using a TiO_2 photocatalyst. Such a higher rate of CO_2 reduction using the ternary composite was observed, because three distinct junctions have been combined in a single composite to boost activity synergistically. The photogenerated charge carriers were separated and transferred more quickly to the multi-junctions built on the surface of well-faceted TiO_2 nanosheets. Also, visible light absorption capability was enhanced with the formation of the heterojunction, which was also responsible for the increased rate of CO_2 reduction [57].

Increased CO_2 reduction to CO with a rate of 27.2 μmol $g^{-1}h^{-1}$ was observed with urchin-like α-Fe_2O_3, and g-C_3N_4 under sunlight irradiation was observed without using any sacrificial reagent or cocatalyst. The inexpensive nature, good thermodynamic stability, and narrow bandgap of 2.2 eV, as well as good capacity to absorb the visible light, render α-Fe_2O_3 a good photocatalyst for the reduction of CO_2. However, its photocatalytic activity was hindered due to the smaller migration length of the holes, generation of the electrons of low energy from the conduction band, and also lesser segregation of the photogenerated excitons. Coupling of the α-Fe_2O_3 with g-C_3N_4 created the Z-scheme heterojunction owing to which the reduction capability of the composites was enhanced. Moreover, the reduction of CO_2 was increased as the adsorption of CO_2 was aided by the basic sites and the 3D urchin-like structure of α-Fe_2O_3. Also, the segregation of the excitons was facilitated by the Z-scheme characteristics of the heterojunction photocatalyst [13].

It has been estimated that 109.5 μmol g^{-1} and 19.2 μmol g^{-1} of methanol and acetaldehyde were produced respectively when CO_2 reduction was carried out photocatalytically over MoS_2 sheets under UV light irradiation. MoS_2 sheets possess a conduction band edge value below the potential necessitated to reduce CO_2 to form methanol and acetaldehyde. On absorption of a sufficient amount of UV light, electrons hop from the valence to conduction band level of the MoS_2 sheets, thereby reducing their capacity to recombine with the holes. Hence, a significant reduction of CO_2 occurs to form the valuable products, as represented in Figure 8.1a. The major product was found to be methanol along with a substantial amount of acetaldehyde in 0.5 M $NaHCO_3$, as depicted in Figure 8.1b. The production rate of 27.4 and 2.2 μmol g^{-1}, and the highest yield of 109.6 and 8.9 μmol g^{-1}, were found for methanol and acetaldehyde respectively. On the other hand, an increased amount of acetaldehyde was produced in comparison to methanol when 0.5 M NaOH was used to carry out CO_2 reduction, as represented in Figure 8.1c. Production rates of 11.2 and

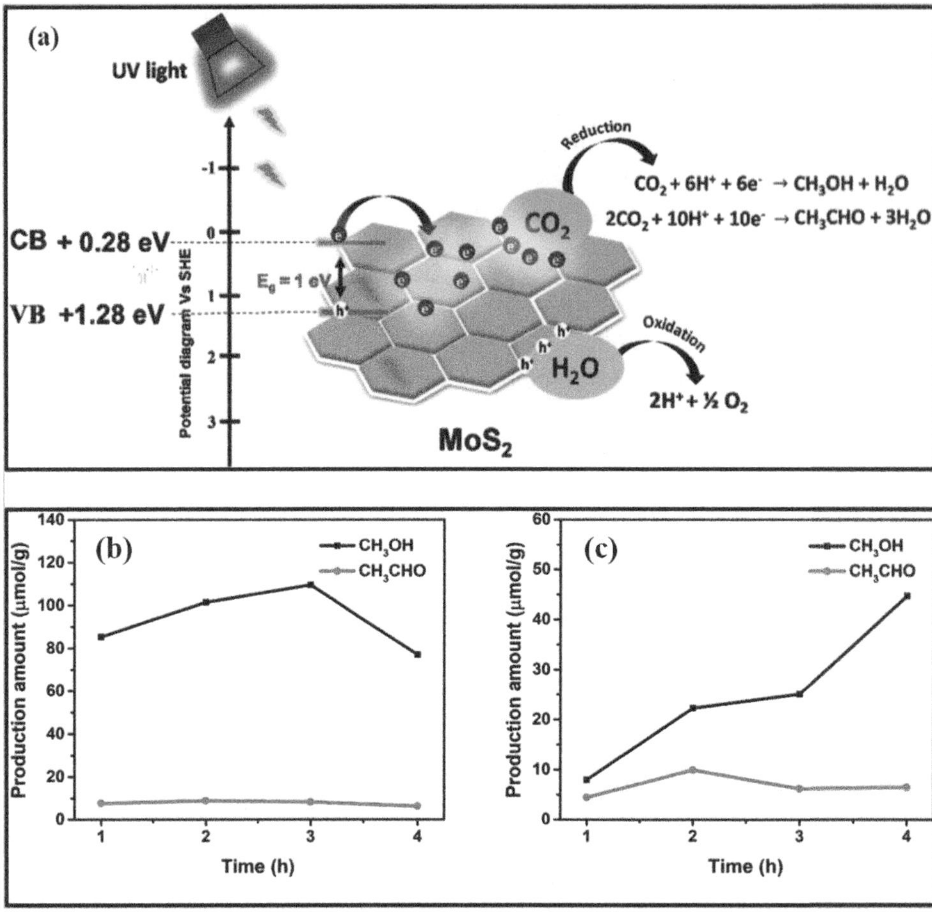

Figure 8.1 (a) Mechanism of the reduction of CO_2 over MoS_2 sheets by the photocatalytic process; and (b) and (c) products formed by the reduction of CO_2 over MoS_2 sheets in 0.5 M $NaHCO_3$ and 0.5 M NaOH, respectively. (Reproduced from [58] with permission from Elsevier).

2.5 μmol g^{-1}, and highest yields of 44.7 and 9.9 μmol g^{-1} were found for methanol and acetaldehyde respectively in 0.5 M NaOH. The selectivity of production of methanol in 0.5 M $NaHCO_3$ may be due to the involvement of HCO_3^- ions in reducing the recombination of the photogenerated charges and so initiating a series of events that began with the production of formate anions, then formyl anions, and lastly methanol. However, an increase in the acetaldehyde production in 0.5 M NaOH may be due to the inclusion of OH^- ions, which possess the ability to scavenge the holes, thereby reducing the recombination rate. Hence, the formation of acetaldehyde occurs via dimerization of the two carbon species owing to accretion of the higher number of photogenerated electrons on MoS_2 sheets [58].

The CO_2 reduction rate of 2.98 and 1.31 μmol g^{-1} for CH_4 and CO respectively have been mentioned for CdS/$BiVO_4$ under the illumination of visible light for 5 h, while the yield of only 8.73 and 1.95 μmol g^{-1} for CH_4 and CO respectively was achieved by using $BiVO_4$ alone. Due to the higher activity toward water splitting and organic degradation, $BiVO_4$ has sparked a lot of interest as a visible light absorbing photocatalyst. CO_2 reduction coupled with water oxidation can be carried out owing to the occurrence of the optimum band edges in $BiVO_4$. However, lower charge migration capability, higher recombination of the excitons, and lower segregation capacity limits the efficacy of $BiVO_4$. Photocatalytic activity of $BiVO_4$ was enhanced by tuning it with CdS, as it contains the appropriate conduction band potential of −0.6 eV for the CO_2 reduction. Also, the construction of a Z-scheme heterojunction was responsible of enhanced CO_2 reduction. The heterojunction improved the carrier charge mobility, reduced the recombination of the charges, and enhanced their segregation [8].

Photocatalytic CO_2 reduction is regarded as an efficient method, as it is a safe, inexpensive, and environmental-friendly method [59]. Moreover, it does not require electricity, reactions occur under milder conditions, and a higher yield of products is obtained [60]. However, difficulty in the product selectivity, complexity in the electron transfer process, and trouble in the partition of the catalyst from the products after the occurrence of the reaction, limit the efficacy of the photocatalytic CO_2 reduction process [61]. This is where the photoelectrocatalytic process comes into play, since it has the ability to overcome the limits of the photocatalytic process.

8.3.5 Photoelectrocatalytic Reduction of CO_2

The combination of electrocatalysis and photocatalysis or photoelectrocatalytic process is based on a unified set of principles [62]. A set of ideas have acquired a lot of attraction for the oxidation of organic compounds with great success, decrease of inorganic ions, inactivation of microbes, and reduction of CO_2, as well as for the formation of hydrogen [63]. The ultimate objective for effective CO_2 reduction is the photoelectrocatalytic configuration, which comprises of photoanode and photocathode of an n-type and an p-type semiconductor respectively without external voltage [64]. Using two n-type and p-type semiconductors with comparable bandgaps to construct a heterojunction, the bias will be totally eliminated. Redox reactions occur by photogenerated electrons and holes left at photocathode and the photoanode respectively after the reintegration of the electrons at the photoanode and the holes at photocathode occur due to the higher overpotential [65]. Various heterocatalysts used for carrying out the photoelectrocatalytic process by the different scientists are described below.

Photoelectrocatalytic CO_2 reduction using $CuFe_2O_4$ hybridized with GO has been reported in the literature. Production of methanol with the yield of 28.8 μmol $L^{-1}cm^{-2}$, Faradaic efficiency of 87 %, and quantum efficacy of 20.5%, were observed by the abatement of CO_2 under visible light illumination. $CuFe_2O_4$ offers a good platform for CO_2 reduction as it possesses a low bandgap and good structural stability. However, higher electron–hole recombination and lower visible light absorption capacity limits its efficacy, which can be enhanced by hybridizing it with the wide bandgap semiconductor. Here, GO acts as electron acceptor and site for reduction of CO_2. Conductive network for the transfer of photogenerated charge carriers forms between GO and $CuFe_2O_4$ (photocathode), which facilitates the reduction of CO_2 [66].

The abatement of CO_2 to hydrocarbons having the production rate of 62.4 μM $cm^{-2}h^{-1}$ was observed with a TiO_2@ZnO/FTO heterojunction under illumination by sunlight. The selectivity of the formation of hydrocarbons like methanol, acetic acid, formic acid, etc., and rate of the reduction of the CO_2, was affected by the deposition of the metal including Ag, Pt, Au, and Ni. The activity of the photoelectrocatalytic reduction was augmented by the high transfer of the electrons and lower conduction band potential [42].

Reduction of CO_2 has been carried out photoelectrochemically over a $CuO/g\text{-}C_3N_4$/carbon paper photocatalyst synthesized by hydrothermal processes and methanol was formed as the product with the highest yield of 25.10 $\mu ML^{-1}cm^{-2}$. The mechanism of the photoelectrocatalytic CO_2 reduction involves the direct transfer of the electrons and holes from one semiconductor to the other for the production of methanol. Equations 8.9–8.11 represent the steps for the conversion of CO_2 into methanol occurring on the surface of the photocatalyst.

$$CuO/g-C_3N_4 + h\nu \rightarrow e^- + h^+ \tag{8.9}$$

$$H_2O + 2h^+ \rightarrow 2H^+ + 1/2O_2 + 2e^- \tag{8.10}$$

$$CO_2 + 6H^+ + 6e^- \rightarrow CH_3OH + H_2O \tag{8.11}$$

$CuO/g\text{-}C_3N_4$/carbon paper photocathodes produced higher methanol output of 25.1 mol Lcm^{-2}, while on $g\text{-}C_3N_4$/carbon paper and CuO/carbon paper photocathodes, 4.1 and 2.6 times respectively, the lower formation of methanol was observed, as shown in Figure 8.2a. The Faradaic and quantum efficacy of 75.0 and 8.9% respectively was observed by using a $CuO/g\text{-}C_3N_4$/carbon paper photocathode (Figure 8.2b), which was higher than that produced on bare $g\text{-}C_3N_4$. The $CuO/g\text{-}C_3N_4$/carbon paper photocathode exhibit a higher yield of methanol by the reduction of CO_2 than bare CuO and $g\text{-}C_3N_4$, due to the greater ability to absorb irradiated visible light, which leads to efficacious production and charge transfer of the photogenerated charge carriers for CO_2 photo-electrocatalytic conversion [67].

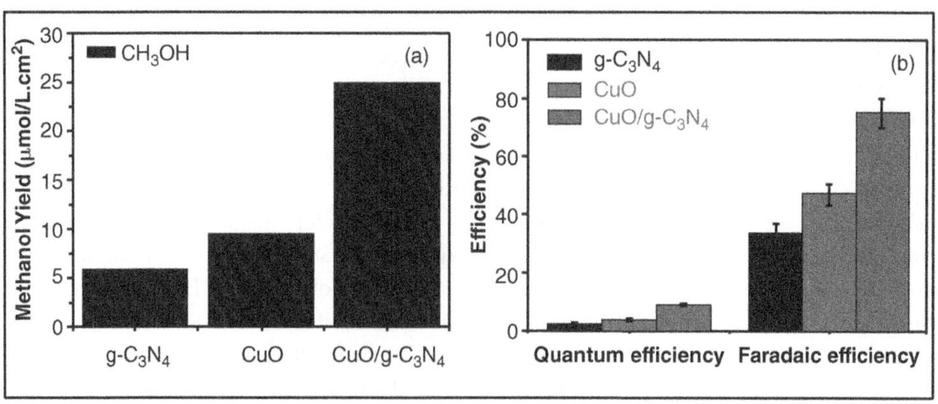

Figure 8.2 (a) Yield of the methanol produced from the photocathodes by photo-electrocatalytic CO_2 reduction by using a $CuO/g\text{-}C_3N_4$ photocatalyst; and (b) quantum and the Faradaic efficacy produced from the photocathode. (Reproduced from [67], with permission from Elsevier).

It is documented in the literature that ethanol was formed as the product of the reduction of the CO_2, possessing a rate of 11.4 μM $h^{-1}cm^{-2}$, and Faradaic efficacy of 80%, with the Bi_2WO_6/BiOCl heterojunctions synthesized by the hydrothermal method grown *in situ* onto an F-SnO_2 transparent conductive glass (FTO) under irradiation by sunlight. With the lattice shift from (101) to (112) for BiOCl, the heterojunctions of the composite preserve an excellent 2D layered structure, increasing the efficiency of photoelectron–hole separation and thus escalating the CO_2 reduction activity [68, 209–217].

Owing to its inexpensive nature, availability of higher product yield as well as selectivity, more options for the catalysts, and CO_2 reduction through PEC, is more effective [69]. This strategy allows for additional chances to reduce the drawbacks of both technologies while maximizing their benefits for effective CO_2 conversion [70]. In PEC, the reintegration rate of the excitons becomes reduced in contrast to photocatalysis as the direct transfer of the electrons takes place and the promotion of band bending occurs due to the applied bias [71]. Moreover, with suitable external bias, photocatalysts with unfavorable band locations can be utilized for CO_2 reduction and H_2O oxidation [72]. Further, the utilization of fossil fuels as a power source for electrocatalysis is no longer necessary because of the integration of photocatalytic and electrocatalytic processes [73]. However, complex design of the system limits its efficiency.

Merits and demerits of the electrocatalytic, photocatalytic, and photo-electrocatalytic methods are tabularized in Table 8.1. Also, CO_2 reduction rate of diverse photocatalysts, products formed, as well as yield of the products by different methods, are summarized in Table 8.2.

Table 8.1 Merits and demerits of electrocatalytic, photocatalytic, and photo-electrocatalytic method.

Methods	Advantages	Disadvantages
Electrocatalytic	• Cleaner method • Selective products achieved	• Need of ambient pressure, electrocatalyst, temperature • HER as competitive side reaction reducing Faradaic efficiency of CO_2 reduction
Photocatalytic	• Safe • Inexpensive • Environmental-friendly method • Does not require electricity • Reactions occur at lower temperatures • A higher yield of the products is obtained	• Difficulty in the product selectivity • Complexity in the electron transfer process • Trouble in the partition of the catalyst from the products after the occurrence of the reaction
Photoelectrocatalytic	• Inexpensive nature • Availability of higher product yield • Higher selectivity of products • Offers more options for the catalysts	• Complex design

Table 8.2 The CO_2 reduction rate of different photocatalysts by different methods.

Method	Photocatalyst	Product	Yield	References
Electrocatalytic	Bi NPs/Bi_2O_3 NSs	Formate	–	[42]
	Mn complex catalyst/ MWCNTs/ K^+	CO	75.5 μmol	[43]
	ZnO/Cu_2O film	CH_3OH	315.656 μmol cm^{-2}	[44]
Photocatalytic	TiO_2-MnO_x-Pt	CH_4 CH_3OH	104 μmol m^{-2} 91 μmol m^{-2}	[57]
	α-Fe_2O_3/g-C_3N_4	CO	27.2 μmol $g^{-1}h^{-1}$	[13]
	MoS_2 sheets	CH_3OH CH_3CHO	109.5 μmol g^{-1} 19.2 μmol g^{-1}	[58]
	CdS/$BiVO_4$	CH_4 CO	2.98 μmol g^{-1} 1.31 μmol g^{-1}	[8]
Photo-electrocatalytic	$CuFe_2O_4$/GO	CH_3OH	28.8 μmol $L^{-1}cm^{-2}$	[66]
	TiO_2@ZnO/FTO	Hydrocarbons	62.4 μM $cm^{-2}h^{-1}$	[42]
	CuO/g-C_3N_4/ carbon paper	CH_3OH	25.10 μM $L^{-1}cm^{-2}$	[67]
	Bi_2WO_6/BiOCl/ F-SnO_2	CH_3CH_2OH	11.4 μM $h^{-1}cm^{-2}$	[68]

8.4 Features of Optimum Photocatalyst

A semiconductor capable of absorbing an optimum amount of light is required to carry out the reduction of CO_2. Ability to possess adequate redox potential, lower recombining ability of the excitons, higher migration ability, and optimum amount of surface active sites are some key features that a optimum photocatalyst should possess [74]. Moreover, a good photocatalyst should be inexpensive, reusable, give higher yield, be abundant, possess narrow bandgap, and should be non-toxic. However, low efficiency is the most significant constraint that photocatalysts face. The cause is photogenerated electrons and holes recombining at a rapid pace. After illumination, electrons hop from valence to conduction band levels of a single photocatalyst [54]. Then the reintegration of the photogenerated excitons occurs owing to the tendency of Columbic forces to withdraw photogenerated charge carriers together at the surface of a photocatalyst. Aside from quick recombination, other important characteristics for an excellent photocatalyst include broad sunlight absorption and adequate redox capacity. These two requirements, unfortunately, are incompatible for a single photocatalyst. As a result of this, heterojunction photocatalysts are considered. Photocatalysts with heterojunctions are made by joining two semiconductor photocatalysts together [55]. The ability to reduce the recombining capability of the charge carriers renders the heterojunction photocatalysts efficacious for the splitting of water and reduction of CO_2. However, heterojunctions formed by 2D nanomaterials are more efficacious than those formed by the other semiconducting materials. The higher the surface area for harvesting the sunlight, the greater the number of surface active sites for initiating redox reactions offered by the 2D nanomaterials, owing to their layered structure due to which they are regarded as efficacious. Also, carrier charge recombination is minimized due to the tendency of the atomic thickness of 2D materials to reduce the carrier's migration distance from the interior to the surface. All these properties render 2D materials more efficient than the other semiconducting materials [75].

8.5 Conclusions and Future Outlook

The emission of detrimental greenhouse gases, as well as rising energy demands, are met by formation of value-added fuels by the conversion of CO_2. CO_2 is now widely acknowledged as a safe, affordable, and abundant source of carbon for use in fuels and chemicals. Despite the fact that generating fuels from CO_2 appears to need significantly more energy than existing fuels, CO_2 reduction through the photo-electrocatalytic technique appears to be the most feasible and dependable way for industrial-scale high-value CO_2 conversion and use. This is because the photo-electrocatalytic process has the ability to overcome the limitations of all other processes and only the complex design of the reactor limits its efficacy. The use of novel developing technological approaches as well as heterogeneous photocatalysts that reduce the particular cost per function is drawing increased research attention and industrial applications for value-added chemicals synthesis. The development of the heterojunction reduces the recombining ability of the photogenerated charge carriers, promotes their migration, is capable of absorbing an ample amount of light for carrying out reduction, and possesses surface active sites. The development of multifunctional photocatalysts capable of absorbing a sufficient amount of sunlight besides the advancement in the hybrid systems, promises a bright future in industrial-scale CO_2 reduction into value-added fuels.

References

1 Singla, S., Sharma, S., and Basu, S. (2021). MoS_2/WO_3 heterojunction with the intensified photocatalytic performance for decomposition of organic pollutants under the broad array of solar light. *Journal of Cleaner Production* 324: 129290. doi:10.1016/j.jclepro.2021.129290.

2 Sharma, S., Basu, S., Shetti, N.P. et al. (2021). Microplastics in the environment: Occurrence, perils, and eradication. *Chemical Engineering Journal* 408: 127317. doi:10.1016/j.cej.2020.127317.

3 Singla, S., Sharma, S., and Basu, S. et al. (2020). Graphene/graphitic carbon nitride-based ternary nanohybrids: Synthesis methods, properties, and applications for photocatalytic hydrogen production. *FlatChem* 24: 100200. doi:10.1016/j.flatc.2020.100200.

4 Singla, S., Sharma, S., and Basu, S. (2021). Photocatalytic water splitting hydrogen production via environmental benign carbon based nanomaterials. *International Journal of Hydrogen Energy* 46(68): 33696–33717. doi:10.1016/j.ijhydene.2021.07.187.

5 Kumar, N., Ray, S.S., Gusain, R. et al. (2020). Polypyrrole-promoted RGO–MoS_2 nanocomposites for enhanced photocatalytic conversion of CO_2 and H_2O to CO, CH_4, and H_2 products. *ACS Applied Energy Materials* 3(10): 9897–9909. doi:10.1021/acsaem.0c01602.

6 Wang, S., Zhan, J., Chen, K. et al. (2020). Potassium-doped g-C_3N_4 achieving efficient visible-light-driven CO_2 reduction. *ACS Sustainable Chemistry and Engineering* 8(22): 8214–8222. doi: 10.1021/acssuschemeng.0c01151.

7 Lee, J., Kim, S.K., and Sohn, Y. (2018). Understanding photocatalytic coupled-dye degradation, and photoelectrocatalytic water splitting and CO_2 reduction over WO_3/MoO_3 hybrid nanostructures. *Journal of Industrial and Engineering Chemistry* 62: 362–374. doi:10.1016/j.jiec.2018.01.016.

8 Wei, Z.-H., Wang, Y., Li, Y. et al. (2018). Enhanced photocatalytic CO_2 reduction activity of Z-scheme CdS/$BiVO^4$ nanocomposite with thinner $BiVO^4$ nanosheets. *Journal of CO_2 Utilization* 28: 15–25. doi: 10.1016/j.jcou.2018.09.008.

9 Xu, Y., Wang, S., Yang, J. et al. (2018b). Highly efficient photoelectrocatalytic reduction of CO_2 on the Ti_3C_2/g-C_3N_4 heterojunction with rich Ti^{3+} and pyri-N species. *Journal of Materials Chemistry A* 6(31): 15213–15220. doi: 10.1039/c8ta03315c.

10 Kim, C., Cho, K.M., Sagaf, A.A. et al. (2018). Z-scheme photocatalytic CO_2 conversion on three-dimensional $BiVO_4$/carbon-coated Cu_2O nanowire arrays under visible light. *ACS Catalysis* 8(5): 4170–4177. doi:10.1021/acscatal.8b00003.

11 Kim, J.H., Magesh, G., Kang, H.J. et al. (2015). Carbonate-coordinated cobalt co-catalyzed $BiVO_4/WO_3$ composite photoanode tailored for CO_2 reduction to fuels. *Nano Energy* 15: 153–163. doi:10.1016/j.nanoen.2015.04.022.

12 Li, M., Zhang, L., Wu, M. et al. (2016). Mesostructured CeO_2/g-C_3N_4 nanocomposites: Remarkably enhanced photocatalytic activity for CO_2 reduction by mutual component activations. *Nano Energy* 19: 145–155. doi:10.1016/j.nanoen.2015.11.010.

13 Jiang, Z., Wan, W., Li, H. et al. (2018b). A hierarchical Z-scheme α-Fe_2O_3/g-C_3N_4 hybrid for enhanced photocatalytic CO_2 reduction. *Advanced Materials* 30(10): 1–9. doi:10.1002/adma.201706108.

14 Cortes, M.A.L.R.M., Hamilton, S.M.J.W.J., Sharma, P.K. et al. (2020). Photoelectrochemical reduction of CO_2 with TiN. *Materials Science in Semiconductor Processing* 108: 104900. doi:10.1016/j.mssp.2019.104900.

15 Liu, L., Wang, S., Huang, H. et al. (2020b). Surface sites engineering on semiconductors to boost photocatalytic CO_2 reduction. *Nano Energy* 75: 104959. doi:10.1016/j.nanoen.2020.104959.

16 Schneider, J., Jia, H., Muckerman, J.T. et al. (2012). Thermodynamics and kinetics of CO_2, CO, and H^+ binding to the metal centre of CO_2 reduction catalysts. *Chem. Soc. Rev.* 41(6): 2036–2051. doi:10.1039/C1CS15278E.

17 Ma, Y., Wang, X., Jia, W. et al. (2014). Titanium dioxide-based nanomaterials for photocatalytic fuel generations. *Chemical Reviews* 114(19): 9987–10043. doi:10.1021/cr500008u.

18 Cho, H., Kim, W.D., Lee, K. et al. (2018). Selectivity of photoelectrochemical CO_2 reduction modulated with electron transfer from size-tunable quantized energy states of CdSe nanocrystals. *Applied Surface Science* 429: 2–8. doi:10.1016/j.apsusc.2017.08.137.

19 Liu, J.-Y., Gong, X.Q., Le, R. et al. (2020a). (Photo)Electrocatalytic CO_2 reduction at the defective anatase TiO_2 (101) surface. *ACS Catalysis* 10(7): 4048–4058. doi:10.1021/acscatal.0c00947.

20 Mei, D., Zhu, X., He, Y.L. et al. (2014). Plasma-assisted conversion of CO_2 in a dielectric barrier discharge reactor: Understanding the effect of packing materials. *Plasma Sources Science and Technology* 24(1): 015011. doi:10.1088/0963-0252/24/1/015011.

21 Galadima, A. and Muraza, O. (2019). Catalytic thermal conversion of CO_2 into fuels: Perspective and challenges. *Renewable and Sustainable Energy Reviews* 115: 109333. doi:10.1016/j.rser.2019.109333.

22 Dibenedetto, A., Stufano, P., Macyk, W. et al. (2012). Hybrid technologies for an enhanced carbon recycling based on the enzymatic reduction of CO_2 to methanol in water: Chemical and photochemical NADH regeneration. *ChemSusChem* 5(2): 373–378. doi:10.1002/cssc.201100484.

23 Luo, J., Meyer, A.S., Mateui, R.V. et al. (2015). Cascade catalysis in membranes with enzyme immobilization for multi-enzymatic conversion of CO_2 to methanol. *New Biotechnology* 32(3): 319–327. doi:10.1016/j.nbt.2015.02.006.

24 Ren, S., Chen, R., Wu, R. et al. (2021). Enzymatic characteristics of immobilized carbonic anhydrase and its applications in CO_2 conversion. *Colloids and Surfaces B: Biointerfaces* 204: 111779. doi:10.1016/j.colsurfb.2021.111779.

25 Kim, S., Kim, M.K., Lee, S.H. et al. (2014). Conversion of CO_2 to formate in an electroenzymatic cell using *Candida boidinii* formate dehydrogenase. *Journal of Molecular Catalysis B: Enzymatic* 102: 9–15. doi:10.1016/j.molcatb.2014.01.007.

26 Zhang, Z., Vasiliu, T., Li, F. et al. (2021b). Electrochemically driven efficient enzymatic conversion of CO_2 to formic acid with artificial cofactors. *Journal of CO_2 Utilization* 52: 101679. doi:10.1016/j.jcou.2021.101679.

27 Sun, Q., Jiang, Y., Jiang, Z. et al. (2009). Green and efficient conversion of CO_2 to methanol by biomimetic coimmobilization of three dehydrogenases in protamine-templated titania. *Industrial & Engineering Chemistry Research* 48(9): 4210–4215. doi:10.1021/ie801931j.

28 Zhai, T., Wang, C., Gu, F. et al. (2020). Dopamine/polyethylenimine-modified silica for enzyme immobilization and strengthening of enzymatic CO_2 conversion. *ACS Sustainable Chemistry & Engineering* 8(40): 15250–15257. doi:10.1021/acssuschemeng.0c04975.

29 Xu, S.W., Lu, Y., Li, J. et al. (2006). Efficient conversion of CO_2 to methanol catalyzed by three dehydrogenases Co-encapsulated in an alginate–silica (ALG–SiO_2) hybrid gel. *Industrial & Engineering Chemistry Research* 45(13): 4567–4573. doi:10.1021/ie051407l.

30 Kim, Y., Huan, K., Ahn, D. et al. (2019). Efficient aluminum catalysts for the chemical conversion of CO_2 into cyclic carbonates at room temperature and atmospheric CO_2 pressure. *ChemSusChem* 12(18): 4211–4220. doi:10.1002/cssc.201901661.

31 Lan, J., Qu, Y., Zhang, X. et al. (2020a). A novel water-stable MOF Zn(Py)(Atz) as heterogeneous catalyst for chemical conversion of CO_2 with various epoxides under mild conditions. *Journal of CO_2 Utilization* 35: 216–224. doi:10.1016/j.jcou.2019.09.019.

32 Brzostek, K.S., Terlecki, M., Sokolowski, K. et al. (2017). Chemical fixation and conversion of CO_2 into cyclic and cage-type metal carbonates. *Coordination Chemistry Reviews* 334: 199–231. doi:10.1016/j.ccr.2016.10.008.

33 Tian, J.-S., Cai, F., Wang, J.W. et al. (2008). Environmentally benign chemical conversion of CO_2 into organic carbonates catalyzed by phosphonium salts. *Phosphorus, Sulfur, and Silicon and the Related Elements* 183(2–3): 494–498. doi:10.1080/10426500701761540.

34 Vieira, M.O., Aquino, A.S., Schutz, M.K. et al. (2017). Chemical conversion of CO_2: Evaluation of different ionic liquids as catalysts in dimethyl carbonate synthesis. *Energy Procedia* 114: 7141–7149. doi:10.1016/j.egypro.2017.03.1876.

35 Wang, H.-H., Hou, H., Li, Y.Z. et al. (2017). Porous MOF with highly efficient selectivity and chemical conversion for CO_2. *ACS Applied Materials & Interfaces* 9(21): 17969–17976. doi:10.1021/acsami.7b03835.

36 Chen, D.-M. and Zhang, X.-J. (2019). A polyhedron-based metal-organic framework with a rare hexanuclear Co(II) cluster for selective sorption and chemical conversion for CO_2. *Journal of Solid State Chemistry* 278: 120906. doi:10.1016/j.jssc.2019.120906.

37 Lan, Y., Niu, G., Wang, F. et al. (2020b). SnO_2-modified two-dimensional CuO for enhanced electrochemical reduction of CO_2 to C_2H_4. *ACS Applied Materials & Interfaces* 12(32): 36128–36136. doi:10.1021/acsami.0c09240.

38 Han, N., Wang, Y., Ma, L. et al. (2017). Supported cobalt polyphthalocyanine for high-performance electrocatalytic CO_2 reduction. *Chem* 3(4): 652–664. doi:10.1016/j.chempr.2017.08.002.

39 Trindell, J.A., Clausmeyer, J., and Crooks, R.M. (2017). Size stability and H_2/CO selectivity for Au nanoparticles during electrocatalytic CO_2 reduction. *Journal of the American Chemical Society* 139(45): 16161–16167. doi:10.1021/jacs.7b06775.

40 Rosen, J., Hutchings, G.S., Lu, Q. et al. (2015). Electrodeposited Zn dendrites with enhanced CO selectivity for electrocatalytic CO_2 reduction. *ACS Catalysis* 5(8): 4586–4591. doi:10.1021/acscatal.5b00922.

41 Li, X., Xi, S., Sun, L. et al. (2020). Isolated FeN_4 sites for efficient electrocatalytic CO_2 reduction. *Advanced Science* 7(17): 2001545. doi:10.1002/advs.202001545.

42 Li, L., Ma, D.K., Qifei, S. et al. (2019). Bi nanoparticles/Bi_2O_3 nanosheets with abundant grain boundaries for efficient electrocatalytic CO_2 reduction. *Electrochimica Acta* 298: 580–586. doi:10.1016/j.electacta.2018.12.116.

43 Sato, S., Saita, K., Sekizawa, K. et al. (2018). Low-energy electrocatalytic CO_2 reduction in water over Mn-complex catalyst electrode aided by a nanocarbon support and K^+ cations. *ACS Catalysis* 8(5): 4452–4458. doi:10.1021/acscatal.8b01068.

44 Zhang, W., Zhou, Q., Li, J.Q.N. et al. (2021a). Promotion of electrocatalytic CO_2 reduction on Cu_2O film by ZnO nanoparticles. *Reaction Kinetics, Mechanisms and Catalysis.* 134: 243–257. doi:10.1007/s11144-021-02047-z.

45 Jiang, B., Zhang, X.G., Jiang, K. et al. (2018a). Boosting formate production in electrocatalytic CO_2 reduction over wide potential window on Pd surfaces. *Journal of the American Chemical Society* 140(8): 2880–2889. doi:10.1021/jacs.7b12506.

46 Han, P., Yu, X., Yuan, D. et al. (2019). Defective graphene for electrocatalytic CO_2 reduction. *Journal of Colloid and Interface Science* 534: 332–337. doi:10.1016/j.jcis.2018.09.036.

47 Luo, W., Zhang, J., Li, M. et al. (2019). Boosting CO production in electrocatalytic CO_2 reduction on highly porous Zn catalysts. *ACS Catalysis* 9(5): 3783–3791. doi:10.1021/acscatal.8b05109.

48 Guo, Y., Shi, W., Yang, H. et al. (2019). Cooperative stabilization of the [Pyridinium-CO_2-Co] adduct on a metal–organic layer enhances electrocatalytic CO_2 reduction. *Journal of the American Chemical Society* 141(44): 17875–17883. doi:10.1021/jacs.9b09227.

49 Ji, L., Chang, L., Zhang, Y. et al. (2019). Electrocatalytic CO_2 reduction to alcohols with high selectivity over a two-dimensional $Fe_2P_2S_6$ nanosheet. *ACS Catalysis* 9(11): 9721–9725. doi:10.1021/acscatal.9b03180.

50 Mo, Z., Zhu, X., Jiang, Z. et al. (2019). Porous nitrogen-rich g-C_3N_4 nanotubes for efficient photocatalytic CO_2 reduction. *Applied Catalysis B: Environmental* 256: 117854. doi:10.1016/j.apcatb.2019.117854.

51 Zhang, J., Huang, G., Zeng, J. et al. (2019). SnS_2 nanosheets coupled with 2D ultrathin MoS_2 nanolayers as face-to-face 2D/2D heterojunction photocatalysts with excellent

photocatalytic and photoelectrochemical activities. *Journal of Alloys & Compounds* 775: 726–735. doi:10.1016/j.jallcom.2018.10.159.

52 Xu, Y., You, Y., Huang, Y. et al. (2020). Bi_4NbO_8Cl nanosheets coupled with g-C_3N_4 as 2D/2D heterojunction for photocatalytic degradation and CO_2 reduction. *Journal of Hazardous Materials* 381: 121159. doi:10.1016/j.jhazmat.2019.121159.

53 Xu, D., Cheng, B., Wang, W. et al. (2018a). Ag_2CrO_4/g-C_3N_4/graphene oxide ternary nanocomposite Z-scheme photocatalyst with enhanced CO_2 reduction activity. *Applied Catalysis B: Environmental* 231: 368–380. doi:10.1016/j.apcatb.2018.03.036.

54 Nie, N., Zhang, L., Wu, J. et al. (2018). Self-assembled hierarchical direct Z-scheme g-C_3N_4/ZnO microspheres with enhanced photocatalytic CO_2 reduction performance. *Applied Surface Science* 441: 12–22. doi:10.1016/j.apsusc.2018.01.193.

55 Rather, R.A., Khan, M., and Lo, I.M.C. (2018). High charge transfer response of g-C_3N_4/Ag/AgCl/$BiVO_4$ microstructure for the selective photocatalytic reduction of CO_2 to CH_4 under alkali activation. *Journal of Catalysis* 366: 28–36. doi:10.1016/j.jcat.2018.07.027.

56 Wang, Y., Zhang, Z., Zhang, L. et al. (2018). Visible-light driven overall conversion of CO_2 and H_2O to CH_4 and O_2 on 3D-SiC@2D-MoS_2 heterostructure. *Journal of the American Chemical Society* 140(44): 14595–14598. doi:10.1021/jacs.8b09344.

57 Meng, A., Zhang, L., Cheng, B. et al. (2019). TiO_2–MnO_x–Pt hybrid multiheterojunction film photocatalyst with enhanced photocatalytic CO_2-reduction activity. *ACS Applied Materials & Interfaces* 11(6): 5581–5589. doi:10.1021/acsami.8b02552.

58 Geioushy, R.A. (2019). Insights into two-dimensional MoS_2 sheets for enhanced CO_2 photoreduction to C1 and C2 hydrocarbon products. *Materials Research Bulletin* 118: 110499. doi:10.1016/j.materresbull.2019.110499.

59 Xu, Y.-F., Sheikh, S.M.E., Hegazy, I.M. et al. (2017). A $CsPbBr_3$ perovskite quantum dot/graphene oxide composite for photocatalytic CO_2 reduction. *Journal of the American Chemical Society* 139(16): 5660–5663. doi:10.1021/jacs.7b00489.

60 Fu, J., Zhu, B., Jiang, C. et al. (2017). Hierarchical porous O-doped g-C_3N_4 with enhanced photocatalytic CO_2 reduction activity. *Small* 13(15): 1603938. doi:10.1002/smll.201603938.

61 Sun, D., Fu, Y., Liu, W. et al. (2013). Studies on photocatalytic CO_2 reduction over NH_2-Uio-66(Zr) and its derivatives: Towards a better understanding of photocatalysis on metal-organic frameworks. *Chemistry – A European Journal* 19(42): 14279–14285. doi:10.1002/chem.201301728.

62 Foster, B.M., Paris, A.R., Frick, J.J. et al. (2020). Catalytic mismatching of $CuInSe_2$ and Ni_3Al demonstrates selective photoelectrochemical CO_2 reduction to methanol. *ACS Applied Energy Materials* 3(1): 109–113. doi:10.1021/acsaem.9b01441.

63 Ahmed, A.M., Mohamed, F., Ashraf, A.M. et al. (2020). Enhanced photoelectrochemical water splitting activity of carbon nanotubes@TiO_2 nanoribbons in different electrolytes. *Chemosphere* 238: 124554. doi:10.1016/j.chemosphere.2019.124554.

64 Butson, J.D., Narangari, P.R., Lysevich, M. et al. (2019). InGaAsP as a promising narrow bandgap semiconductor for photoelectrochemical water splitting. *ACS Applied Materials & Interfaces* 11(28): 25236–25242. doi:10.1021/acsami.9b06656.

65 Duchene, J.S., Taglibue, G., Welch, A.J. et al. (2018). Hot hole collection and photoelectrochemical CO_2 reduction with plasmonic Au/p-GaN photocathodes. *Nano Letters* 18(4): 2545–2550. doi:10.1021/acs.nanolett.8b00241.

66 Karim, K.M.R., Tarek, M., Ong, H.R. et al. (2019). Photo-electrocatalytic reduction of carbon dioxide to methanol using $CuFe_2O_4$ modified with graphene oxide under visible light irradiation. *Industrial & Engineering Chemistry Research* 58(2): 563–572. doi:10.1021/acs.iecr.8b03569.

67 Jiang, X.X., Hu, X.D., Tarek, M. et al. (2020). Tailoring the properties of g-C_3N_4 with CuO for enhanced photoelectrocatalytic CO_2 reduction to methanol. *Journal of CO_2 Utilization* 40: 101222. doi:10.1016/j.jcou.2020.101222.

68 Wang, J., Wei, Y., Yang, B. et al. (2019). *In situ* grown heterojunction of Bi_2WO_6/BiOCl for efficient photoelectrocatalytic CO_2 reduction. *Journal of Catalysis* 377: 209–217. doi:10.1016/j.jcat.2019.06.007.

69 Dong, Y., Nie, R., Wang, J. et al. (2019). Photo-electrocatalytic CO_2 reduction based on metalloporphyrin-modified TiO_2 photocathode. *Chinese Journal of Catalysis* 40(8): 1222–1230. doi:10.1016/S1872-2067(19)63375-9.

70 Han, B., Wang, J., Yan, C. et al. (2018). The photoelectrocatalytic CO_2 reduction on TiO_2@ZnO heterojunction by tuning the conduction band potential. *Electrochimica Acta* 285: 23–29. doi:10.1016/j.electacta.2018.07.216.

71 Shen, Q., Chen, Z., Huang, X. et al. (2015). High-yield and selective photoelectrocatalytic reduction of CO_2 to formate by metallic copper decorated Co_3O_4 nanotube arrays. *Environmental Science & Technology* 49(9): 5828–5835. doi:10.1021/acs.est.5b00066.

72 Brito, J.F.D., Genovese, C., Tavella, F. et al. (2019). CO_2 reduction of hybrid Cu_2O–Cu/Gas diffusion layer electrodes and their integration in a Cu-based photoelectrocatalytic cell. *ChemSusChem* 12(18): 4274–4284. doi:10.1002/cssc.201901352.

73 Cheng, J., Zhang, M., Wu, G. et al. (2014). Photoelectrocatalytic reduction of CO_2 into chemicals using Pt-modified reduced graphene oxide combined with Pt-modified TiO_2 nanotubes. *Environmental Science & Technology* 48(12): 7076–7084. doi:10.1021/es500364g.

74 Liang, M., Borjigin, T., Zhang, Y. et al. (2019). Controlled assemble of hollow heterostructured g-C_3N_4@CeO_2 with rich oxygen vacancies for enhanced photocatalytic CO_2 reduction. *Applied Catalysis B: Environmental* 243: 566–575. doi:10.1016/j.apcatb.2018.11.010.

75 Kumar, N., Kumar, S., Gusaine, R. et al. (2020). Polypyrrole-promoted rGO–MoS_2 nanocomposites for enhanced photocatalytic conversion of CO_2 and H_2O to CO, CH_4, and H_2 products. *ACS Applied Energy Materials* 3(10): 9897–9909. doi:10.1021/acsaem.0c01602.

9

MXenes for CO_2 Reduction and H_2 Generation

N. Usha Kiran,[1,2] Laxmidhar Besra,[1,2], and Sriparna Chatterjee[1]

[1] *Materials Chemistry Department, CSIR-Institute of Minerals and Materials Technology, Bhubaneswar 751013, Odisha, India*
[2] *Academy of Scientific and Innovative Research (AcSIR), Ghaziabad 201002, Uttar Pradesh, India*

9.1 Introduction to MXenes

During the past few years, there has been substantial growth in global warming and energy consumption as two challenging issues due to the rapid growth of global population and industrial sector [1]. The traditional energy sources such as coal, gas, and oil are the major contributors to the emission of carbon into the atmosphere, leading to environmental degradation [2, 3]. The International Energy Agency [4] at Paris reported that by 2050 the global energy consumption would be increased by ~30%, with subsequent emission of carbon into the atmosphere reaching ~36% Gt yr^{-1}. Therefore, it is absolutely essential to harvest energy from renewable and sustainable energy sources such as fossil fuels, nuclear energy, solar energy, etc. [5, 6]. Among the emerging sources, the burning of fossil fuel releases toxic gases such as NO_2, SO_2, and CO_2, which contributes to severe environmental devastation [7]. Carbon dioxide (CO_2), being chemically inert, acts as a primary pollutant released from many different sources by human activities, which contributes to the greenhouse effect and global warming [8]. Therefore, the establishment of carbon capture and storage technologies are very important to protect the environment. Among known methods for reducing the level of atmospheric CO_2, transport and storage of CO_2 is carefully implemented to minimize the risk of leakage [9]. Another approach is to reduce CO_2 into high value hydrocarbons such as CO, HCOOH, CH_4, CH_3OH, C_2H_3, and C_2H_5OH, which can be deployed as clean fuels [10–12]. Due to its excellent chemical reaction and stability, the CO_2 reduction reaction (CO_2RR) is a viable method to mitigate CO_2 with great promise [13]. However, in particular, the CO_2 reduction reaction mechanism is a complicated process under standard conditions and requires a large number of elementary reactions for producing various products. Considering the production of methane, the most favorable pathway of subsequent steps includes [14]:

$$CO_2 \rightarrow HCO_2 \rightarrow H_2CO_2 \rightarrow H_2COOH \rightarrow HCHO \rightarrow CH_3OH \rightarrow CH_4$$

Hydrogen (H_2) energy is identified as a clean, sustainable, and renewable energy source among all the other available fuel sources, which can be technically produced by photocatalytic water splitting under water and sunlight [15, 16]. This process releases O_2 into the atmosphere. Three main strategies are required to produce H_2: from fossil fuels, biofuels, and water splitting [17]. H_2 evolution from fossil fuel includes fermentation of marsh gas or organic waste, natural gas pyrolysis, and coal vaporization. However, it consumes a

Green Energy Harvesting: Materials for Hydrogen Generation and Carbon Dioxide Reduction,
First Edition. Edited by Pooja Devi.

large amount of energy during the chemical reaction and releases CO_2 as a final product which may add to the greenhouse effect, resulting in further environmental problems.

$$CH_4 + H_2O \rightarrow 3H_2 + CO \quad \Delta H^0 = 206 \text{ kJ/mol} \tag{9.1}$$

$$CO + H_2O \rightarrow CO_2 + H_2 \quad \Delta H^0 = -41 \text{ KJ/mol} \tag{9.2}$$

H_2 evolution from biofuels involves gasification of biomass, fermentation of organic matter by bacteria, and photosynthetically, etc. [18]. Although it has little harmful impact on the environment, large-scale production of H_2 using this technique is costly. The water-splitting-based H_2 production is an eco-friendly approach which utilizes waste water as the source [19]. Among the three classified processes of water splitting (photochemical, thermochemical, and electrochemical), intense research has been performed on photochemical H_2 evolution as an efficient technique for both academic and industrial interests [20]. Currently, photo- and electrocatalytic water splitting contributes to only 4% of global H_2 production. This is mainly because of the lack of efficient and stable catalysts for H_2 evolution reaction (HER) and CO_2 reduction reaction (CO_2RR). At present, the most efficient noble metals, namely Au, Pt, Pd, and Ir, are being used, which are extremely expensive and hence find limited practical applications in water splitting. On the other hand, non-precious materials, which were explored as catalysts for this reaction are low cost with excellent catalytic activity, which include transition metal sulfides, phosphor-sulfides, phosphides, etc. [21–24]. In search of adequate catalysts, two-dimensional (2D) materials have emerged as a special class that show remarkable efficiency due to their tunable functionalities.

Since the isolation of single-layer 2D graphene [25] in 2004, there has been a flurry of research on development of several other 2D nanostructured materials such as transition metal dichalcogenides (TMDs) [26], hexagonal boron nitride (h-BN) [27], layered metal oxides (TMOs) [28], layered hydroxides [29], and metal organic frameworks (MOFs) [30]. etc. in the field of photocatalytic H_2 production and CO_2 reduction. Although most of these materials are of purely academic interest, most of the others hold tremendous applications due to their fascinating properties that are not found in their bulk precursors [31–33]. However, the poor non-magnetic properties of layered TMDs and the lack of bandgap in graphene limit their use in photocatalytic applications. With the advancement in synthetic procedures, many 2D nanomaterials beyond graphene have been successfully synthesized such as silicene [34], germanene, [31, 33], and phosphorene [34, 35]. One of the most recent additions to the 2D family of nanomaterials are transition metal carbides, nitrides, and carbonitrides, known as MXenes, which were discovered by Prof Yury Gogotsi and his team in 2011 at Drexel University, USA [36–38]. These ternary carbides possess the general chemical formula of $M_{n+1} X_n T_x$. Here "M" stands for an early transition metal (Sc, Ti, V, Zr, Cr, Hf, Ta, Nb, etc.), "X" represents carbon and/or nitrogen, and "T_x" represents surface terminating groups such as oxide (-O), hydroxyl (-OH), fluoride (-F), etc. Due to these functional groups, MXenes are highly hydrophilic in nature and possess high electrical conductivity for which they appear as popular materials for electrochemical applications including energy storage [39], electromagnetic interference shielding [40], water purification [41], H_2 storage [42, 43], and medicine [44], etc. $Ti_3C_2T_X$ MXene is the first that was reported in 2011. Some more examples of MXenes include $Nb_4C_3T_X$, Ti_2CT_X, and Ti_3CNT_X [45], etc. Unlike other 2D materials, including graphene that only exists in one or a few chemical compositions,

MXenes can be made of many different compositions and structures. This led to the formation of about 100 stoichiometric MXene compositions and an unlimited solid solution which offer unique combination of properties due to tunable M or X elements and functional groups [46]. Certain amalgamations of transition metals forming ordered MXenes are energetically more stable as compared to solid solutions. Density functional theory (DFT) calculations have predicted the formation of more than 30 different ordered MXenes to date [47]. Aside from the surface chemistry of MXenes, which is different from the surface of graphene, the mechanical and electrical properties of MXenes are similar. $Ti_3C_2T_X$ MXene exhibits an electrical conductivity of ~10,000 S cm^{-1} with high thermal stability up to ~1000°C and optical transparency of ~97% [48]. Ta_2CT_X MXenes possess an impressive Young's modulus of ~800 GPa [which is very close to the micro-mechanically exfoliated graphene (~1 TPa)]. The thermal conductivity of Sc_2CF_2 MXene shows ~722 W m^{-1} K^{-1}, which is so far the best in comparison to other MXene compositions [49], and the Hf_2CO_2 MXene [50] shows a very low thermal expansion coefficient of 6.094×10^{-6} K^{-1}. Besides these extraordinary properties, they also exhibit exceptional catalytic activity for which many replace noble metal catalysts in photocatalytic and electrocatalytic H_2 evolution reactions and CO_2 reduction reactions in terms of industrialized use.

9.1.1 Methods of Synthesis of MXenes

In 2011, researchers at Drexel University discovered a class of new 2D carbides, nitrides, and carbonitrides of transition metals, termed "MXenes." These can be synthesized from their corresponding layered precursor MAX phases, specifically by etching certain atomic "A" layers (Group-13 and 14 elements) [51]. The general chemical formula of the MAX phase is $M_{n+1}AX_n$. Here, "M" is an early transition metal (Ti, Sc, Zr, V, Hf, Cr, Ta, Nb, etc.), "A" is Group 13 and 14 elements, and "X" is carbon and/or nitrogen. Depending on their "n" value, MAX phases are mainly classified into three categories: M_2AX type, M_3AX_2 type, and M_4AX_3 type [52]. These interstitial metal carbides and nitrides can also be considered as nano-laminar composites, as they possess a layered micro-structure with a hexagonal crystal lattice system where the mono-atomic "A"-elements are interleaved in between. Soon after the synthesis of the first MXene $Ti_3C_2T_X$ from its corresponding Ti_3AlC_2 MAX phase, they have rapidly been used in numerous promising applications because of their excellent mechanical properties and metallic conductivity, tuneable optical properties, and unique in-plane anisotropic structure. [51, 53, 54]. The compact layered MAX phases are considered as the only precursors for the synthesis of extraordinary 2D MXenes, exhibiting immense applications in the field of energy storage [52], electromagnetic interference shielding [40, 55], electrolysis [56, 57], medicine [58], advanced composite structures [59], water purification [60], etc. In general, there are two different methods for the synthesis of 2D MXenes. One is the bottom-up approach, which includes the chemical vapor deposition (CVD) technique that produces high-quality films on various substrates. However, this approach is not suitable enough to produce single-layer MXene nanosheets [61]. The second approach is the top-down approach, which involves mechanical and chemical exfoliation of layered solids. The former method was used to separate graphene layers from graphite by scotch adhesive tape, as graphene layers are stuck together by a weak van der Waal's force of attraction in graphite [62]. However, it is difficult to mechanically exfoliate

MXene layers from their precursor MAX phase, as "M" atoms are covalently bonded to "A" atoms. As noted above, MXenes can be formed only by selectively etching "A" atoms (Group-13 and 14 elements) from their corresponding precursor MAX phase by chemical exfoliation using different etchants.

The pioneering synthesis of MXenes reported in the literature involves many different protocols [63] and it must be noted that one synthesis method may be suitable for one application but not for another. On synthesis of MXenes, one must be aware of the resulting properties of the end product to develop the best material for a particular application. In general, synthesis of the MXene may take hours to days, depending on two major factors, namely concentration of the etchant and etching temperature. Many previous reports in the literature used hydrofluoric acid (HF) as the etchant for the massive production of monolayer and multi-layer MXenes [64]. The different etching methods are described below.

9.1.1.1 HF Etching

Synthesis of $Ti_3C_2T_X$ MXenes begins with chemical exfoliation of the Ti_3AlC_2 MAX phase with HF at room temperature. The material obtained after etching is terminated multi-layered MXene powder, where 2D layers are held together by H-bonds bonds and/or the van der Waals force of attraction. After etching, the powder is required to be washed several times via centrifugation until the pH attains neutrality (~6) to remove residual acid. Then, the multi-layered MXene flakes are collected via vacuum filtration and dried. The proposed mechanism in conversion of MAX to MXene is described below:

$$Ti_3AlC_2 + 3HF \rightarrow Ti_3C_2 + AlF_3 + 3/2H_2 \tag{9.3}$$

$$Ti_3C_2 + 2H_2O \rightarrow Ti_3C_2(OH)_2 + H_2 \tag{9.4}$$

$$Ti_3C_2 + 2HF \rightarrow Ti_3C_2F_2 + H_2 \tag{9.5}$$

Reaction 1 represents the removal of Al atoms from the Ti_3AlC_2 MAX phase, producing Ti_3C_2 MXene. The Ti atoms, being highly reactive, form Ti-OH and Ti-F bonds on reaction with water and HF (Reactions 2 and 3) [65]. Multi-layered MXene has strong inter-layer interactions and because of the effect of inter-layer distances on several applications, many researchers have explored intercalants like tetrabutylammonium hydroxide (TBAOH) [66], dimethyl sulfoxide (DMSO) [67], isopropylamine (i-PrA) [68], tetramethylammonium hydroxide (TMAOH) [69], etc. followed by ultrasonic treatment to obtain aqueous colloidal solution of de-laminated MXene which is extremely stable.

Depending on the bond energy of the M–A bond in various MAX phases, significant differences in etching conditions are implemented in the synthesis of many MXene members. Alhabeb et al. reported $Ti_3C_2T_X$ MXene synthesis from the Ti_3SiC_2 MAX phase by oxidant-assisted selective etching of Si atoms [71]. Recent studies revealed that $Zr_3Al_3C_5$ and Mo_2Ga_2C can also act as precursor material for the synthesis of 2D MXenes, although they do not belong to the MAX family [70, 71]. The quality of MXene flakes majorly depends on different etching parameters such as particle size, temperature, concentration of the etchant, and etching time. Although strong etchant and increase in temperature can significantly shorten the etching time, if these parameters are not properly controlled, the destruction of layered structure and properties is unavoidable. However, HF procedures have a limitation in scaling-up processes which includes hazardous chemistry solutions and do not produce monolayer MXene materials.

9.1.1.2 *In-Situ* HF Etching

To avoid using unnecessarily high HF concentrations, Ghidiu et al. [72] explored the reaction of the Ti_3AlC_2 MAX phase with a mixture of LiF and HCl solution to generate HF *in situ*, which ultimately produces $Ti_3C_2T_X$ MXene flakes. The reaction mechanism involves:

$$LiF(aq) + HCl(aq) \rightarrow HF(aq) + LiCl(aq) \tag{9.6}$$

The advantages of this procedure are: i) mild corrosive environment of HF in comparision to highly corrosive media on direct addition; ii) requires no sonication or low sonication time for de-lamination of MXene flakes; iii) high yield with single- or few-layered MXenes; iv) less defective MXene sheets; and v) highly flexible clay-like MXenes. In order to enhance MXene exfoliation, the environmentally benign etchant, LiF/HCl is used to facilitate intercalation of Li^+ ions during etching. This weakens the inter-layer interaction and increases the gap between the MXene nanosheets. The change in molar ratio of the reactants results in a mild route which produces de-laminated, high-quality MXene nanosheets with high yield. This method is known as the minimally intensive layer delamination (MILD) method. Currently, this method is being most extensively used to synthesize high-quality MXenes, particularly in the field of energy storage and opto-electronics.

9.1.1.3 Bifluoride-Based Etchants

HF salts such as ammonium bifluoride (NH_4HF_2) have been used by Halim et al. [73] in 2014 to etch the Ti_3AlC_2 MAX phase at room temperature or elevated temperature for a few hours to easily prepare $Ti_3C_2T_X$ MXenes. This method is devoid of hazardous HF and the larger d-spacing after drying is attributed to intercalation of NH_4^+ ions in between the MXene layers. Unlike the HF-etched MXenes, MXenes obtained by this method requires a longer time and higher temperature vacuum drying to completely remove the intercalated water molecules.

9.1.1.4 Molten Salt Etching

Due to unsuccessful production of a nitride-based MXene using HF or HCl-LiF as etchants, the Ti_4N_3-MXene was first synthesized in 2016 by the reaction of the Ti_4AlN_3 MAX phase with molten fluoride salt at 550°C in argon atmosphere [74]. Recently, a salt templated method was implemented to produce Mo_2NT_X and V_2NT_X by amination of MXene carbides, Mo_2CT_X and V_2CT_X [75]. This is due to the less cohesive and high formation energies of $M_{n+1}N_n$ class of compounds. The de-lamination process using TBAOH was carried out successfully to obtain single or few-layered MXene nanosheets. Some recent reports described the synthesis of Cl-terminated Ti_2C and Ti_3C_2 MXene flakes from their corresponding Al-based MAX phase in a Lewis acidic melt, i.e., $ZnCl_2$ [76].

9.1.1.5 Fluoride-Free Etching

The MXene obtained using a fluoride containing etchant has a higher fraction of fluoride (–F) terminating moieties due to which the electrochemical performance of the material becomes reduced. Most of the previously reported literature follows the HF-related method to produce

MXene flakes successfully by etching the acidic "A" atoms from the MAX phase. Since HF is a corrosive chemical, other researchers have tried alkali etching but no one has achieved high-quality fluoride-free MXene [77]. Li et al. [78] explored the hydrothermal alkaline etching method using NaOH as an etchant, where it formed the Al containing oxides and hydroxides, resulting in up to 92% yield of fluoride-free high pure $Ti_3C_2T_X$ MXenes. Yang et al. explained the synthesis of $Ti_3C_2T_X$ MXenes using the electrochemical etching method [79]. This kind of method can avoid harsh etching conditions and corrosive fluoride-containing compounds.

9.1.2 Structure and Elemental Distribution

Ternary MAX phases have a hexagonal crystal lattice system which belongs to the point group of $P6_3/mmc$ symmetry, with two formula units per unit-cell. They have the general chemical formula of $M_{n+1}AX_n$. Here, "M" is a transition metal (V, Sc, Ti, Zr, Cr, Ta, Nb, Hf, etc.), "A" is a post-transition metal, and "X" is nitrogen and/or carbon. The crystal structure of the MAX phase consists of an edge-sharing M_6X octahedral layer interleaved with "A" elements. The difference between the various phases as mentioned earlier (211, 312, and 413 phase) lies in the different number of transition metal (M) layers separating the post-transition metal (A) layers. In 2017, ordering between "M" elements was discovered, which has been noted as in-plane MAX (*i*-MAX) and out-of-plane MAX (*o*-MAX) phases to make a distinction between the two groups of materials [80]. The ordering of "M" layers in the *o*-MAX phase is such that two M' layers are sandwiched between either one or two layers of M" in each M-X block, where M' and M" are early transition metals. Similarly, the in-plane ordering in the MAX phase (*i*-MAX) reveals that the M' atoms are arranged in a hexagonal arrangement with M" atoms situated at the center of the hexagon, where M' and M" represents the early transition metals. Only six *o*-MAX phases have been successfully discovered to date in which the first *o*-MAX phase, $(Cr_{2/3}Ti_{1/3})_3AlC_2$, was discovered by Liu et al. on heating a mixture of Cr_2AlC and TiC at 1500°C in an argon atmosphere [81]. The *i*-MAX phases are the most recent addition to the MXene family, with 31 *i*-MAX phases discovered so far [82]. The occurrence of these MAX phases is significant because it enables the synthesis of MXenes with surface terminations that would otherwise be impossible to produce. For example, MXene with Cr in the outer layer is not conceivable due to the complete dissolution of Cr_2AlC during etching. Hence, $(Cr_{2/3}Ti_{1/3})_3AlC_2$ is the only MAX phase that produces Cr-containing MXene flakes.

MXenes are arranged in a hexagonal crystal lattice system like their precursor MAX phase, where "X" atoms occupy the octahedral site. However, the atomic arrangement can be altered with the change in stoichiometric parameter "n." The M_2X compounds have ABAB stacking with a hexagonal close packing (hcp) system, whereas M_3X_2 and M_4X_3 compounds have ABCABC ordering with face-centred cubic (fcc) stacking. This atomic ordering is very important for the synthesis of stable MXenes with high "n" indexes. For example, Mo_2C is stable, whereas Mo_3C_2 and Mo_4C_3 are unstable due to the ABCABC ordering of "Mo" and "C" atoms [83]. Surface terminating groups such as –OH, –F, and –O are seen on MXene nanosheets produced using aqueous HF solutions. The functionalization of the MXene surface with different terminating groups depends on the influence of different synthesis routes, delamination conditions, post-synthesis treatment, and storage. Many studies have analyzed the configurations and effects of different functional groups. Computational calculations found that most stable conformations are usually achieved when the terminations (T) are in a different atomic position. As a result, ABCABC ordering is created with its neighboring "M," "X," and "T" atoms, respectively. However, some exceptions revealed that "T" atoms are positioned directly on the "X" ones to enhance their electronic interaction. Based on transmission electron microscopy (TEM), NMR spectroscopy, electron energy-loss spectroscopy (EELS), and neutron scattering measurements, it was confirmed that on $Ti_3C_2T_X$ and V_2CT_X MXenes, the surface terminating groups are distributed randomly rather than in specific atomic positions [84, 85]. The chemical composition and elemental stoichiometric ratio can be identified by using X-ray photoelectron spectroscopy (XPS). The presence of H_2 bonding between –F and –O atoms of one MXene surface and –OH group of other surfaces, as well as van der Waals interaction of –O and –F atoms between two surfaces, explains the interaction between the two MXene nanolayers. Intercalated water creates a strong H-bond bond with the –O and –OH terminations of MXene nanosheets. The –O and –OH groups on the surface of the $Ti_3C_2T_X$ MXene plays an important role in photocatalytic H_2 generation and CO_2 reduction [42, 86].

9.1.3 Modified MXene

MXene's diverse compositions and structure have resulted in the development of a growing range of 2D nanomaterials. To understand MXene's structure with new possible stable compounds, these can be found in six possible structures depending on the number of "M" atomic species contained in the MXene nanosheet: i) vacancy-free mono "M" element MXene (Mo_2C, Ti_3C_2, and Ti_4N_3); ii) solid-solution MXenes [$(Cr,V)_3C_2$ and $(Ti,V)_3C_2$]; iii) ordered out-of-plane double transition metal MXenes in which one transition metal is present at the external layer while the other occupies the center (Mo_2TiC and $Mo_2Ti_2C_3$); iv) ordered in-plane double transition metal MXenes where the "M" atoms are ordered in the basal plane [$(Mo_{2/3}Y_{1/3})_2AlC$]; v) ordered divacancy MXenes ($Mo_{1.33}C$, $W_{1.33}C$); and vi) vacancies randomly distributed MXenes ($Nb_{1.33}C$) [64]. In these categories, traditional mono transition metal MXenes are the most studied MXenes to date, which were discussed earlier. Although many possible combinations of "M" and "X" elements are possible, only 14 mono-M MXenes have been synthesized. MXenes with more than one M-element contain solid solutions in which two different transition metals are randomly arranged. Unlike ordered MXenes, solid-solution MXenes and their precursor MAX phase contain a large range of stoichiometric ratio of M':M", where M' and M" are two different early transition metals. This kind of MXene reported the highest order of MXenes ($M_4C_3T_X$ and $M_5C_4T_X$) to date, along with the degree of substitution wherever possible [87, 88]. Some of the key factors to further enhance the diversity of the solid-solution MAX phase include:

1) Amalgamation of small amounts of Mn and Fe into V_2AlC, Cr_2GeC, and Cr_2AlC has led to the development of magnetic properties which may be ideal precursors for the synthesis of magnetic MXenes [89].
2) Elements like Bi, Cu, and Au are the newest elements to be incorporated onto the A-sites to form $Zr_2(Al_{0.42}Bi_{0.58})C$, $Ti_3(Al_{1-x}Cu_x)C_2$, and $(Cr_{1-x}Mn_x)_2AuC$ MAX precursors, which may lead to symmetry reduction [90–92].
3) In addition to this, more complex phases with five elements on the M and A sites have been discovered in the $(Ti_{0.95}Zr_{0.05})_3(Si_{0.9}Al_{0.1})C_2$ system [93].

Double transition-metal (DTM) MXenes are further classified into two categories, namely, in-plane and out-of-plane ordered MXenes depending on the atomic position of "M" atoms in specific sites. In-plane ordered DTM MXenes have a general chemical formula of $M'_{4/3}M''_{2/3}XT_x$ and are only observed in the thinnest M_2XT_X MXene structure. Each M-layer of this MXene is occupied by two different transition metals M′ and M″, where M′ is typically Cr, W, Mo, V, or, Mn and M″ is Zr, Y, or Sc. These in-plane ordered DTM MXenes are derived from ordered *i*-MAX phases. For example, an elemental mixture of Mo, Y, Al, and C sintered at 1500°C for 20 h in an argon atmosphere results in formation of an in-plane ordered $Mo_{4/3}Y_{2/3}AlC$ MAX phase [94]. The "Al" atoms are then selectively etched onto a milder etching condition such as low HF concentration or shorter etching time to form the corresponding DTM MXene. The second type of MXene, known as the out-of-plane ordered DTM MXene, has a general chemical formula of $M'_2M''X_2T_X$ or $M'_2M''_2X_3T_X$, in which inner layers of M″ transition metals are sandwiched by outer M′ transition metal layers [95]. The structural conformation of transition metals in ordered out-of-plane DTM MXenes involves the ordering in separate atomic planes, whereas in in-plane DTM MXenes, the ordering of transition metals is in an atomic plane. All the theoretically predicted and experimentally synthesized DTM MXenes are carbides only. However, the nitride- and carbonitride-based DTM MXene have not been explored. Since MXenes are derived from their precursor MAX phase, it is the composition and structure of the DTM MAX phase that directs the composition and structure of the DTM MXene. More than 20 DTM MAX phases (in-plane and out-of-plane) have been synthesized, the majority of which have not been selectively etched to their corresponding MXenes. So far, Mo_2ScAlC_2, Cr_2VAlC_2, Mo_2TiAlC_2, Cr_2TiAlC_2, Ti_2ZrAlC_2, $Cr_2V_2AlC_3$, and $Mo_2Ti_2AlC_3$ are the only experimentally synthesized out-of-plane MAX phases from which corresponding MXenes can be derived on etching with retention of the structural ordering. The discovery of ordered divacancy MXenes, such as $Mo_{4/3}CT_X$ and $W_{4/3}CT_X$ MXenes with randomly distributed vacancies, depends on the etching conditions of single and double "M" element MAX phase precursors. With longer etching time and higher HF concentration, M″- and A-layers can be etched out, forming di-vacancy ordered MXenes with a structure similar to that of MXene derived from the single "M" element MAX phase. The precursor of divacancy ordered $Mo_{4/3}CT_X$ MXene is the $Mo_{4/3}Y_{2/3}AlC$ MAX phase, which has been etched in 48 wt% HF for 12 to 72 h at room temperature [94].

$Ti_3C_2T_X$, the first ever identified MXene has received so much attention that it has accounted for more than 70% of all MXene studies to date. By changing the ratios of "M" or "X" elements, at least 100 stoichiometric MXene compositions and limitless solid solutions offer unique combinations of intrinsic properties. The addition of more complexity in compositions towards the development of high-entropy MXenes with extraordinary properties could open the door to a variety of different applications.

9.2 Properties/Characteristics of MXenes

The rich surface chemistry of MXenes led to several computational studies, exploring a unique combination of properties which includes high electrical conductivity, high Young's modulus, functionalized surface that makes MXene hydrophilic in nature with tunable band-gap, high thermal stability, high negative zeta-potential forming a stable colloidal aqueous solution, and efficient absorption of electromagnetic waves. It is noteworthy that the tunable surface and high metallic conductivity of MXenes distinguish them from other 2D nanomaterials [51]. Depending on different applications, their properties may be tuned through: i) composition (change in atomic position of different "M" and "X" elements); ii) surface functionalization (through thermal and chemical treatment); and iii) structural modification. Some major properties of MXenes family are described below.

9.2.1 Electrical Properties

Theoretical studies of 2D MXenes show its incomparable electrical properties ranging from pure metallic to semi-conducting, depending on alteration of functional groups and transition metals. Some of the MXenes containing heavier transition metals (Cr, W, Mo, etc.) are predicted to be topological insulators. For example, although $Ti_3C_2T_X$ shows purely metallic behavior, Mo-incorporated MXene ($Mo_2TiC_2T_X$) shows semiconductor-like properties [96]. The conductive metal carbide core in the middle of the MXene structure plays an important role in their excellent conductivity [97]. The electrical conductivity of MXene depends upon many factors such as surface functionalization, concentration of defects, nano-flakes size, and contact resistance between the flakes [36, 37, 51]. Experimentally, it was found that the electrical conductivity of MXene discs evaluates similarly to multi-layered graphene nanosheets (resistance of 22 to 339 Ω), but higher than reduced graphene oxide (rGO) and carbon nanotubes [98]. Further, it was observed that the resistivity values increase with increase in number of layers and presence of surface terminations. Because of this, the electrical conductivity calculated by computational simulations generally show higher values than the ones observed experimentally [99]. MXene has good on-sheet conductivity and has poor conductivity in between the sheets. The measured electrical conductivity of $Ti_3C_2T_X$ MXene varies from 850 to 9880 S cm^{-1}, which depends on concentration defects, lateral size of nanosheets, surface functionalization, and d-spacing between the MXene nanoflakes, etc. [100]. Wang et al. [48] reported an increase in conductivity of nearly three times by roasting $Ti_3C_2T_X$ MXene nanosheets at 600°C for 1 hr. In many cases, the electrical conductivity of MXenes depends upon surface modification through thermal and alkaline treatments. A low concentration of defects with larger flake size are more likely to be obtained under mild etching conditions, which results in higher conductivity. The increase in conductivity can also be attributed to removal of functional groups and intercalated molecules upon post-treatment. The effect of external environment on the $Ti_3C_2T_X$ MXene nanosheet for a longer exposure time may lead to oxidation of the MXene flakes, which significantly deteriorates conductivity [101].

9.2.2 Mechanical Properties

Since metal carbides and metal nitrides have the strongest bonds, mechanical properties of MXenes have also received great attention. Density functional theory (DFT) [102] and molecular dynamic (MD) [103] simulations predicted that M_2X MXenes are stronger than their M_3X_2 and M_4X_3 counterparts [51]. Experimental results revealed that despite 2 to 4 times lower tensile strength as compared to graphene, MXenes possess high bending stiffness which tend to be used as reinforcers in composites. Due to the presence of surface terminations, MXenes have better interaction with polymeric matrixes than graphene to form composites. Further studies have validated the mechanical properties by making a cylinder of 5 μm thick $Ti_3C_2T_X$ MXene film which can support ~ 4000 times its own weight. These films can be further strengthened by making a composite of $Ti_3C_2T_X$ MXene with 10 wt% (PVA) polyvinyl alcohol, which can hold ~15,000 times its own weight [104]. Many polymers such as PVA, polypyrrole, and polydimethyl siloxane (PDMS), etc. have been introduced into the MXene nanosheets to enhance the mechanical, electrochemical, and wear resistance of the material [105, 106]. Other than polymers, MXene/CNT hybrids have also been prepared which show better electrochemical performance [107]. Recently, a rare negative Poisson's ratio of W_2C MXene was predicted, which may find applications in fracture components in aircrafts and automobiles [108].

There are several mechanical testing methods for characterization of bulk materials. However, determining the mechanical characteristics of 2D nanomaterials remains difficult. The experimental properties of 2D nanomaterials were determined by a nanoindentation technique which was recently used to calculate the experimental Young's modulus of the $Ti_3C_2T_X$ monolayer MXene (333 ± 30 GPa). This value is almost 60% higher than graphene oxide and MoS_2. Despite the difficulties associated with measurement techniques, the presence of vacancies in MXene structure and weak composite interface still hinders MXene's mechanical evaluation. Thus, new experimental studies should be focused more on development of a defect-free MXene structure with different functionalization groups for theoretical and experimental evaluation of mechanical properties.

9.2.3 Optical Properties

The UV and visible light absorption are very important for photocatalytic, photovoltaic, and optoelectronic electrode devices. Experimental studies on thin films of $Ti_3C_2T_X$ MXene could absorb light in the UV-Vis region of 300 to 500 nm and a thickness of 5 nm film shows transmittance up to 91.2% [109]. $Ti_3C_2T_X$ MXenes intercalated with NH_4HF_2, TMAOH possess a transmittance of 90%, while hydrazine, urea, and DMSO reduces the film transmittance. The least transparent film is of the Ti_3AlC_2 MAX phase, which has a transparency of 30%. Hence, the transmittance value of MXene films depend upon its thickness and ion intercalation. Further, a broad and strong absorption band at around 700–800 nm has been found, which is crucial for photothermal therapy applications [110]. DFT simulations revealed that the presence of functional groups also effects the optical properties of 2D materials. In the visible region, –F and –OH terminated MXenes reduce absorption and enhance the reflectivity as compared to the pristine MXene. Recently, it was observed that lower absorbance values are obtained with reduction in lateral size of the MXene flakes [100]. Due to high metallic conductivity and optical transparency in the visible region, MXenes act as potential candidates for flexible transparent electrode application. Because of their strong UV reflectivity, they can be used as an anti-ultraviolet rays coating material. Further, it was observed that the outstanding light-to-heat conversion efficiency (~100%) of MXene nanosheets is very useful for biomedical and water evaporation applications [98].

9.2.4 Magnetic Properties

The spin-polarized DFT calculations indicate that many pristine and surface functionalized MXenes are non-magnetic at their ground state due to the strong covalent bonding between the transition metal "M" and the non-metal "X" element as well as the functional groups attached to them. However, by applying external strain, the covalent bond releases d-electrons which favors magnetism in non-magnetic systems. For instance, monolayers of V_2N and V_2C are anti-ferromagnetic and non-magnetic, respectively. By applying tensile stress, a large magnetic moment can be induced in both the monolayered MXenes [111]. On surface functionalization, V_2C becomes an anti-ferromagnetic semi-conductor. Although magnetism disappears in the presence of surface terminating groups, some pristine MXenes such as Ti_4C_3, Fe_2C, Cr_2C, Ti_3CN, and Zr_2C, etc. are predicted to be intrinsically magnetic, showing ferromagnetic and anti-ferromagnetic properties [74, 112, 113]. For example, Ti_3CNT_X and $Ti_4C_3T_X$ MXenes are non-magnetic with the presence of surface terminations, whereas the termination-free Mn_2NT_X MXene is ferromagnetic in nature [64]. Since Cr_2N has a larger number of valence electrons than Cr_2C, Cr_2N is anti-ferromagnetic whereas Cr_2C is a half-metallic ferromagnetic at their ground states. However, surface functionalized Cr_2CT_X and Cr_2NT_X MXenes possess ferromagnetism at room temperature with –F and –OH groups attached [113]. It has also been discovered that the energy difference between magnetic and non-magnetic Cr-based MXenes is sufficiently large that they could retain their magnetism at ambient temperature. Hence, these MXenes could be used as a promising material for spintronic devices. More recently, the magnetic properties of DTM MXenes have been predicted which shows non-magnetic, anti-ferromagnetic, or ferromagnetic behavior, depending on M″ and T elements. For example, $Cr_2TiC_2O_2$ is non-magnetic, whereas $Cr_2TiC_2(OH)_2$ and $Cr_2TiC_2F_2$, and $Cr_2VC_2(OH)_2$ and $Cr_2VC_2O_2$, are anti-ferromagnetic and ferromagnetic, respectively. The reported magnetic moments are only theoretical predictions and are not yet established experimentally, because of limited synthesis and lack of surface chemistry of MXene compounds.

9.2.5 Others

Thermal conductivity and thermal expansion co-efficient properties of MXenes are essential for electronic and energy-related devices. Simulation studies predicted that MXenes possess lower thermal expansion co-efficients and higher thermal conductivities than the phosphorene and MoS_2 monolayer. The predicted thermal conductivities of Ti_2CO_2, Hf_2CO_2, and Sc_2CF_2 vary from 22 to 472 $Wm^{-1}K^{-1}$ at room temperature [49, 50]. Similar to other 2D materials, MXenes are anisotropic and present two high symmetry routes; the *x*-axis coincides with the zig-zag direction while the *y*-axis is parallel to the armchair direction. The thermal conductivity of MXenes increases with increase in flake length [49]. It was discovered that the thermal conductivity of –O terminated MXenes increases as the atomic number of the metal "M" increases. The thermal conductivity of $Ti_3C_2T_X$ MXenes has evaluated until now and thus, other MXene materials need to be explored soon.

The contact angle measurement of $Ti_3C_2T_X$ MXene film (35°) indicates that the $Ti_3C_2T_X$ MXene surface is highly hydrophilic in nature. This hydrophilicity is due to the presence of surface terminating groups such as –OH, –O, and –F, etc., which allow aqueous electrolytes to infiltrate and render MXene non-conductive in wet environments. By removing the functional groups from MXene's surface and increasing the roughness of the film surface, it shows good hydrophobicity and could be applicable to wet environments without any degradation of the film. The contact angle of MXene foam [114] initiated by hydrazine is 94° and the 3D macroporous $Ti_3C_2T_X$ MXene film [115] is 135°.

9.3 Role of MXenes for CO_2 Reduction

Economic development is progressively decoupling the energy consumption as well as CO_2 emissions, and concurrently momentous development is presently being made in decarbonization towards producing electricity. However, significant challenges remain to be overcome for full phase decarbonization. Since CO_2 [which has a formal oxidation state of (+4)] is the most oxidized form of carbon, converting CO_2 requires the transfer of electrons to carbon, resulting in a more energetic product. Basically, a set of reactions by which CO_2 is reduced to a most favorable and feasible product is often called CO_2 reduction. These reactions are known as CO_2 hydrogenation [116, 117]. CO_2 hydrogenation can be carried out by thermo-catalytic reaction at high temperature and pressure. For example, a pressure of ~100 bar and temperature of 250°C are typical conditions for methanol synthesis. In comparison to this, electrocatalytic and photocatalytic methods can be regarded as more acceptable methods as they are environmental-friendly in nature. Over the past few decades, the electrochemical conversion of CO_2 to CO via CO_2RR has been investigated on various catalysts but only a few materials have come up as potential catalysts which can be scaled up to an efficient industrial level. Many precious metals such as gold and silver, etc. have been designated as the potential catalysts for electrochemical reduction of CO_2 to CO because of their low overpotential values with high Faradaic efficiency (FE). However, these materials cannot be used at an industrial level as they are highly expensive and not earth-abundant [118, 119]. Late transition metals such as copper were particularly used as liquid electrolytes in understanding the reaction mechanism of CO_2RR because of their unusual ability to convert CO_2 to multi-carbon products [120]. However, the activity and selectivity of these transition metal catalysts seem to be limited despite intense optimization.

In order to improve CO_2RR activity, many promising ways has been implemented by the scientific research community in exploring new catalyst material systems which allow stabilization of intermediates with different scaling relations. Discovered in 2011 by Yury Gogotsi et al., MXenes are a fascinating class of 2D materials with a wide range of applications such as energy storage, batteries, electromagnetic shielding, and catalysis. They can be represented by three diverse formulas such as: M_2X, M_3X_2, and M_4X_3 ("M" is an early transition metal and "X" is nitrogen and/or carbon) and are available in three dissimilar forms such as: i) metal elements on their own (Ex: Ti_2C and Nb_4C_3); ii) a solid solution of two unlike coupled metal elements (Ex: $(Ti,V)_3C_2$ and $(Cr,V)_3C_2$); and iii) ordered double transition metal elements with one transition metal on the outer layer and the other "M" layer in the middle (Ex: The exterior "M" layers of $Mo_2Ti_2C_2$ and $Mo_2Ti_2C_3$ are Mo, while the center "M" layers are Ti), as shown in Figure 9.1a [51]. Basically, electrochemical and photocatalytic methods are the two major methods to carry out CO_2 reduction reaction (CO_2RR). Between the two, the second one is likely to be more important as it is well known that photocatalytic reactions are comparatively more environmentally sustainable among all the other methods. This section explains a smooth representation of basic and mechanistic fundamental scientific ideas of CO_2 reduction and photocatalytic approach of MXene nano-catalyst toward CO_2 reduction. In addition, the process used for this application has more importance in order to meet environmental sustainability [121].

The basic surprising properties of MXenes are their superior electronic conductivity, electronegativity, and high hydrophilicity which generates from the metallic and surface terminating groups, providing a suitable platform for catalytic CO_2 reduction [116]. Indeed, DFT calculations showed that –O terminated MXenes (i.e., Ti_3CO_2 and W_2CO_2) have a low overpotential and strong selectivity for electrocatalytic CO_2 to HCOOH conversion. The –O terminating groups that aid in the stabilization of intermediates via –H coordination give rise to such a reaction pathway. As presented in Figure 9.1a, among various forms of MXenes, the Ti-based MXenes have been extensively used by a wide range of research groups throughout the world to check their photocatalytic efficiency toward CO_2 reduction, because of their high electrical conductivity and low Fermi level [117]. Further, Ti containing MXenes enable electron–hole separation when connected to a semiconductor-based photocatalyst. In addition, computational calculations revealed that Ti containing MXenes with oxygen vacancies are advantageous for photocatalytic CO_2 reduction with excellent selectivity.

MXenes with greater conductivity and –OH surface terminations are beneficial for CO_2 molecule adsorption, activation, and reduction. Moreover, the heterojunction of pristine MXenes with other semiconductor can enhance the charge separation ability and active surface area which directly impact the photocatalytic activity of CO_2 reduction. A 2D/2D Ti_3C_2/Bi_2WO_6 heterojunction was described to build an efficient material with significant solar absorption, electron separation, and reductive capability for CO_2 conversion [122]. In this regard, Li et al. published a significant finding in 2017 on the Cr_3C_2 and Mo_3C_2 MXene for selective CO_2 to CH_4 conversion using density functional theory calculations [123]. The findings imply that –OH and –O terminated MXenes are preferable for selective CO_2 to CH_4 conversion because they significantly reduce the overall reaction energy than bare materials. Similarly, Zhang et al. used density functional theory calculations to calculate the CO_2RR in the presence of $Ti_3C_2O_2$, V_2CO_2, and Ti_2CO_2, and found that Ti_2CO_2 MXene is a potential photocatalyst for CO_2 reduction [124]. Moreover, Li et al. has reported the photocatalytic CO_2 reduction by using eight different types of bare MXenes having the formula M_3C_2 (no surface terminations) and concluded that Mo_3C_2 and Cr_3C_2 have promising photocatalytic output toward the conversion of CO_2 to CH_4 [98, 116]. Handoko et al.has repeated the same reaction by using 19 different –O containing MXenes and concluded the least negative CO_2 reduction potential for W_2CO_2 and Ti_2CO_2 (–0.35 and –0.52 V, respectively) [125]. In 2018, Zeng et al. reported the photocatalytic CO_2 reduction by using Cu_2O nanowire by MXene quantum dots. The energy-level diagram of Ti_3C_2 QDs/Cu_2O NWs/Cu and Ti_3C_2 sheets/Cu_2O NWs/Cu heterostructures are shown in Figure 9.1e. The outcome of this analysis shows that pristine Cu_2O nanowire is activated in the presence of visible light to form a corresponding hole and electron at their respective valence band and conduction band. As the Fermi level of Ti_3C_2 MXene is placed below that of pristine Cu_2O nanowire (Figure 9.1c) the electrons are trapped by the Ti_3C_2 MXene co-catalyst for further catalytic conversion of CO_2 to methanol (CH_3OH) [126]. Yang et al. [127] reported the synthesis of an ultrathin 2D/2D Ti_3C_2/g-C_3N_4 heterojunction, which shows enhanced

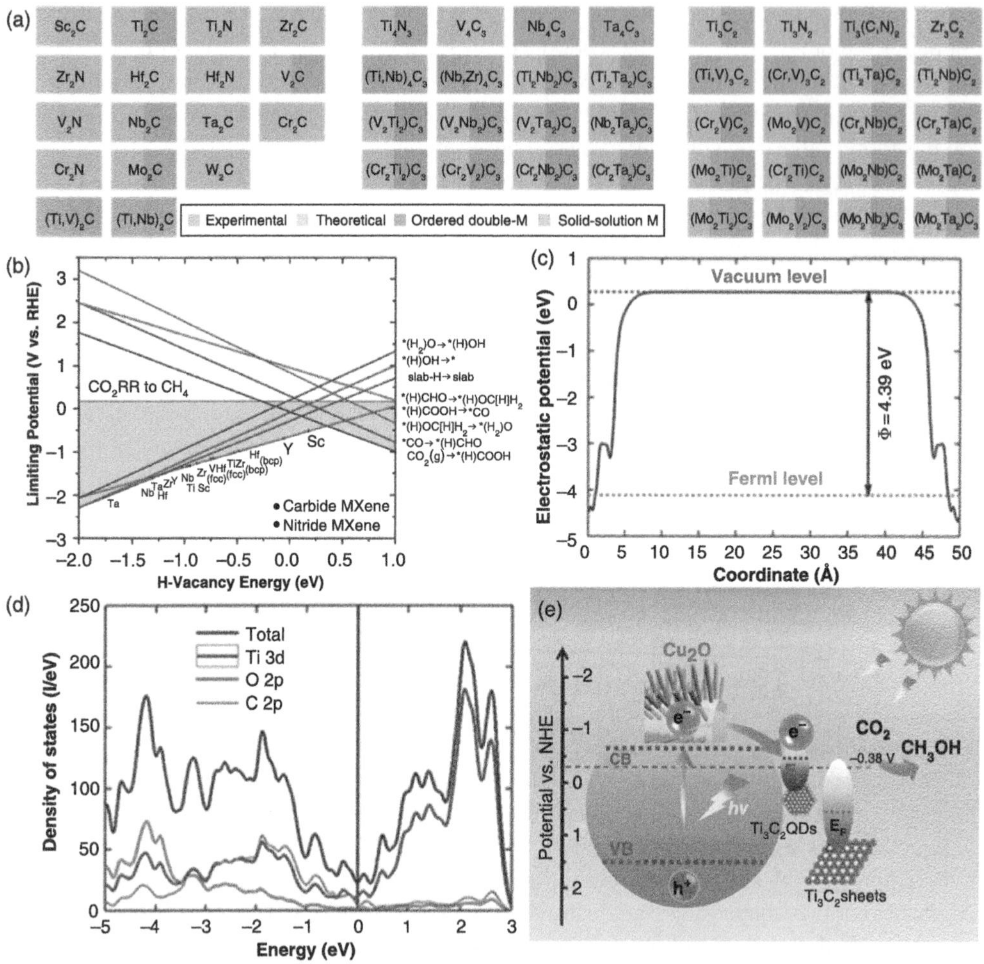

Figure 9.1 (a) MXenes reported so far have at least three different formulas: M_2X, M_3X_2, and M_4X_3, where "M" is an early transition metal and "X" is carbon and/or nitrogen; (b) Volcano curve of –OH terminated MXenes. Each line represents an elementary step described on the right side. The potential-limiting step for each H-vacancy energy (EH_{vac}) is the elementary step represented by the bottom line with the most negative limiting potential. The different MXenes are marked with dots and arranged according to their EH_{vac} value. The equilibrium potential for CO_2RR to CH_4 is 0.17 V vs. RHE; (c) and (d) calculated Fermi level and density of states (DOS) of –O terminated Ti_3C_2 QD model; and (e) energy-level diagram of Ti_3C_2, QDs/Cu_2O, NWs/Cu, and Ti_3C_2 sheets/Cu_2O and NWs/Cu heterostructures. Reproduced permission from [118]. American Chemical Society.

photocatalytic performance with an 8.1 times higher CO_2 conversion rate than that of pure g-C_3N_4. Ye et al.reported that surface alkalinization of the Ti_3C_2 MXene shows enhanced photocatalytic activity due to superior electrical conductivity, charge carrier separation ability, and maximum CO_2 adsorption. It shows dramatical increase in photocatalytic activity with 3- and 277-times higher evolution rates of CO and CH_4, respectively than that of titania. In addition, a number of similar reported works and their practical outcome are comprised in Table 9.1 [127–134].

Table 9.1 Photocatalytic reduction of CO_2 by different MXene composites.

Sl. No.	Photocatalyst	Reaction Condition	Light Source		Desired product yield		References
			Power	Lamp	Product	Yield	
1	2D-2D $g\text{-}C_3N_4/Ti_3C_2$	$NaHCO_3 + H_2SO_4 \rightarrow CO_2$ + water vapor	300 W	Xe	CO CH_4	5.19 $\mu mol\ g^{-1}h^{-1}$ 0.044 $\mu mol\ g^{-1}h^{-1}$	[127]
2	$g\text{-}C_3N_4/$ alkalized Ti_3C_2	Water vapor	300 W	Xe	CO CH_4	11.21 $\mu mol\ g^{-1}h^{-1}$ 0.269 $\mu mol\ g^{-1}h^{-1}$	[128]
3	TiO_2/Ti_3C_2	$NaHCO_3 + HCl \rightarrow CO_2$ + water vapor	300 W	Xe	CO CH_4	0.66 $\mu mol\ g^{-1-}h^{1}$ 0.22 $\mu mol\ g^{-1}h^{-1}$	[129]
4	$TiO_2/$ alkalized Ti_3C_2	Water vapor	300 W	Xe	CO CH_4	11.74 $\mu mol\ g^{-1}h^{-1}$ 16.61 $\mu mol\ g^{-1}h^{-1}$	[86]
5	$g\text{-}C_3N_4/Ti_3AlC_2/TiO$	Water and methanol	150 $mW\ cm^{-2}$	UV light	CO CH_4 H_2	4.97 $mmol\ g^{-1}h^{-1}$ 0.64 $mmol\ g^{-1}h^{-1}$ 91.9 $mmol\ g^{-1}h^{-1}$	[130]
6	CeO_2/Ti_3C_2	$NaHCO_3 + HCl \rightarrow CO_2$ + water vapor	300 W	Xe	CO	40.2 $\mu mol\ m^{-2}h^{-1}$	[131]
7	Cu_2O/Ti_3C_2 Sheets/Cu	Water vapor	300 W	Xe	CH_3OH	11.98 $ppm\ cm^{-2}h^{-1}$	[126]
8	Bi_2WO_6/Ti_3C_2	Water vapor	300 W	Xe	CH_3OH	25.77 $ppm\ cm^{-2}h^{-1}$	[132]
9	$CsPbBr_3/Ti_3C_2$	Ethyl acetate	300 W	Xe	CO CH_4	26.32 $\mu mol\ g^{-1}h^{-1}$ 7.25 $\mu mol\ g^{-1}h^{-1}$	[133]
10	Co-Co LDH/ Ti_3C_2	$MeCN/H_2O/$ triethanolamine (TEOA) and $[Ru(bpy)_3]Cl_2 \cdot 6H_2O$	300 W	Xe	CO	6.248 $\mu mol\ g^{-1}h^{-1}$	[134]
11	TiO_2/Ti_3C_2	Water vapor	300 W	Xe	CH_4	0.22 $\mu mol\ g^{-1}h^{-1}$	[129]

This section reveals the catalytic properties of MXenes during CO_2RR and, at the same time, provides a new outlook in optimizing the catalytic performance of MXenes.

9.4 Role of MXenes for H_2 Generation

The unabated exploitation of non-renewable resources with added constrictions in the existing energy reserves has led to increased environmental pollution as well as a severe energy crisis globally. In order to meet the worldwide energy demands along with maintaining an ecological balance, a green shift has been observed towards discovering and utilizing the potential of the renewable sources of energy [6]. In the league of clean and green sources, H_2 as a fuel has garnered serious attention in R&D ventures due to its abounding sources, high calorific value, zero pollution emission, energy density, recyclability, and no greenhouse gas emissions among others. Out of the many available processes for H_2 evolution reactions (HER), like decomposition of formic acid, conversion of biomass, water-gas shift reaction, and oxidation of hydrocarbons, the simplest yet most widely used approach for production of H_2 gas is the process of electrolysis of water [18]. During the water-splitting reaction, the H_2 ions after being reduced tend to combine, which results in the production of H_2 gas. However, the limitation of the electrolysis process is the energy loss due to the overpotential of the said reaction. Thus, for an efficient H_2 evolution reaction, a suitable catalyst is required which substantially reduces the overpotential, making the reaction feasible. Photocatalysis and electrocatalysis are two types of catalysis that are key aspects of the chemical processes in water splitting, which are necessary for harvesting all-pervasive kinds of desired quantity [20]. The underlying principle in the process of photocatalysis is the use of specific semiconductors which respond to incident light and accelerates the reaction rate. When incident light strikes the surface of semiconductors, particularly "photocatalysts," the valence band (VB) electrons get excited to the conduction band (CB), creating holes in the VB. These free electrons possess strong reduction affinities toward the valence state of target elements. Once the photon strikes the surface of the photocatalyst, photogenerated electron–hole pairs are produced which participate in the subsequent redox reactions, thus lowering the activation energy. On the other hand, electrocatalysis is a process where the reaction rate is enhanced by speeding up charge transfer between the electrodes and the electrolyte surface under the influence of an electric field. A typical electrocatalyst works by lowering the activation energy of an electrochemical process, which is either by reducing the oxidation/reduction potential or by facilitating specific chemical reactions at the electrode–electrolyte interface [135]. Despite promising results of H_2 evolution by both processes, photocatalysis has certain advantages over the electrocatalysis process like cost efficacy, sustainability, extended cycles per unit catalyst used, and recyclability etc. and has proven to have extensive prospects in the areas of clean energy, ecological remediation, and several other fields.

For an efficient and enhanced HER reaction to occur, the design and composition of the photocatalyst is crucial. Noble metal-based catalysts like gold (Au), palladium (Pd), and platinum (Pt) have made significant progress and have stood out as highly effective catalysts for HER. However, the increasing demand for metal-based electrocatalysts are limited as they are highly expensive and are limited resources. The discovery of 2D nanostructures is a significant achievement for the expansion and development of the energy industry [136]. Graphene, being the first 2D material consisting of single or few layers in its bulk structure graphite, has an extensive use in various energy conversion reactions owing to its unique electro-photocatalytic activity. It has an extraordinary surface with a large specific area which acts as a catalyst [137]. However, its usage in photocatalytic applications is limited because of very poor optical absorbance due to lack of bandgap. Many other 2D nanomaterials such as g-C_3N_4 and transition metal dichalcogenides (TMDs) [138], etc., and some non-precious electrode materials such as oxides [139], sulfides [140], and phosphides, [141] etc., show high performance catalytic activity when used as HER catalysts. However, there are still two problems: (1) poor intrinsic conductivity which hinders charge transfer, leading to a sharp decline in HER catalytic activity, and (2) electrochemical reaction instability of non-precious metal catalysts in aqueous electrolytes. With the advances of nanostructured catalyst materials for HER activity, transition metal carbide, nitrides, and carbonitrides, so-called MXenes have shown excellent performance like high catalytic activity, natural abundance, economically viability, and ease of commercialization. These are a class of layered smart materials with exceptional properties such as: (a) high electrical conductivity which facilitates efficient charge-carrier transfer; (b) large surface area with -OH and -O terminations can establish a strong correlation with a variety of non-precious metals and other nanostructured materials' surface; (c) excellent chemical stability and adjustable structure in aqueous electrolyte; (d) high hydrophilicity because of the surface terminations with excellent stability of the suspension; and (e) strong redox activity of the transition metals exposed at the edges of MXenes in comparison to carbon materials. In recent years, much experimental and theoretical research has been carried out on the development of MXene-based electrocatalysts. They have been optimized intrinsically and extrinsically via doping of metal atoms, hybridization, fabrication of nanostructures, and nanohybrids, and tunable surface, etc. Theoretical calculations predicted that 2D MXenes possess essential requirements for H_2 evolution reaction (HER) due to the presence of surface terminating groups such as -O and -OH, which encourages the charge transfer and transportation. Moreover, the MXene surface modified with oxygen atoms offers active sites due to the weak H–MXene interaction which assists in H_2 evolution. The exceptional catalytic activity toward HER has rendered MXenes as the most promising substitute to the noble metal catalysts in terms of industrialized use [142].

Seventeen different kinds of -O terminated MXenes were explored by Ling et al.[143] using DFT calculations for HER activity. They showed that the MXene surface terminated with -O atoms can absorb H* species, which serves as the active sites for HER activity. In 2016, Seh et al. found that the Mo_2CT_X MXene synthesized from the Mo_2Ga_2C MAX phase exhibits improvement in HER activity and stability compared to Ti_2CT_X MXenes. The Mo_2CT_X MXene exhibits an overpotential of 283 mV at a current density of 10 mA cm^{-2}, which is considered as lower than that of the Ti_2CT_X MXene (609 mV), suggesting an outperformed candidate for HER activity [56]. Recently, DFT calculations revealed that Cr-based MXenes such as Cr_2C and Cr_2CO_2 possess the valuable conductive ability for electron transfer to generate H_2. These preliminary results explore the application of Cr-based MXenes for an effective HER [144]. In another investigation, it was found that the MXenes containing surface di-vacancies exhibit less HER activity than the single metal MXenes. For example, the HER activity of the $Mo_{1.33}CT_X$ MXene is lower than that of the Mo_2CT_X MXene. This change could be due to the modifications in both co-ordination geometries and the electronic properties of the $Mo_{1.33}CT_X$ MXene structure [145]. Wang et al. reported a hydrothermally synthesized 2D transitional metal carbide as a co-catalyst with rutile TiO_2(TiO_2/composite) for H_2 production from water splitting under visible light [146]. A 200 W Hg lamp was used as the light source with a wavelength (λ) higher than 400 nm. Using methanol as the sacrificial reagent, the catalyst shows an efficiency of 17.8 μmol $h^{-1}g^{-1}$ for HER, which is nearly 4 times more than the pristine value. Yao et al. [147] demonstrated the synthesis of black phosphorous quantum dots (BPQDs)@TiO_2 NPs on the surface of a Ti_3C_2 MXene nanosheet. The H_2 production rate was reported to be 684.5 μmol $h^{-1}g^{-1}$ under visible light using 25% TEOA as the reagent with a 300 W Xe lamp of wavelength (λ) higher than 420 nm. In comparison to Ti_3C_2@TiO_2 and BPQDs/Ti_3C_2@TiO_2 samples, the increment factor was 11.35 times higher. Li et al. reported a 3D porous framework of Ti_3C_2/TiO_2 nanoflowers as high-performance photocatalysts for water splitting, free from any noble metal. This multi-layered system has an H_2 generation activity of 783.11 mol $h^{-1}g^{-1}$, which is 6 times more than the original, utilizing methanol as the reagent. The close proximity between Ti_3C_2 MXene and TiO_2 results in a synergetic effect and Schottky junction which enhances the overall reaction [148].

Apart from metal oxide, synthesis and development of hybrid transition metal chalcogenides with MXene has generated some exciting results. In 2017, Ran et al. reported a rational strategy for integrating Ti_3C_2 NPs with CdS via a hydrothermal method which is highly efficient for visible-light photocatalytic HER [42]. Figure 9.2 shows the representation of the photocatalytic activity of the MXene composites. Lactic acid was used as the sacrificial reagent with a 300 W Xe lamp ($\lambda > 420$ nm). With a quantum efficiency of 40.1% the H_2 production was reported to have a super value of 14,342 μmol $h^{-1}g^{-1}$. This exceptional performance is attributed to the electrical conductivity, favorable Fermi level position, and H_2 evolution capacity of Ti_3C_2 MXene nanosheets. Li et al. reported a unique 2D/2D/2D structure with *in-situ* development of TiO_2 nanoparticles on a highly conductive Ti_3C_2 MXene nanosheet followed by loading of 15% MoS_2. The MoS_2 nanosheets were deposited on (101) facets of TiO_2 with exposed highly-active (001) facets. The co-exposed (101) and (001) facets form a suitable heterojunction with the TiO_2, facilitating transfer and separation of charge carriers. This results in a highly active water-splitting reaction for H_2 production of 6425.297 μmol $g^{-1}h^{-1}$, which is 7.15 times more than the original. Triethanolamine (TEOA) has been used as the sacrificial reagent with a 300 W Xe lamp [149]. Ternary metal sulfides have attracted attention as co-catalysts due to suitable bandgap, strong absorption in the visible region, and high chemical stability. Recently, Cheng et al. [150] reported the synthesis of $CdLa_2S_4$ anchored on the surface of 2D Ti_3C_2 nanosheets. This novel composite attained a maximum H_2-evolution rate of

Figure 9.2 (a) A typical high-angle annular dark-field (HAADF) image of CT2.5 and the six different points (O_1, O_2, O_3, O_4, O_5, and O_6) for EDX analysis [scale bar: 200 nm]; (b) the EDX spectrum at O_3 point in figure "a" [scale bar: 2 nm]; (c) the high-resolution TEM image near O_3 point in figure "a" [scale bar: 500 nm]; (d) and (e) a typical SEM image of CT2.5 and its corresponding EDX spectrum; (f–h) the high-resolution XPS spectra of Ti 2p, O 1s, and F 1s for CT2.5; (i) a comparison of the photocatalytic H_2-production activities of CT0, CT0.05, CT0.1, CT2.5, CT5, CT7.5, Ti_3C_2 NPs, Pt–CdS, NiS–CdS, Ni–CdS, and MoS_2–CdS. The error bars are defined as SD (standard deviation); (j) ultraviolet-visible diffuse reflectance spectra of CT0, CT2.5, and Ti_3C_2-E. The inset shows the colors of all the samples as well as the UV-visible absorbance spectrum and image of the Ti_3C_2 NPs aqueous solution; (k) time-resolved PL spectra of CT0 and CT2.5; (l) EIS Nyquist plots of CT0 and CT2.5 electrodes measured under the open-circle potential and visible-light irradiation in 0.5 M potassium phosphate buffer (pH = 7) solution. The inset shows the transient photocurrent responses of CT0 and CT2.5 electrodes in 0.2 M Na_2S and 0.04 M Na_2SO_3 mixed aqueous solution under visible-light irradiation; (m) the influence of the co-catalyst's surface area on the photocatalytic activity. The error bars are defined as SD (standard deviation); (n) the influence of the co-catalyst's surface containing F to O atomic ratio on the photocatalytic activity. The error bars are defined as SD (standard deviation); (o) the charge separation and transfer in the CdS/Ti_3C_2 system under visible-light irradiation. Red and blue spheres denote photo-induced electrons and holes, respectively; (p) proposed mechanism for photocatalytic H_2 production in the CdS/Ti_3C_2 system under visible-light illumination. Green sphere denotes H^+ ions. White, grey, red, yellow, cyan, and gold spheres denote H, C, O, S, Ti, and Cd atoms, respectively [42] / Springer Nature / CC BY-4.0.

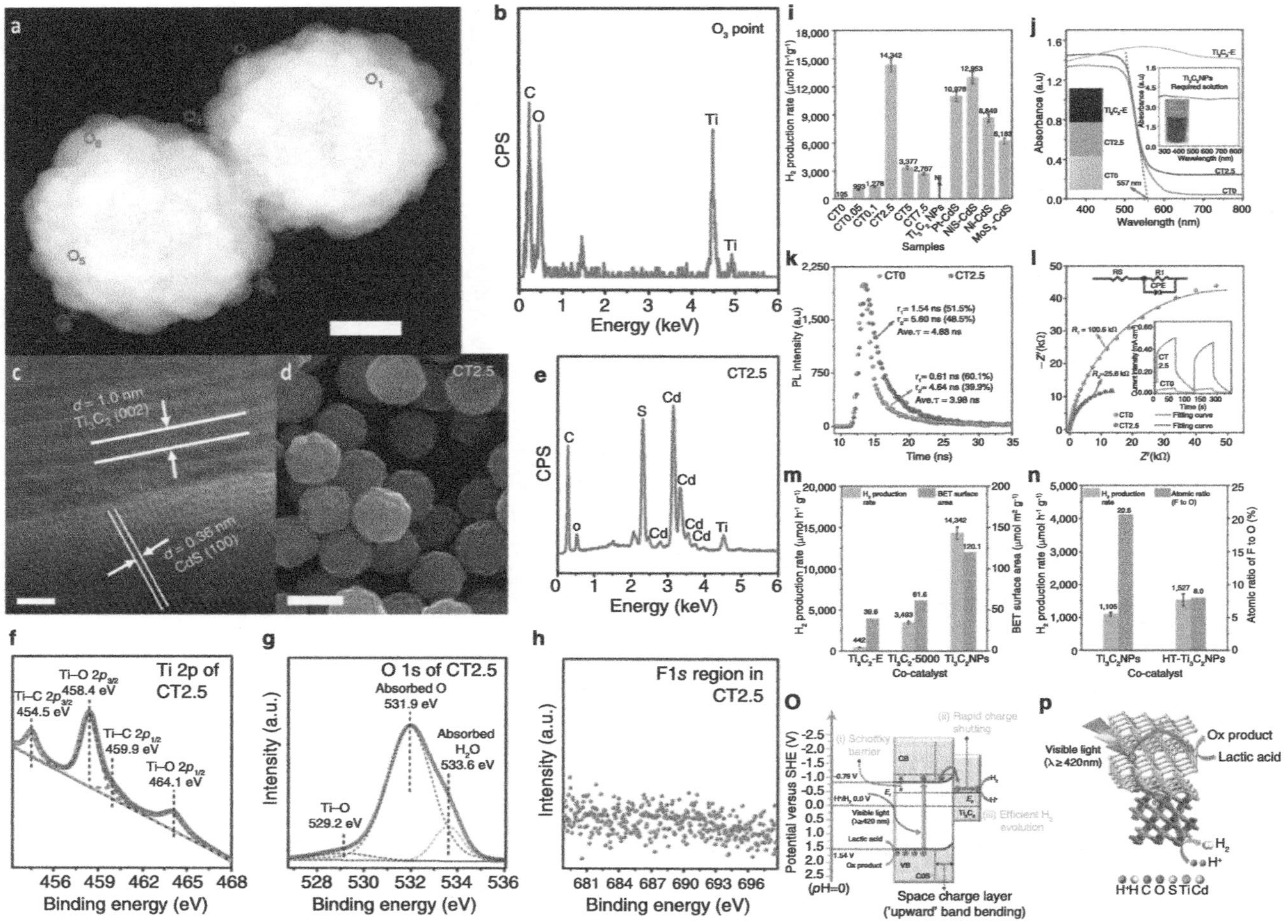

a
b
O3 point
CPS
Energy (keV)
c
d = 1.0 nm
Ti3C2 (002)
d = 0.38 nm
CdS (100)
d
CT2.5
e
CT2.5
f
Ti 2p of CT2.5
Ti–C 2p3/2 454.5 eV
Ti–O 2p3/2 458.4 eV
Ti–C 2p1/2 459.9 eV
Ti–O 2p1/2 464.1 eV
Binding energy (eV)
g
O 1s of CT2.5
Absorbed O 531.9 eV
Absorbed H2O 533.6 eV
Ti–O 529.2 eV
Intensity (a.u.)
h
F1s region in CT2.5
i
H2 production rate (μmol h−1g−1)
Samples
j
Absorbance (a.u)
Wavelength (nm)
k
PL intensity (a.u)
Time (ns)
l
−Z″(kΩ)
Z′(kΩ)
m
H2 production rate (μmol h−1 g−1)
BET surface area (μmol m2 g−1)
Co-catalyst
n
H2 production rate (μmol h−1 g−1)
Atomic ratio of F to O (%)
Co-catalyst
O
Potential versus SHE (V)
(pH=0)
Space charge layer
('upward' band bending)
p
Visible light
(λ ≥ 420nm)
Ox product
Lactic acid
H2
H+

11,182.4 μmol g^{-1} h^{-1}, which is 13.4 times greater than pristine MXene at 420 nm. Among various other semiconductor photocatalysts with layered structures, g-C_3N_4 is a distinct polymeric semiconductor which exhibits excellent properties like tunable bandgap, and amplified visible light response along with remarkable thermal and chemical stability. Although pristine g-C_3N_4 suffers from limited H_2 gas production due to drawbacks like insufficient active sites and rapid recombination of the photoexcited carrier, addition of noble metals into the layers of g-C_3N_4 are very likely to enhance the charge separation efficiencies and increase the number of active sites to lessen the HER energy barriers. Lin et al. [151] developed a 2D-2D structural Schottky junction of oxygen-doped g-C_3N_4 and Ti_3C_2 MXene. The junction resulted due to the collective *in-situ* electrostatic assembly of the negative end of Ti_3C_2 MXene and the positive end of O-doped g-C_3N_4, respectively, producing H_2 gas of 25,124 μmol g^{-1} h^{-1}, which is almost 2 times more than the pristine value. The advanced photocatalytic performance was ascribed to the synergistic interaction of the 2D interfacial interaction of the Schottky junction, where the MXene nanosheets are acceptor of photogenerated electrons and enhance the charge separation efficiency by increasing catalytic reaction kinetics. Li et al. recently reported the successful ionothermal synthesis of the Schottky heterojunction of heptazine-based crystalline carbon nitride (HCN) and Ti_3C_2 MXene. The composite showed the highest photocatalytic HER activity of 4225 μmol $g^{-1}h^{-1}$, which is about 8 times higher than that of bulk CN and twice that of pristine HCN [152]. The well-constructed Schottky heterojunction, crystalline polymeric CN, and excellent conductivity of Ti_3C_2 MXene contributed towards the enhanced photocatalytic activity. Table 9.2 gives a comparative collection of MXene-based photocatalysts for H_2 evolution reaction (HER) [153–170].

In this section, we rationally summarized and discussed various structural and compositional modifications of MXene using metal oxides, dichalcogenides, nitrides, and noble metals for application as catalyst and co-catalyst for enhanced HER activity.

Table 9.2 Comparative collection of MXene-based photocatalysts for H_2 evolution reaction.

Sl. No	Photocatalyst	Light source	Reagent	Monolayer/ Multilayer	H_2 production (μmol $h^{-1}g^{-1}$)	Improvement factor	Reference
1	$Ti_3C_2T_X/TiO_2$ nanoflowers	300 W Xe lamp	Methanol (20%)	Multilayer	783.11	6 times	[148]
2	Black phosphorus quantum dots/ $Ti_3C_2@TiO_2$	300 W Xe lamp	TEOA (25%)	Multilayer	684.5	11.35 times	[149]
3	$Ti_3C_2T_X$/rutile TiO_2	200 WHg lamp	Methanol (25%)	Monolayer	17.8	4 times	[146]
4	Au/MoS_2/ Ti_3C_2	—	Methanol	Multilayer	12,000	—	[153]
5	MXene@Au @CdS	300 W Xe lamp	0.35 mol L^{-1} Na_2S and 0.25 mol L^{-1} Na_2SO_3	Monolayer	17,070.43	1.85 times	[154]

(*Continued*)

Table 9.2 (Continued)

Sl. No	Photocatalyst	Light source	Reagent	Monolayer/ Multilayer	H_2 production ($\mu mol\ h^{-1} g^{-1}$)	Improvement factor	Reference
6	$CdS/Ti_3C_2T_X$	300 W Xe lamp	Lactic acid	——	14,342	135.59 times	[42]
7	Sulfur-doped Carbon/ TiO_2	300 W Xe lamp	Methanol (10%)	Multilayer	333	——	[155]
8	ZnS/Ti_3C_2	300 W Xe lamp	Lactic acid (20%)	Multilayer	502.6	4 times	[156]
9	1D CdS nanorod/2D $Ti_3C_2T_X$ MXene	300 W Xe lamp	Lactic acid (10%)	Monolayer	2407	6.68 time	[157]
10	MoS_2/Ti_3C_2	300 W Xe lamp	Methanol (30%)	Multilayer	6144	2.33 times	[158]
11	$CdLa_2S_4/Ti_3C_2$	300 W Xe lamp	0.25 M Na_2SO_3 and 0.35 M Na_2S	Single layer	11,182.4	13.4 times	[150]
12	$Mo_xS@TiO_2@Ti_3C_2T_X$	300 W Xe lamp	TEOA	Multilayer	10,505.8	5.99 times	[159]
13	$CdS@\ Ti_3C_2\ @\ CoO$	300 W Xe lamp	——	Monolayer	134.46	1.75 times	[160]
14	$TiO_2/\ Ti_3C_2T_X\ /\ CoS_X$	300 W Xe lamp	Methanol (20%)	Multilayer	950	5.8 times	[161]
15	$Ti_3C_2\ (TiO_2)@CdS/MoS_2$	300 W Xe lamp	lactic acid (20%)	Multilayer	344.74	3.76 times	[162]
16	$Ti_3C_2/MoS_2/TiO_2$	300 W Xe lamp	TEOA	Multilayer	6425.297	7.15 times	[149]
17	ZnS/Ti_3C_2	300 W Xe lamp	Lactic acid (20%)	Multilayer	502.6	4 times	[156]
18	$Zn_2In_2S_5/Ti_3C_2T_X$	300 W Xe lamp	0.25 M Na_2SO_3/ 0.35 M Na_2S/ H_2PtCl_6	Multilayer	2596.76	1.97 times	[163]
19	$Ti_3C_2/Pt/g\text{-}C_3N_4$	300 W Xe lamp	$g\text{-}C_3N_4$ TEOA (10%)	Monolayer	5000	15 times	[164]
20	$C\text{-}TiO_2/g\text{-}C_3N_4$	300 W Xe lamp	TEOA (10%)	Multilayer	1409	8 times	[165]
21	Ti_3C_2/ O-doped $g\text{-}C_3N_4$	300 W Xe lamp	TEOA	Multilayer	25,124	1.8 times	[151]

(Continued)

Table 9.2 (Continued)

Sl. No	Photocatalyst	Light source	Reagent	Monolayer/ Multilayer	H_2 production ($\mu mol\ h^{-1} g^{-1}$)	Improvement factor	Reference
22	$Ti_3C_2T_X$/ g-C_3N_4	300 W Xe lamp	TEOA (15%)	——	5111.8	25.97 times	[166]
23	2D/3D g-C_3N_4/ $Ti_3C_2T_X$	300 W Xe lamp	TEOA (10%)	Multilayer	116.2	6.64 times	[167]
24	2D/2D Ti_3C_2/g-C_3N_4	200WHg lamp	TEOA (10%)	Monolayer	72.3	10.18 times	[168]
25	d-Ti_3C_2/TiO_2/ g-C_3N_4	300 W Xe lamp	TEOA (10%)	Monolayer	1620	12.15 times	[169]
26	Ti_3C_2/porous MOFs (UiO-66-NH_2)	350 W Xe lamp	0.1 M Na_2S and 0.1 M Na_2SO_3	Monolayer	204	8 times	[170]
27	(HCN)/Ti_3C_2 Schottky heterojunction	LED lamps	10% TEOA	Multilayer	4225	8 times	[152]

9.5 Challenges in Using MXene

Two-dimensional MXenes have gained immense applications in the field of H_2 evolution and CO_2 reduction reactions because of their tunable surface chemistry, defects on the surface, hydrophilicity, greater interlayer spacing, and excellent EM wave absorbing properties. However, these 2D MXenes are prone to oxidation and chemical degradation, and their photocatalytic and electrocatalytic efficacy is insecure, depending on the environment. Hence, the stability of the MXene plays an important role in photocatalytic and electrocatalytic HER and CO_2RR. Many researchers have discovered that the fabrication of the MXene-based nanocomposite catalyst can circumvent this issue and significantly promote HER and CO_2RR performance as compared to the pristine catalyst [171, 172]. Although MXenes and MXene-based nanocomposites show excellent optical response and broad catalytic activity in clean energy conversion application, the catalytic pathways and working mechanisms still remain mysterious which warrants further investigation. Moreover, high catalytic activity, stability, and selectivity is needed for the long-term commercialization of MXenes and MXene-based nanocomposites in catalysis application. This could be achieved by the amalgamation of theoretical simulations and *in-situ* measurements for the evolution of active sites on the catalysts.

MXenes, the potentially largest class of 2D material, have the real potential to be used at the industrial level in the near future. Despite the several efforts made by the scientific community to watch out for MXenes from laboratory to industry, most of them have ended up in small quantities because they are expensive. Ultimately, their precursor MAX phases are expensive too, have scalability issues, and are infeasible to market materials. For example, it is possible to synthesize Sc- and V-based MXenes, but due to the cost and reactivity of the metal, they are hardly ever found in the synthesized MXene family. In addition, MAX phases are considered as the only precursors for the development of MXenes and all the conventional techniques are used for the synthesis of the MAX phase are either very complex, expensive, time-consuming, power-consuming, need high levels of sophistication, and specifically need high temperatures (>1200°C) in an oxygen-free environment. In this regard, we have recently innovated a very simple and cost-effective "Flash sintering" technique to synthesize different MAX phases with desired composition in any atmosphere (vacuum, inert, or air) in an extremely short time (<30 s). This scientific innovation was filed for Indian patent on 2 June 2021 and PCT (Application No. PCT/IN2022/050494, dated 27 May 2022) as well. [173]. This will be a breakthrough in bulk synthesis of high-quality MAX phases economically, in a short time to cater to the demand as a precursor material for the synthesis of extraordinary 2D MXenes, consequently leading to a technology of great commercial value with global impact.

References

1 Lee, R. (2011). The outlook for population growth. *Science* 333(6042): 569–573.

2 Leung, D.Y., Caramanna, G., and Maroto-Valer, M.M. (2014). An overview of current status of carbon dioxide capture and storage technologies. *Renewable & Sustainable Energy Reviews* 39: 426–443.

3 Kumar, P., Boukherroub, R., and Shankar, K. (2018). Sunlight-driven water-splitting using two-dimensional carbon based semiconductors. *Journal of Materials Chemistry A* 6(27): 12876–12931.

4 Jestin-Fleury, N. (1994). International energy agency. World energy outlook. *Politique étrangère* 59(2): 564–565.

5 Rafiee, A., Khalilpour, K.R., Milani, D. et al. (2018). Trends in CO_2 conversion and utilization: A review from process systems perspective. *Journal of Environmental Chemical Engineering* 6(5): 5771–5794.

6 Kåberger, T. (2018). Progress of renewable electricity replacing fossil fuels. *Global Energy Interconnection* 1(1): 48–52.

7 Perera, F. (2018). Pollution from fossil-fuel combustion is the leading environmental threat to global pediatric health and equity: Solutions exist. *International Journal of Environmental Research & Public Health* 15(1): 16.

8 Anderson, T.R., Hawkins, E., and Jones, P.D. (2016). CO_2, the greenhouse effect and global warming: From the pioneering work of Arrhenius and callendar to today's earth system models. *Endeavour* 40(3): 178–187.

9 Maginn, E.J. (2010). What to do with CO_2. *The Journal of Physical Chemistry Letters* 1(24): 3478–3479.

10 Nitopi, S., Bertheussen, E., Scott, S.B. et al. (2019). Progress and perspectives of electrochemical CO_2 reduction on copper in aqueous electrolyte. *Chemical Reviews* 119(12): 7610–7672.

11 Halmann, M. (1978). Photoelectrochemical reduction of aqueous carbon dioxide on p-type gallium phosphide in liquid junction solar cells. *Nature* 275(5676): 115–116.

12 Maina, J.W., Pozo-Gonzalo, C., Kong, L. et al. (2017). Metal organic framework-based catalysts for CO_2 conversion. *Materials Horizons* 4(3): 345–361.

13 Varela, A.S., Ranjbar Sahraie, N., Steinberg, J. et al. (2015). Metal-doped nitrogenated carbon as an efficient catalyst for direct CO_2 electro reduction to CO and hydrocarbons. *Angewandte Chemie International Edition* 54(37): 10758–10762.

14 Yin, Z., Palmore, G.T., and Sun, S. (2019). Electrochemical reduction of CO_2 catalyzed by metal nanocatalysts. *Trends in Chemistry* 1(8): 739–750.

15 Carmo, M., Fritz, D.L., Mergel, J. et al. (2013). A comprehensive review on PEM water electrolysis. *International Journal of Hydrogen Energy* 38(12): 4901–4934.

16 Maeda, K. and Domen, K. (2010). Photocatalytic water splitting: Recent progress and future challenges. *The Journal of Physical Chemistry Letters* 1(18): 2655–2661.

17 Balat, M. (2008). Possible methods for hydrogen production. *Energy Sources, Part A: Recovery, Utilization, & Environmental Effects* 31(1): 39–50.

18 Ren, N., Cao, G., Wang, A. et al. (2008). Dark fermentation of xylose and glucose mix using isolated *Thermoanaerobacterium thermosaccharolyticum* W16. *International Journal of Hydrogen Energy* 33(21): 6124–6132.

19. Chen, Z., Duan, X., Wei, W. et al. (2019). Recent advances in transition metal-based electrocatalysts for alkaline hydrogen evolution. *Journal of Materials Chemistry A* 7(25): 14971–15005.

20 Zhao, W., Chen, Z., Yang, X. et al. (2020). Recent advances in photocatalytic hydrogen evolution with high-performance catalysts without precious metals. *Renewable & Sustainable Energy Reviews* 132: 110040.

21 Tekalgne, M.A., Van Nguyen, K., Nguyen, D.L. et al. (2020). Hierarchical molybdenum disulfide on carbon nanotube–reduced graphene oxide composite paper as efficient catalysts for hydrogen evolution reaction. *Journal of Alloys & Compounds* 823: 153897.

22 Nguyen, T.P., Nguyen, D.L., Nguyen, V.H. et al. (2020). Facile synthesis of WS_2 hollow spheres and their hydrogen evolution reaction performance. *Applied Surface Science* 505: 144574.

23 Hasani, A., Van Le, Q., Tekalgne, M. et al. (2019). Fabrication of a WS_2/p-Si heterostructure photocathode using direct hybrid thermolysis. *ACS Applied Materials & Interfaces* 11(33): 29910–29916.

24 Nguyen, T.P., Van Le, Q., Choi, S. et al. (2018). Surface extension of MeS_2 (Me=Mo or W) nanosheets by embedding MeS_x for hydrogen evolution reaction. *Electrochimica Acta* 292: 136–141.

25 Novoselov, K.S. and Geim, A.K. (2007). The rise of graphene. *Nature Materials* 6(3): 183–191.

26 You, J., Hossain, M.D., and Luo, Z. (2018). Synthesis of 2D transition metal dichalcogenides by chemical vapor deposition with controlled layer number and morphology. *Nano Convergence* 5(1): 1–3.

27 O'Connor, T.E. (1962). Synthesis of boron nitride. *Journal of the American Chemical Society* 84(9): 1753–1754.

28 Kalantar-Zadeh, K., Ou, J.Z., Daeneke, T. et al. (2016). Two-dimensional and layered transition metal oxides. *Applied Materials Today* 5: 73–89.

29 Lu, X., Xue, H., Gong, H. et al. (2020). 2D layered double hydroxide nanosheets and their derivatives toward efficient oxygen evolution reaction. *Nano-Micro Letters* 12(1): 1–32.

30 Baumann, A.E., Burns, D.A., Liu, B. et al. (2019). Metal-organic framework functionalization and design strategies for advanced electrochemical energy storage devices. *Communications Chemistry* 2(1): 1–4.

31 Dávila, M.E., Xian, L., Cahangirov, S. et al. (2014). Germanene: A novel two-dimensional germanium allotrope akin to graphene and silicene. *New Journal of Physics* 16(9): 095002.

32 Liu, H., Neal, A.T., Zhu, Z. et al. (2014). Phosphorene: An unexplored 2D semiconductor with a high hole mobility. *ACS Nano* 8(4): 4033–4041.

33 Cahangirov, S., Topsaka, M., Aktürk, E. et al. (2009). Two-and one-dimensional honeycomb structures of silicon and germanium. *Physical Review Letters* 102(23): 236804.

34 Lalmi, B., Oughaddou, H., Enriquez, H. et al. (2010). Epitaxial growth of a silicene sheet. *Applied Physics Letters* 97(22): 223109.

35 Li, L., Yu, Y., Ye, G.J. et al. Black phosphorus field-effect transistors. *Nature Nanotechnology* 9(5): 372–377.

36 Naguib, M., Kurtoglu, M., Presser, V. et al. (2011). Two-dimensional nanocrystals produced by exfoliation of Ti_3AlC_2. *Advanced Materials* 23(37): 4248–4253.

37 Naguib, M., Mashtalir, O., Carle, J. et al. (2012). Two-dimensional transition metal carbides. *ACS Nano* 6(2): 1322–1331.

38 Naguib, M., Halim, J., Lu, J. et al. (2013). New two-dimensional niobium and vanadium carbides as promising materials for Li-ion batteries. *Journal of the American Chemical Society* 135(43): 15966–15969.

39 Yang, J., Bao, W., Jaumaux, P. et al. (2019). MXene-based composites: Synthesis and applications in rechargeable batteries and supercapacitors. *Advanced Materials Interfaces* 6(8): 1802004.

40 Shahzad, F., Alhabeb, M., Hatter, C.B. et al. (2016). Electromagnetic interference shielding with 2D transition metal carbides (MXenes). *Science* 353(6304): 1137–1140.

41 Zhang, Q., Teng, J., Zou, G. et al. (2016). Efficient phosphate sequestration for water purification by unique sandwich-like MXene/magnetic iron oxide nanocomposites. *Nanoscale* 8(13): 7085–7093.

42 Ran, J., Gao, G., Li, F.T. et al. (2017). Ti_3C_2 MXene co-catalyst on metal sulfide photo-absorbers for enhanced visible-light photocatalytic hydrogen production. *Nature Communications* 8(1): 1–0.

43 Su, T., Peng, R., Hood, Z.D. et al. (2018). One-step synthesis of $Nb_2O_5/C/Nb_2C$ (MXene) composites and their use as photocatalysts for hydrogen evolution. *ChemSusChem* 11(4): 688–699.

44 Rafieerad, A., Yan, W., Sequiera, G.L. et al. (2019). Application of Ti_3C_2 MXene quantum dots for immunomodulation and regenerative medicine. *Advanced Healthcare Materials* 8(16): 1900569.

45 Zhu, J., Wang, M., Lyu, M. et al. (2018). Two-dimensional titanium carbonitride MXene for high-performance sodium ion batteries. *ACS Applied Nano Materials* 1(12): 6854–6863.

46 Gogotsi, Y. and Anasori, B. (2019). The rise of MXenes. *ACS Nano* 13(8): 8491–8494.

47 Anasori, B., Xie, Y., Beidaghi, M. et al. (2015). Two-dimensional, ordered, double transition metals carbides (MXenes). *ACS Nano* 9(10): 9507–9516.

48 Wang, H., Wu, Y., Zhang, J. et al. (2015). Enhancement of the electrical properties of MXene Ti_3C_2 nanosheets by post-treatments of alkalization and calcination. *Materials Letters* 160: 537–540.

49 Zha, X.H., Zhou, J., Zhou, Y. et al. (2016). Promising electron mobility and high thermal conductivity in Sc_2CT_2 (T= F, OH) MXenes. *Nanoscale* 8(11): 6110–6117.

50 Zha, X.H., Huang, Q., He, J. et al. (2016). The thermal and electrical properties of the promising semiconductor MXene Hf_2CO_2. *Scientific Reports* 6(1): 1–0.

51 Anasori, B., Lukatskaya, M.R., and Gogotsi, Y. (2017). 2D metal carbides and nitrides (MXenes) for energy storage. *Nature Reviews Materials* 2(2): 1–7.

52 Sokol, M., Natu, V., Kota, S. et al. (2019). On the chemical diversity of the MAX phases. *Trends in Chemistry* 1(2): 210–223.

53 Sang, X., Xie, Y., Lin, M.W. et al. (2016). Atomic defects in monolayer titanium carbide ($Ti_3C_2T_X$) MXene. *ACS Nano* 10(10): 9193–9200.

54 Dillon, A.D., Ghidiu, M.J., Krick, A.L. et al. (2016). Highly conductive optical quality solution-processed films of 2D titanium carbide. *Advanced Functional Materials* 26(23): 4162–4168.

55 Iqbal, A., Kwon, J., Kim, M.K. et al. (2021). MXenes for electromagnetic interference shielding: Experimental and theoretical perspectives. *Materials Today Advances* 9: 100124.

56 Seh, Z.W., Fredrickson, K.D., Anasori, B. et al. (2016). Two-dimensional molybdenum carbide (MXene) as an efficient electrocatalyst for hydrogen evolution. *ACS Energy Letters* 1(3): 589–594.

57 Fan, C., Zhang, P., Wang, R. et al. (2021). Applications of two-dimensional material-MXene for proton exchange membrane fuel cells (PEMFCs) and water electrolysis. *Current Nanoscience* 17(1): 2–13.

58 Sreenilayam, S.P., Ahad, I.U., Nicolosi, V. et al. (2021). MXene materials based printed flexible devices for healthcare, biomedical and energy storage applications. *Materials Today* 43: 99–131.

59 Wyatt, B.C., Nemani, S.K., Desai, K. et al. (2021). High-temperature stability and phase transformations of titanium carbide (Ti_3C_2TX) MXene. *Journal of Physics: Condensed Matter* 33(22): 224002.

60 Ren, C.E., Hatzel, K.B., Alhabeb, M. et al. (2015). Charge-and size-selective ion sieving through $Ti_3C_2T_X$ MXene membranes. *The Journal of Physical Chemistry Letters* 6(20): 4026–4031.

61 Xu, C., Wang, L., Liu, Z. et al. (2015). Large-area high-quality 2D ultrathin Mo_2C superconducting crystals. *Nature Materials* 14(11): 1135–1141.

62 Novoselov, K.S., Geim, A.K., Morozov, S.V. et al. (2004). Electric field effect in atomically thin carbon films. *Science* 306(5696): 666–669.

63 Alhabeb, M., Maleski, K., Anasori, B. et al. (2017). Guidelines for synthesis and processing of two-dimensional titanium carbide ($Ti_3C_2T_X$ MXene). *Chemistry of Materials* 29(18): 7633–7644.

64 Ronchi, R.M., Arantes, J.T., and Santos, S.F. (2019). Synthesis, structure, properties and applications of MXenes: Current status and perspectives. *Ceramics International* 45(15): 18167–18188.

65 Karlsson, L.H., Birch, J., Halim, J. et al. (2015). Atomically resolved structural and chemical investigation of single MXene sheets. *Nano Letters* 15(8): 4955–4960.

66 Naguib, M., Unocic, R.R., Armstrong, B.L. et al. (2015). Large-scale delamination of multi-layers transition metal carbides and carbonitrides "MXenes". *Dalton Transactions* 44(20): 9353–9358.

67 Naguib, M., Mashtalir, O., Lukatskaya, M.R. et al. (2014). One-step synthesis of nanocrystalline transition metal oxides on thin sheets of disordered graphitic carbon by oxidation of MXenes. *Chemical Communications* 50(56): 7420–7423.

68 Mashtalir, O., Lukatskaya, M.R., Zhao, M.Q. et al. (2015). Amine-assisted delamination of Nb_2C MXene for Li-ion energy storage devices. *Advanced Materials* 27(23): 3501–3506.

69 Alhabeb, M., Maleski, K., Mathis, T.S. et al. (2018). Selective etching of silicon from Ti_3SiC_2 (MAX) to obtain 2D titanium carbide (MXene). *Angewandte Chemie* 130(19): 5542–5546.

70 Zhou, J., Zha, X., Chen, F.Y. et al. (2016). A two-dimensional zirconium carbide by selective etching of Al_3C_3 from nanolaminated $Zr_3Al_3C_5$. *Angewandte Chemie International Edition* 55(16): 5008–5013.

71 Meshkian, R., Näslund, LÅ, Halim, J. et al. (2015). Synthesis of two-dimensional molybdenum carbide, Mo_2C, from the gallium based atomic laminate Mo_2Ga_2C. *Scripta Materialia* 108: 147–150.

72 Ghidiu, M., Lukatskaya, M.R., Zhao, M.Q. et al. (2014). Conductive two-dimensional titanium carbide "clay" with high volumetric capacitance. *Nature* 516(7529): 78–81.

73 Halim, J., Lukatskaya, M.R., Cook, K.M. et al. (2014). Transparent conductive two-dimensional titanium carbide epitaxial thin films. *Chemistry of Materials* 26(7): 2374–2381.

74 Urbankowski, P., Anasori, B., Makaryan, T. et al. (2016). Synthesis of two-dimensional titanium nitride Ti_4N_3 (MXene). *Nanoscale* 8(22): 11385–11391.

75 Urbankowski, P., Anasori, B., Hantanasirisakul, K. et al. (2017). 2D molybdenum and vanadium nitrides synthesized by ammoniation of 2D transition metal carbides (MXenes). *Nanoscale* 9(45): 17722–17730.

76 Li, M., Lu, J., Luo, K. et al. (2019). Element replacement approach by reaction with Lewis acidic molten salts to synthesize nanolaminated MAX phases and MXenes. *Journal of the American Chemical Society* 141(11): 4730–4737.

77 Xuan, J., Wang, Z., Chen, Y. et al. (2016). Organic-base-driven intercalation and delamination for the production of functionalized titanium carbide nanosheets with superior photothermal therapeutic performance. *Angewandte Chemie* 128(47): 14789–14794.

78 Li, T., Yao, L., Liu, Q. et al. (2018). Fluorine-free synthesis of high-purity Ti_3C2T_x (T= OH, O) via alkali treatment. *Angewandte Chemie International Edition* 57(21): 6115–6119.

79 Yang, S., Zhang, P., Wang, F. et al. (2018). Fluoride-free synthesis of two-dimensional titanium carbide (MXene) using a binary aqueous system. *Angewandte Chemie* 130(47): 15717–15721.

80 Rosen, J., Dahlqvist, M., Tao, Q. et al. (2019). In- and out-of-plane ordered MAX phases and their MXene derivatives. In: *2D Metal Carbides and Nitrides (MXenes).* (eds Anasori, B. and Gogotsi, Y.), pp. 37–52. Cham: Springer.

81 Liu, Z., Wu, E., Wang, J. et al. (2014). Crystal structure and formation mechanism of $(Cr_2/_3Ti_1/_3)$ $3AlC_2$ MAX phase. *Acta Materialia* 73: 186–93.

82 Tao, Q., Lu, J., Dahlqvist, M. et al. (2019). Atomically layered and ordered rare-earth i-MAX phases: A new class of magnetic quaternary compounds. *Chemistry of Materials* 31(7): 2476–2485.

83 Halim, J., Kota, S., Lukatskaya, M.R. et al. (2016). Synthesis and characterization of 2D molybdenum carbide (MXene). *Advanced Functional Materials* 26(18): 3118–3127.

84 Wesolowski, D.J., Wang, H.W., Page, K.L. et al. (2015). Resolving the structure of $Ti_3C_2T_X$ MXenes through multilevel structural modeling of the atomic pair distribution function. *Chemistry of Materials* 28: 1.

85 Harris, K.J., Bugnet, M., Naguib, M. et al. (2015). Direct measurement of surface termination groups and their connectivity in the 2D MXene V_2CT_x using NMR spectroscopy. *The Journal of Physical Chemistry C* 119(24): 13713–13720.

86 Han, X., Wang, X., Liu, E. et al. (2018). Boosting the photocatalytic activity of P25 for carbon dioxide reduction by using a surface-alkalinized titanium carbide MXene as cocatalyst. *ChemSusChem* 11(10): 1606–1611.

87 Yang, J., Naguib, M., Ghidiu, M. et al. (2016). Two-dimensional Nb-based M_4C_3 solid solutions (MXenes). *Journal of the American Ceramic Society* 99(2): 660–666.

88 Deysher, G., Shuck, C.E., Hantanasirisakul, K. et al. (2019). Synthesis of Mo_4VAlC_4 MAX phase and two-dimensional Mo_4VC_4 MXene with five atomic layers of transition metals. *ACS Nano* 14(1): 204–217.

89 Hamm, C.M., Dürrschnabel, M., Molina-Luna, L. et al. (2018). Structural, magnetic and electrical transport properties of non-conventionally prepared MAX phases V_2AlC and $(V/Mn)_2AlC$. *Materials Chemistry Frontiers* 2(3): 483–490.

90 Lai, C.C., Fashandi, H., Lu, J. et al. (2017). Phase formation of nanolaminated Mo_2AuC and Mo_2 $(Au_{1-x}Ga_x)$ 2C by a substitutional reaction within Au-capped Mo_2 GaC and Mo_2 Ga_2C thin films. *Nanoscale* 9(45): 17681–17687.

91 Horlait, D., Middleburgh, S.C., Chroneos, A. et al. (2016). Synthesis and DFT investigation of new bismuth-containing MAX phases. *Scientific Reports* 6(1): 1–9.

92 Nechiche, M., Cabioc'h, T., Caspi, E.N. et al. (2017). Evidence for symmetry reduction in $Ti_3(Al_{1-\delta}Cu_{\delta})C_2$ MAX phase solid solutions. *Inorganic Chemistry* 56(23): 14388–14395.

93 Monnet, I. (2013). Saturation of irradiation damage in $(Ti,Zr)_3(Si,Al)C_2$ compounds. *Journal of Nuclear Materials* 433(1–3): 534–537.

94 Persson, I., El Ghazaly, A., Tao, Q. et al. (2018). Tailoring structure, composition, and energy storage properties of MXenes from selective etching of in-plane, chemically ordered MAX phases. *Small* 14(17): 1703676.

95 Hong, W., Wyatt, B.C., Nemani, S.K. et al. (2020). Double transition-metal MXenes: Atomistic design of two-dimensional carbides and nitrides. *MRS Bulletin* 45(10): 850–861.

96 Anasori, B., Shi, C., Moon, E.J. et al. (2016). Control of electronic properties of 2D carbides (MXenes) by manipulating their transition metal layers. *Nanoscale Horizons* 1(3): 227–234.

97 Xie, Y. and Kent, P.R. (2013). Hybrid density functional study of structural and electronic properties of functionalized $Ti_{n+1}X_n$ (X=C,N) monolayers. *Physical Review B* 87(23): 235441.

98 Li, R., Zhang, L., Shi, L. et al. (2017). MXene Ti_3C_2: An effective 2D light-to-heat conversion material. *ACS Nano* 11(4): 3752–3759.

99 Mashtalir, O., Naguib, M., Mochalin, V.N. et al. (2013). Intercalation and delamination of layered carbides and carbonitrides. *Nature Communications* 4(1): 1–7.

100 Maleski, K., Ren, C.E., Zhao, M.Q. et al. Size-dependent physical and electrochemical properties of two-dimensional MXene flakes. *ACS Applied Materials & Interfaces* 10(29): 24491–24498.

101 Lipatov, A., Alhabeb, M., Lukatskaya, M.R. et al. (2016). Effect of synthesis on quality, electronic properties and environmental stability of individual monolayer Ti_3C_2 MXene flakes. *Advanced Electronic Materials* 2(12): 1600255.

102 Kurtoglu, M., Naguib, M., Gogotsi, Y. et al. (2012). First principles study of two-dimensional early transition metal carbides. *MRS Communications* 2(4): 133–137.

103 Borysiuk, V.N., Mochalin, V.N., and Gogotsi, Y. (2015). Molecular dynamic study of the mechanical properties of two-dimensional titanium carbides $Tin^{+1}Cn$ (MXenes). *Nanotechnology* 26(26): 265705.

104 Ling, Z., Ren, C.E., Zhao, M.Q. et al. (2014). Flexible and conductive MXene films and nanocomposites with high capacitance. *Proceedings of the National Academy of Sciences* 111(47): 16676–16681.

105 Zhang, H., Wang, L., Chen, Q. et al. (2016). Preparation, mechanical and anti-friction performance of MXene/polymer composites. *Materials & Design* 92: 682–689.

106 Wu, X., Hao, L., Zhang, J. et al. (2016). Polymer-$Ti_3C_2T_x$ composite membranes to overcome the trade-off in solvent resistant nanofiltration for alcohol-based system. *Journal of Membrane Science* 515: 175–188.

107 Liu, Y., Wang, W., Ying, Y. et al. (2015). Binder-free layered Ti_3C_2/CNTs nanocomposite anodes with enhanced capacity and long-cycle life for lithium-ion batteries. *Dalton Transactions* 44(16): 7123–7126.

108 Wu, D., Wang, S., Zhang, S. et al. (2018). Highly negative Poisson's ratio in a flexible two-dimensional tungsten carbide monolayer. *Physical Chemistry Chemical Physics* 20(28): 18924–18930.

109 Hantanasirisakul, K., Zhao, M.Q., Urbankowski, P. et al. (2016). Fabrication of Ti_3C_2TX MXene transparent thin films with tunable optoelectronic properties. *Advanced Electronic Materials* 2(6): 1600050.

110 Xing, C., Chen, S., Liang, X. et al. (2018). Two-dimensional MXene (Ti_3C_2)-integrated cellulose hydrogels: Toward smart three-dimensional network nanoplatforms exhibiting light-induced swelling and bimodal photothermal/chemotherapy anticancer activity. *ACS Applied Materials & Interfaces* 10(33): 27631–27643.

111 Gao, G., Ding, G., Li, J. et al. (2016). Monolayer MXenes: Promising half-metals and spin gapless semiconductors. *Nanoscale* 8(16): 8986–8994.

112 Wu, F., Luo, K., Huang, C. et al. (2015). Theoretical understanding of magnetic and electronic structures of Ti_3C_2 monolayer and its derivatives. *Solid State Communications* 222: 9–13.

113 Meng, Q., Ma, J., Zhang, Y. et al. (2018). Theoretical investigation of zirconium carbide MXenes as prospective high-capacity anode materials for Na-ion batteries. *Journal of Materials Chemistry A* 6(28): 13652–3660.

114 Liu, J., Zhang, H.B., Sun, R. et al. (2017). Hydrophobic, flexible, and lightweight MXene foams for high-performance electromagnetic-interference shielding. *Advanced Materials* 29(38): 1702367.

115 Zhao, M.Q., Xie, X., Ren, C.E. et al. (2017). Hollow MXene spheres and 3D macroporous MXene frameworks for Na-ion storage. *Advanced Materials* 29(37): 1702410.

116 Chen, H., Handoko, A.D., Xiao, J. et al. Catalytic effect on CO_2 electroreduction by hydroxyl-terminated two-dimensional MXenes. *ACS Applied Materials & Interfaces* 11(40): 36571–36579.

117 Nguyen, T.P., Nguyen, D.M., Le, H.K. et al. MXenes: Applications in electrocatalytic, photocatalytic hydrogen evolution reaction and CO_2 reduction. *Molecular Catalysis* 486: 110850.

118 Salehi-Khojin, A., Jhong, H.R., Rosen, B.A. et al. Nanoparticle silver catalysts that show enhanced activity for carbon dioxide electrolysis. *The Journal of Physical Chemistry C* 117(4): 1627–1632.

119 Hori, Y., Wakebe, H., Tsukamoto, T. et al. Electrocatalytic process of CO selectivity in electrochemical reduction of CO_2 at metal electrodes in aqueous media. *Electrochimica Acta* 39(11–12): 1833–1839.

120 Huang, Y., Handoko, A.D., Hirunsit, P. et al. Electrochemical reduction of CO_2 using copper single-crystal surfaces: Effects of CO* coverage on the selective formation of ethylene. *ACS Catalysis* 7(3): 1749–1756.

121 Zhao, Y., Que, M., Chen, J. et al. (2020). MXenes as co-catalysts for the solar-driven photocatalytic reduction of CO_2. *Journal of Materials Chemistry C* 8(46): 16258–16281.

122 Wang, Z. (2021). A review on MXene: Synthesis, properties and applications on alkali metal ion batteries. *In IOP Conference Series: Earth and Environmental Science* 714(4): 042030.

123 Li, N., Chen, X., Ong, W.J. et al. (2017). Understanding of electrochemical mechanisms for CO_2 capture and conversion into hydrocarbon fuels in transition-metal carbides (MXenes). *ACS Nano* 11(11): 10825–10833.

124 Zhang, X., Zhang, Z., Li, J. et al. (2017). Ti_2CO_2 MXene: A highly active and selective photocatalyst for CO_2 reduction. *Journal of Materials Chemistry A* 5(25): 12899–12903.

125 Handoko, A.D., Khoo, K.H., Tan, T.L. (2018). et al. Establishing new scaling relations on two-dimensional MXenes for CO_2 electroreduction. *Journal of Materials Chemistry A* 6(44): 21885–21890.

126 Zeng, Z., Yan, Y., Chen, J. et al. (2019). Boosting the photocatalytic ability of Cu_2O nanowires for CO_2 conversion by MXene quantum dots. *Advanced Functional Materials* 29(2): 1806500.

127 Yang, C., Tan, Q., Li, Q. et al. (2020). 2D/2D Ti_3C_2 MXene/g-C_3N_4 nanosheets heterojunction for high efficient CO_2 reduction photocatalyst: Dual effects of urea. *Applied Catalysis B: Environmental* 268: 118738.

128 Tang, Q., Sun, Z., Deng, S. et al. (2020). Decorating g-C_3N_4 with alkalinized Ti_3C_2 MXene for promoted photocatalytic CO_2 reduction performance. *Journal of Colloid and Interface Science* 564: 406–417.

129 Low, J., Zhang, L., Tong, T. et al. (2018). TiO_2/MXene Ti_3C_2 composite with excellent photocatalytic CO_2 reduction activity. *Journal of Catalysis* 361: 255–266.

130 Tahir, M. (2020). Construction of a stable two-dimensional MAX supported protonated graphitic carbon nitride (pg-C_3N_4)/Ti_3AlC_2/TiO_2 Z-scheme multi-heterojunction system for efficient photocatalytic CO_2 reduction through dry reforming of methanol. *Energy & Fuels* 34(3): 3540–3556.

131 Shen, J., Shen, J., Zhang, W. et al. (2019). Built-in electric field induced CeO_2/Ti_3C_2-MXene Schottky-junction for coupled photocatalytic tetracycline degradation and CO_2 reduction. *Ceramics International* 45(18): 24146–24153.

132 Cao, S., Shen, B., Tong, T. et al. (2018). 2D/2D heterojunction of ultrathin MXene/Bi_2WO_6 nanosheets for improved photocatalytic CO_2 reduction. *Advanced Functional Materials* 28(21): 1800136.

133 Pan, A., Ma, X., Huang, S. et al. (2019). $CsPbBr_3$ perovskite nanocrystal grown on MXene nanosheets for enhanced photoelectric detection and photocatalytic CO_2 reduction. *The Journal of Physical Chemistry Letters* 10(21): 6590–6597.

134 Chen, W., Han, B., Xie, Y. et al. (2020). Ultrathin Co-Co LDHs nanosheets assembled vertically on MXene: 3D nanoarrays for boosted visible-light-driven CO_2 reduction. *Chemical Engineering Journal* 391: 123519.

135 Dubouis, N. and Grimaud, A. (2019). The hydrogen evolution reaction: From material to interfacial descriptors. *Chemical Science* 10(40): 9165–9181.

136 Low, J., Cao, S., Yu, J. et al. (2014). Two-dimensional layered composite photocatalysts. *Chemical Communications* 50(74): 10768–10777.

137 Kumar, A. and Xu, Q. (2018). Two-dimensional layered materials as catalyst supports. *ChemNanoMat* 4(1): 28–40.

138 Wang, H., Xiao, X., Liu, S. et al. (2019). Structural and electronic optimization of MoS_2 edges for hydrogen evolution. *Journal of the American Chemical Society* 141(46): 18578–18584.

139 Zhu, Y., Liu, X., Jin, S. et al. (2019). Anionic defect engineering of transition metal oxides for oxygen reduction and evolution reactions. *Journal of Materials Chemistry A* 7(11): 5875–5897.

140 Fu, G. and Lee, J.M. (2019). Ternary metal sulfides for electrocatalytic energy conversion. *Journal of Materials Chemistry A* 7(16): 9386–9405.

141 Parra-Puerto, A., Ng, K.L., Fahy, K. et al. Supported transition metal phosphides: Activity survey for HER, ORR, OER, and corrosion resistance in acid and alkaline electrolytes. *ACS Catalysis* 9(12): 11515–11529.

142 Bai, S., Yang, M., Jiang, J. et al. (2021). Recent advances of MXenes as electrocatalysts for hydrogen evolution reaction. *NPJ 2D Materials & Applications* 5(1): 1–5.

143 Ling, C., Shi, L., Ouyang, Y. et al. (2016). Searching for highly active catalysts for hydrogen evolution reaction based on O-terminated MXenes through a simple descriptor. *Chemistry of Materials* 28(24): 9026–9032.

144 Cheng, Y., Wang, L., Li, Y. et al. (2019). Etching and exfoliation properties of Cr_2AlC into Cr_2CO_2 and the electrocatalytic performances of 2D Cr_2CO_2 MXene. *The Journal of Physical Chemistry C* 123(25): 15629–15636.

145 Intikhab, S., Natu, V., Li, J. et al. (2019). Stoichiometry and surface structure dependence of hydrogen evolution reaction activity and stability of Mo_xC MXenes. *Journal of Catalysis* 371: 325–332.

146 Wang, H., Peng, R., Hood, Z.D. et al. (2016). Titania composites with 2D transition metal carbides as photocatalysts for hydrogen production under visible-light irradiation. *ChemSusChem* 9(12): 1490–1497.

147 Yao, Z., Sun, H., Sui, H. et al. (2020). Construction of $BPQDs/Ti_3C_2@TiO_2$ composites with favorable charge transfer channels for enhanced photocatalytic activity under visible light irradiation. *Nanomaterials* 10(3): 452.

148 Li, Y., Deng, X., Tian, J. et al. (2018). Ti_3C_2 MXene-derived Ti_3C_2/TiO_2 nanoflowers for noble-metal-free photocatalytic overall water splitting. *Applied Materials Today* 13: 217–227.

149 Li, Y., Yin, Z., Ji, G. et al. (2019). 2D/2D/2D heterojunction of Ti_3C_2 MXene/MoS_2 nanosheets/TiO_2 nanosheets with exposed (001) facets toward enhanced photocatalytic hydrogen production activity. *Applied Catalysis B: Environmental* 246: 12–20.

150 Cheng, L., Chen, Q., Li, J. et al. (2020). Boosting the photocatalytic activity of $CdLa_2S_4$ for hydrogen production using Ti_3C_2 MXene as a co-catalyst. *Applied Catalysis B: Environmental* 267: 118379.

151 Lin, P., Shen, J., Yu, X. et al. (2019). Construction of Ti_3C_2 MXene/O-doped g-C_3N_4 2D–2D Schottky-junction for enhanced photocatalytic hydrogen evolution. *Ceramics International* 45(18): 24656–24663.

152 Li, J., Li, J., Wu, C. et al. (2021). Crystalline carbon nitride anchored on MXene as an ordered Schottky heterojunction photocatalyst for enhanced visible-light hydrogen evolution. *Carbon* 179: 387–399.

153 Zhang, J., Liu, M., Wang, Y. et al. (2020). $Au/MoS_2/Ti_3C_2$ composite catalyst for efficient photocatalytic hydrogen evolution. *Cryst Eng Comm* 22(21): 3683–3691.

154 Yin, J., Zhan, F., Jiao, T. et al. (2020). Facile preparation of self-assembled MXene@Au@CdS nanocomposite with enhanced photocatalytic hydrogen production activity. *Science China Materials* 63(11): 2228–2238.

155 Yuan, W., Cheng, L., An, Y. et al. (2018). Laminated hybrid junction of sulfur-doped TiO_2 and a carbon substrate derived from Ti_3C_2 MXenes: Toward highly visible light-driven photocatalytic hydrogen evolution. *Advanced Science* 5(6): 1700870.

156 Tie, L., Yang, S., Yu, C. et al. (2019). *In situ* decoration of ZnS nanoparticles with Ti_3C_2 MXene nanosheets for efficient photocatalytic hydrogen evolution. *Journal of Colloid & Interface Science* 545: 63–70.

157 Xiao, R., Zhao, C., Zou, Z., et al. (2020). *In situ* fabrication of 1D CdS nanorod/2D Ti_3C_2 MXene nanosheet Schottky heterojunction toward enhanced photocatalytic hydrogen evolution. *Applied Catalysis B: Environmental* 268: 118382.

158 Zhang, J., Xing, C., and Shi, F. (2020). MoS_2/Ti_3C_2 heterostructure for efficient visible-light photocatalytic hydrogen generation. *International Journal of Hydrogen Energy* 45(11): 6291–6301.

159 Li, Y., Ding, L., Liang, Z. et al. (2020). Synergetic effect of defects rich MoS_2 and Ti_3C_2 MXene as cocatalysts for enhanced photocatalytic H_2 production activity of TiO_2. *Chemical Engineering Journal* 383: 123178.

160 Ai, Z., Zhang, K., Chang, B. et al. (2020). Construction of CdS@Ti_3C_2@CoO hierarchical tandem PN heterojunction for boosting photocatalytic hydrogen production in pure water. *Chemical Engineering Journal* 383: 123130.

161 Zhao, J.H., Liu, L.W., Li, K. et al. (2019). Conductive Ti_3C_2 and MOF-derived CoS_x boosting the photocatalytic hydrogen production activity of TiO_2. *CrystEngComm* 21(14): 2416–2421.

162 Ai, Z., Shao, Y., Chang, B. et al. (2019). Effective orientation control of photogenerated carrier separation via rational design of a $Ti_3C_2(TiO_2)$@CdS/MoS_2 photocatalytic system. *Applied Catalysis B: Environmental* 242: 202–208.

163 Wang, H., Sun, Y., Wu, Y. et al. (2019). Electrical promotion of spatially photoinduced charge separation via interfacial-built-in quasi-alloying effect in hierarchical $Zn_2In_2S_5$/Ti_3C_2 (O,OH) x hybrids toward efficient photocatalytic hydrogen evolution and environmental remediation. *Applied Catalysis B: Environmental* 245: 290–301.

164 An, X., Wang, W., Wang, J. et al. (2018). The synergetic effects of Ti_3C_2 MXene and Pt as co-catalysts for highly efficient photocatalytic hydrogen evolution over g-C_3N_4. *Physical Chemistry Chemical Physics* 20(16): 11405–11411.

165 Han, X., An, L., Hu, Y. et al. (2020). Ti_3C_2 MXene-derived carbon-doped TiO_2 coupled with g-C_3N_4 as the visible-light photocatalysts for photocatalytic H_2 generation. *Applied Catalysis B: Environmental* 265: 118539.

166 Li, Y., Ding, L., Guo, Y. et al. (2019). Boosting the photocatalytic ability of g-C_3N_4 for hydrogen production by Ti_3C_2 MXene quantum dots. *ACS Applied Materials & Interfaces* 11(44): 41440–44147.

167 Li, J., Zhao, L., Wang, S. et al. (2020). *In situ* fabrication of 2D/3D g-C_3N_4/Ti_3C_2 (MXene) heterojunction for efficient visible-light photocatalytic hydrogen evolution. *Applied Surface Science* 515: 145922.

168 Su, T., Hood, Z.D., Naguib, M. et al. (2019). 2D/2D heterojunction of Ti_3C_2/g-C_3N_4 nanosheets for enhanced photocatalytic hydrogen evolution. *Nanoscale* 11(17): 8138–8149.

169 Zhang, M., Qin, J., Rajendran, S. et al. (2018). Heterostructured d-$Ti_3C_2/TiO_2/g\text{-}C_3N_4$ nanocomposites with enhanced visible-light photocatalytic hydrogen production activity. *Chem Sus Chem* 11(24): 4226–4236.

170 Tian, P., He, X., Zhao, L. et al. (2019). Ti_3C_2 nanosheets modified Zr-MOFs with Schottky junction for boosting photocatalytic HER performance. *Solar Energy* 188: 750–759.

171 Liang, J., Ding, C., Liu, J. et al. (2019). Heterostructure engineering of Co-doped MoS_2 coupled with Mo_2CT x MXene for enhanced hydrogen evolution in alkaline media. *Nanoscale* 11(22): 10992–11000.

172 Jiang, Y., Wu, X., Yan, Y. et al. (2019). Coupling PtNi ultrathin nanowires with MXenes for boosting electrocatalytic hydrogen evolution in both acidic and alkaline solutions. *Small* 15(12): 1805474.

173 Kiran, N.U., Choudhary, B., Chatterjee, S. et al. (2021). *Ultrafast synthesis of MAX phases by reactive flash sintering and a method thereof.* Indian Patent Application No. 202111024506, dated: 2 June 2021; PCT application no.: PCT/IN2022/050494, dated 27 May 2022

10

The Role of Transition Metal-Based Electrocatalyst Toward Efficient Electrochemical Hydrogen Fuel Generation

Tribani Boruah and Ramendra Sundar Dey

Institute of Nano Science and Technology (INST), Sector-81, Mohali 140306, Punjab, India

10.1 Introduction

According to the International Energy Agency's estimate, worldwide energy consumption will increase by over 25% between 2017 and 2040, including annual emissions of CO_2 rising to 35.8 Gt by 2040 [1]. To address environmental deterioration and the increasing utilization of fossil fuel, has increased the demand for renewable and sustainable energy sources in the last few years [1–3]. As a potential prospect for future energy sources, molecular hydrogen (H_2) has been proposed as a viable alternative, since it has the highest gravimetric energy density as compared to other fuels and generates primarily non-polluting water as a combustion product. In addition, hydrogen (H_2) is employed in the generation of ammonia (NH_3) for fertilizer and the refining of petroleum, rendering it a pivotal chemical feedstock in modern industry. Currently, large-scale H_2 production from fossil fuels employs both H_2 and CO_2. Nevertheless, this production strategy not only increases fossil fuel consumption but also contributes to global CO_2 emissions. As a consequence, the research for a sustainable, renewable, and efficient approach for hydrogen production is critically mandated. To implement the hydrogen economy at the terawatt scale, minimal-cost, relatively high hydrogen generation is anticipated. There are several strategies to generate hydrogen, while there has been considerable interest in electrochemical water splitting as a potentially effective approach to hydrogen generation in the past few decades [4, 5].

The hydrogen evolution reaction (HER) possesses great significance from both the fundamental and a practical perspective. Modern electrochemistry relied significantly on studies of hydrogen adsorption before HER. Nevertheless, HER has always been of fundamental concern. This has received progressively further attention in recent decades as a consequence of the rising needs for renewable energy sources as well as the prospect of employing fuel cells as environment-friendly appliances for energy transmission. HER, unlike other evolution phenomena, necessitates a significant overpotential, rendering it imperative to develop adequate electrocatalysts, in particular to maximize operational efficiencies. In terms of overpotential, noble metals like Pt (platinum), Ru (ruthenium), and Pd (palladium) are suitable hydrogen evolution reaction (HER) electrocatalysts. Despite this, the higher cost and unavailability of these materials make them economically unviable alternatives, as well as the pursuit for low-cost electrocatalysts becomes the crucial topic of research. Yet, the expensive price and rarity of such precious metals limit their

Green Energy Harvesting: Materials for Hydrogen Generation and Carbon Dioxide Reduction, First Edition. Edited by Pooja Devi.

widespread application. As a consequence, it is critical to study and develop cost-effective electrocatalysts. It is considered that various metals (Fe, Ni, Co, etc.), as well as non-metals (P, S), are being employed in the development of extremely active electrocatalysts as they are easily accessible [6, 7]. Transition metal (TM) (Ni, Fe, Co, and other transition metals) have been evaluated as electrocatalysts for efficient HER attributed to their enormous abundance and inevitable activity [8–10]. It has been revealed that a wide variety of transition metal oxides (TMO) [10], nitrides (TMN) [11], phosphides (TMP) [12, 13], carbides (TMC) [14, 15], and sulfides (TMS) [16, 17] are extremely effective hydrogen producers. However, it is critical to enhancing the electrical conductivities of the TM-based material employed as HER electrocatalysts, since they are semiconducting and possess minimal electrical conductivities for electrocatalytic performance. Various approaches including carbon conductive composites have been utilized to address this challenge. Nevertheless, increasing the proportion of reactive sites along with electrical conductivity, while maintaining the overpotential requirement for HER to a bare minimum value, appears to be the foremost difficult challenge.

The purpose of this chapter is to provide an overview of current research on transition metal-based compounds to encourage widespread application in the field of electrochemical energy conversion. Here, we have provided a brief overview of current advancements based on transition metal compounds throughout the chapter. Further, the most widely adopted synthetic methodologies for producing these materials are examined. An overview of the most prominent and remarkable developments in the chemical content together with interface structure properties of transition metal compounds, as well as their practical utilization in existing and prospective energy technologies, is presented herein. Eventually, the challenges and research opportunities in this domain are highlighted, as well as techniques for improving multifunctional transition metal compounds.

10.2 Hydrogen Evolution Reaction (HER)

Protons as well as water molecules can be converted to hydrogen atoms (H_2) over the electrode interfaces with a relatively low voltage in electrochemical HER mechanisms, which comprises three probable fundamental pathways. The Volmer reaction originates whenever a proton interacts with an electron to generate an adsorbed hydrogen atom (H*) upon the electrode surface of a material (M) (Figure 10.1) [18]. Primary proton sources in acidic as well as alkaline electrolytes are the hydronium cation (H_3O^+) and the water molecule (H_2O). Thereafter, either the Heyrovsky or Tafel reactions, or both occurs. Then, a further proton combines with an electron to generate H_2, the Heyrovsky reaction. Even during the Tafel step, two nearest H* interact on the electrode surface and form H_2. The HER can be written as:

Acidic medium

$$H_3O(aq) + M + e^- \rightleftarrows M - H + H_2O \text{ Volmer reaction}$$

$$M - H + M - H \rightleftarrows H_2 \uparrow \text{ Tafel reaction}$$

$$H_3O(aq) + M - H + e^- \rightleftarrows M + H_2(g) \uparrow + H_2O(l) \text{ Heyrovsky reaction}$$

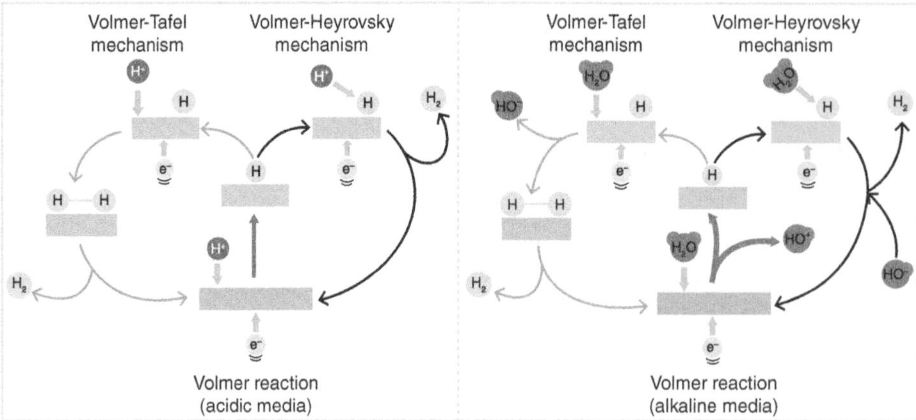

Figure 10.1 Mechanism of HER electrocatalysis in two different media. Reproduced with permission from [18], American Chemical Society (2019).

Basic medium

$$H_2O(l) + M + e^- \rightleftarrows M-H + OH^-(aq)\ \text{Volmer reaction}$$
$$M-H + M-H \rightleftarrows H_2(g)\uparrow\ \text{Tafel reaction}$$
$$M-H + H_2O(l) + e^- \rightleftarrows M + H_2(g)\uparrow + OH^-(aq)\ \text{Heyrovsky reaction}$$

10.3 Parameters Evaluating Hydrogen Evolution Reaction

10.3.1 Overpotential

Overpotential is a term employed to describe the voltage difference between the actual and theoretical thresholds necessary for HER, which typically might be considerably higher than theoretical values. The greater activity of the electrode is associated with the electrocatalyst exhibiting a lower overpotential [19]. Overpotential is divided into three categories: activation, concentration, and resistance. The activation overpotential of the HER catalyst is an inherent characteristic of the catalyst that can be significantly reduced by employing the prominent electrocatalysts, whereas concentration overpotential is induced by the ion concentration differential across the bulk solution and the electrode interface. Overpotential attributed to resistance (junction overpotential) refers to a potential difference between the reference and working electrodes that results in a further drop in voltage that can be addressed through IR compensation to obtain an appropriate overpotential. Accordingly, to examine the real functionality of developed catalysts, it is highly encouraged to load a standard amount of catalyst on a standard GC (glassy carbon) electrode for further analysis in order to get accurate results of our findings.

10.3.2 Tafel Slope/Exchange Current Density

The Tafel equation can be addressed employing polarization curves (overpotential vs. lg|current density|). Thereby, the Tafel equation depicts the linear relationship among overpotential and lg|j|, wherein g is the overpotential, a signifies the Tafel constant, b indicates the Tafel slope, while j denotes current density. The linear relationship is acknowledged whenever the overpotential approaches 59 mV. Catalytic efficiency and probable HER reaction mechanisms can be determined utilizing the Tafel slope. Thus, faster charge transfer kinetics can be achieved when the Tafel slope is lower, which requires a smaller overpotential. To calculate the intrinsic catalytic activity of an electrode material, consider exchange current density (j_0) obtained when g is zero. Rapid electron transfer, as well as greater kinetic performance, is assisted by higher exchange current density. Thus, a lower overpotential with a smaller Tafel slope indicates electrocatalytic behavior exhibiting remarkable exchange current density.

10.3.3 Stability

Stability is a further important parameter to address while evaluating HER electrocatalysts. The following are the two most common approaches for stability measurements: linear sweep voltammetry (LSV) or cyclic voltammetry (CV) measurements, and chronoamperometry (or chronopotentiometry) measurements. With CV or LSV, the electrode is continuously cycled for thousands of cycles, and the catalyst's stability is determined by the degree of overpotential deviation. Chronoamperometry examines the time of an electrocatalyst's current density (or overpotential) at a constant potential voltage

10.3.4 Turnover Frequency (TOF)

Turnover frequency (TOF) is another important metric for evaluating electrocatalytic HER productivity. A catalytically active site's inherent efficiency is determined using the amount of H_2 molecules delivered per active site/second [20]. In the case of solid-state (heterogeneous) catalytic materials, it is hard to determine an accurate TOF value due to the difficulty in precisely determining the HER active sites. However, comparing TOF values of different electrocatalysts can provide some insight into their intrinsic catalytic activity.

10.3.5 Faradaic Efficiency (FE)

The Faradaic efficiency (FE) is a measurable metric employed to describe the effectiveness through which electrons are transferred from an external network to intended products in a reaction mechanism.

10.4 Transition Metal-Based Electrocatalyst for HER

Despite transition metal alloys, transition metal-based compounds (e.g., metal-chalcogenides, metal-phosphides, metal-nitrides, metal-oxides, or metal-sulfides) comprise nonmetallic components that might alter the electronic properties of the catalyst, enhancing hydrogen interaction. Transition metal complexes typically exhibit highly variable crystal architecture and geometry. Consequently, they have been embraced for HER with significant studies as promising substitutes toward Pt-based noble metal HER electrocatalysts. This section highlights transition metal selenides, phosphides, and sufides.

10.5 Transition Metal Selenides

Selenium is a ubiquitous naturally occurring element and essential for human existence. As a consequence of their metalloid characteristics, nano-structured selenides and their derivatives have gained considerable attention during the last 40 years [21]. Meanwhile, with the advancement of technology, varied sizes, configurations, and morphologies of selenides have indeed been examined in a wide range of sectors, notably supercapacitors [22], photocatalytic, thermo-electric materials [23], and electrocatalytic materials [24]. In this section, we will provide a brief understanding of the methodology used to synthesize transition-metal selenides.

10.5.1 Various Synthesis Approaches of Transition Metal Selenides

Several physical and chemical methods and techniques for transition metal selenides have been reported over the decades. In terms of these approaches, hydrothermal/solvothermal is the most widely employed. This process comprises heating metal supplies, Se precursors, and reducing reagents to elevated temperatures and pressures to generate nanomaterials with uniform and varied dimensions. The pressure, temperature, and solvent significantly impact the overall quality [25]. For instance, Hou et al. constructed CNT/$Ni_{0.85}Se$-SnO_2 networks at 140°C for 22 h via reacting Na-selenite and hydrazine hydrate to obtain the Se-precursor. Particles of $Ni_{0.85}Se$-SnO_2 supported by carbon nanotubes are one-third the diameter of bare nanoparticles, which is around 300 nm owing to the dispersion phenomenon [26]. Whereas, the synthesized $Ni_{0.85}Se$ nanoparticles in the range of 10–27 nm at 180°C for 10 h to generate hydrogen evolution electrocatalysts were evaluated by Fu et al. [27]. Interestingly, 2D nanosheets are another a type of material that is widely synthesized employing a hydrothermal approach. Utilizing Na_2SeO_3 as the Se source and water as the solvent [28], developed $Co_{0.85}Se$ nanosheets with a thickness of 30 nm. In addition to the non-metallic doping approach, Qian et al. [29] synthesized Zn-doped $MoSe_2$-NS at 180°C for 20 h, which have delivered remarkable electrocatalytic efficiency for HER in acidic environments. Further, a number of significant ternary selenides nanosheets have been identified and characterized. As an illustration, Chen et al. constructed ultrathin MoFe selenide nanosheets (thickness 20 nanometer) at 180°C for 24 h [30].

Chemical-Vapor-Deposition (CVD) is an adaptable, cost-effective, and efficient methodology of depositing thin layers of material [31]. As a consequence, it typically occurs at the phase surface in between the gas and solid in a tubular furnace tube, which can accompany a higher temperature over conventional hydrothermal technique. In this regard, Gao et al. [32] produced metal-selenides catalysts featuring various nanostructures employing the CVD technique at varying temperatures, including 880°C nanoplates, 940°C nanosheets, and 980°C microparticles. To fabricate $MoSe_2$, Se powder and MoO_2 were deposited discretely in 2 ceramic boats then heated to a set temperature at 25°C min^{-1} underneath mixed gas flow which comprised 80 sccm of Ar and 20 sccm of H_2. Since vacancies can be regulated by modifying growing temperature, overall order of Mo and Se voids follows the nanosheets > nanoplates > microparticles order. By adopting single-step CVD process, Zhou et al. [33] generated an $NiSe_2/Ni$ hybrid foam, where bare Ni-foam as well as Se-powder were placed downstream and upstream of the furnace, and heated to the appropriate temperature at 30°C min^{-1} for 1 h under 600 sccm Ar flow. It is critical to emphasize that the interface of NiSe grew significantly rougher as the growth temperature was increased from 450 to 600°C. At appropriate temperatures, Se vapors can react with target material without affecting its morphological architecture [34].

10.5.2 Types of Transition Metal Selenides

10.5.2.1 Nickel Selenides

Nickel and selenide potentially synthesize various nickel selenides owing to their comparable electronegativity (Ni = 1.9, Se = 2.4), as well as the distinctive valence electronic configuration of Ni ($3d^8 4s^2$). Nickel selenides are a prospective catalytic alternative to precious noble Pt compounds attributed to their diverse tunable valence state, earth-abundance, and minimal cost with higher electrical conductance arising from narrow bandgap [33, 35]. Further, three stable nickel selenide phases exist at room temperature: $Ni_{1-x}Se_{0.15)}$, Ni_3Se_2, and $NiSe_2$ [36]. $Ni_{0.85}Se$ is a prominent material of $Ni_{1-x}Se$ utilized for hydrogen evolution catalyst, possessing a unique electronic arrangement and abundant unsaturated atoms. Utilizing a porous film of hexagonal phase $Ni_{0.85}Se$ grown on a graphite substrate, Wu et al. observed a 200 mV overpotential at 10 mA cm^{-2} with a Tafel slope of 81 mV dec^{-1}, and 48 h stability in basic environments [37]. Tian et al. also synthesized $Ni_{0.85}Se$ nanoparticles over N-doped graphene oxide (NGO), which exhibit a superior TOF value, a greater electrochemical active surface area, as well as a lower charge transfer resistance than $Ni_{0.85}Se$/rGO or pure $Ni_{0.85}Se$. During electrochemical measurements, the $Ni_{0.85}Se$/NGO displayed a 104 mV overpotential at 10 mA cm^{-2} and a 50.7 mV dec^{-1} Tafel slope [27]. Hou et al. likewise demonstrated self-assembled carbon nanotube mixed metal selenides, particularly CNT/$Ni_{0.85}Se$-SnO_2 networks exhibiting a lower Tafel slope of 33.2 mV/dec^{-1}, with high cathodic current density of around 28.4 mA cm^{-2} at −270 mV, as well as outstanding sustainability throughout I-t tests over 20 h. CNTs also promote dispersibility and electronic permeability within nanoparticles, leading to increased SSA (specific surface area) and lower charge-transfer barrier relative to conventional Ni0.85Se as well as Ni0.85Se-SnO_2 [26]. They also developed a Ni_3Se_2 nanowire having a 95 mV over-potential at 50 mA cm^{-2} with excellent durability. Currently, there is no research emphasizing the rich grain boundary with Ni-selenides [38].

$NiSe_2$ is the third-most common nickel selenide catalyst for hydrogen evolution, with a bandgap of zero and a dumbbell-shaped Se_2 molecule placed between two Ni molecules [39]. For example, Chen et al. reported that increasing material irregularity improved the overpotential of $NiSe_2$/NF with many more binding sites to −196 mV for 100 mA cm^{-2} having a Tafel slope of 49 mV dec^{-1} [40]. Further, Wang et al. [41] examined the hydrogen evolution mechanism in acidic environments with a selenium-enriched nickel selenide nanosheet array. The result showed a 117 mV overpotential at 10 mA cm^{-2} and a 32 mV per decade Tafel slope. The effect of post-synthesis strengthening was also examined in this study because of the accumulation of Se-residues on the surface subsequent selenization.

The results demonstrated that Se residue could accelerate charge transmission over the surface and facilitate the effective conversion of H^+ into H_2 bubbles by lowering exchange current density and increasing overpotential.

10.5.2.2 Molybdenum Selenides

Molybdenum is a remarkable component for hydrogen desorption, as revealed by research [42]. Consequently, Mo-selenides have been evaluated as electrocatalysts for H_2 evolution. For instance, Ravikumar et al. reported that nanoflower-like $MoSe_2$ architectures outperformed MoS_2 in acidic HER due to improved conductivity, higher exposed active sites, and greater tensile stress contributing to metallic, as well as greater interlayer atomic proximity [43]. Interestingly, porous $MoSe_x$ films with varying pore sizes were electro-synthesized and found to exhibit a low overpotential of −0.57 V and a low Tafel slope of 118 mV dec^{-1} for HER [44]. To activate amorphous molybdenum selenide under H_2 evolving circumstances, Nguyen et al. [45] performed an electrochemical corrosion technique. This was primarily attributed to the production of molybdenum oxygen sulfide species in the electrochemical system, which aided in the methodology. However, the lower the ΔG_{H^*} barrier, the better the HER. Thus, decreasing ΔG_{H^*} (2 eV) is regarded as the key to increasing Mo-Se activity. This activates the inactive basal plane (0001) of all molybdenum (Mo)-based TM dichalcogenides (TMDs), increasing its electrical conductivity [46]. So far, several approaches have been proposed to promote MoSe conductivity. As an example, Gao et al. revealed that doping $MoSe_2$ with B narrowed the bandgap and enhanced electrical conductivity by allowing electron mobility and charge transfer. With more B atoms in the $MoSe_2$ lattice, the Fermi bands of $MoSe_2$ shifted closer together, and the basal-plane conducting charges extended. It has been illustrated that the synthesized B-doped $MoSe_2$ had excellent catalytic efficiency exhibiting a lower overpotential (84 mV) having a Tafel slope (39 mV dec^{-1}), along with remarkable stability [47]. By utilizing the ball-milling with chemical Li-intercalation approaches, Tan et al. [48] were enabled to create single-layer ternary nanodots (MoSSe) employing a lower overpotential of 140 mV at a current density of 10 mA cm^{-2} as well as a Tafel slope of 40 mV dec^{-1}, where their highly exposed reactive edge locations, higher metallic 1T phase content, and pristine surface contribute to this. The high metallic 1T phase might enable rapid electron/charge transmission throughout HER, whereas the clean interface can minimize surfactant/ligand passivation of active edge areas. The basal plane's catalytic activity is enhanced by the alloying action of sulfur and selenium (Se) atoms, along with its vacancies over MoSSe-nanodots. Whereas $MoSe_2$ nano-dots (NDs), owing to a hetero-dimensional hybrid architecture (HDH) embedded on few-layer $MoSe_2$/NSs ($MoSe_2$-HDH), was adopted as an efficient electrocatalyst for HER by Mao et al., which focused on increasing the active site and electron transfer efficiency. The abundance of edges and flaws at the NS/ND interface enhanced the amount of surface-active sites, allowing unoriented stacking of flake-like NSs on the supporting electrode surface to increase electron transport efficiency. Throughout this instance, the material had an overpotential of 191 mV at 10 mA cm^{-2} with a Tafel slope of 72 mV dec^{-1} [49].

Growing selenides onto a carbon matrix has recently been developed as a viable alternative method of increasing conductivity within materials. Yang et al. showed that $MoSe_2$-NSs supported on rGO exhibited minimal overpotential than MoS_2 and its graphene hybrids [50]. This might be explained by the increased three-dimensional (3D) electrical contact and the abundance of folded edges. Accompanied by the use of a similar materials-macroporous $MoSe_2$–rGO composite, Park et al. [51] were able to generate an overpotential of 0.21 V, which resulted in a Tafel slope of 57%. The synergistic impact of highly conductive rGO (reduced graphene oxides) nanosheets and ultrafine $MoSe_2$ nanocrystals, as well as the crumpled shape with unoccupied nanovoids, was determined to be the cause of the high approachability for electrolyte ions. The presence of the rGO, on the other hand, can hinder the stacking and growth of few-layered $MoSe_2$. Further, nitrogen doping into $MoSe_2$/rGO by Zhang et al. [52] demonstrated that the activity of $MoSe_2$/NG first enhanced and subsequently reduced as the N/C ratio in NG increased. This was attributable to the fact that the interfacial energy barrier between $MoSe_2$ and NG could be minimal at a low N/C ratio, allowing for a mode of tremendous electrochemical activity. However, the energy barrier significantly impeded the passage of electrons from NG to $MoSe_2$ at high concentrations. Consequently, at an ideal doping ratio, N-doped RGO/$MoSe_2$ had substantially higher activity than pure $MoSe_2$ or $MoSe_2$/RGO. On the other hand, Lai et al. [53] described the hierarchical nanoarchitecture of few-layered $MoSe_2$-NSs perpendicularly formed on the carbon-nanotubes (CNTs) for HER electrocatalysis.

To optimize the number of exposed active surface for rapid ion-electrolyte transference, the one-dimensiuonal (1D) CNTs might not only suppress the agglomeration along with restacking of $MoSe_2$-NSs, but also ensure strong electrical interaction between perpendicularly adapted $MoSe_2$ nanosheets because of the higher specific surface area with inherent conductivity [54]. Combining $MoSe_2$-rGO with CNTs, Park et al. [55] were able to achieve an overpotential of 0.24 V exhibiting a current density of 10 mA cm^{-2} with a Tafel slope of 53 mV dec^{-1}. Synergistically, the CNTs with much higher aspect ratio and the rGO nanosheets reduced the formation of MoS_2 nanocrystals in this composite, which had a porous spherical backbone and a spherical CNT core. Molybdenum selenides have been studied extensively, but they have so far only been used in acidic media because of their low stability in all pH electrolytes and the lack of H_2O dissociation in alkaline media [56].

10.5.2.3 Cobalt Selenides

Cobalt selenides have shown a remarkable potential for electrocatalytic HER across the broad pH range due to their abundance on earth and strong electrically conducting. Aside from other cobalt selenides, $CoSe_2$ is the best known, having three distinct crystal architectures: orthorhombic marcasite (o-$CoSe_2$), cubic pyrite (c-CoSe, $t_{2g}{}^6e_g{}^1$), and polymorphic (p-$CoSe_2$) through mixed orthorhombic and cubic phases. In acidic environments, orthorhombic $CoSe_2$ showed an overpotential (174 mV at 10 mA cm^{-2}) with a Tafel slope (37.8 mV dec^{-1}) owing to better H_2 removal rate, reduced electrode polarization, and improving the active sites and electrical conductivity as compared to macroporous CNT [57]. As C-$CoSe_2$ has a stronger Co–Se bond than o-$CoSe_2$, thus the adsorption of H_{ads} on negatively charged selenium-anions and their conversion into H_2-molecules will be favored by this structure. This group presented a c-$CoSe_2$ catalyst with a 190 mV overpotential and an 85 mV dec^{-1} Tafel slope. Further, c-$CoSe_2$ has greater electrical conductivity [58]. A similar c-$CoSe_2$ from Co concentered metalorganic frameworks (MOFs) was explored by Gao et al. [59] for HER catalytic activity. The porous MOF cages exposed efficient integrated $CoSe_2$ active areas while limiting catalyst agglomeration throughout the catalytic process. Addressing the mixed phase of cubic pyrite as well as marcasite, it was reported that a $CoSe_2$

crystal exhibiting exceptional catalytic activity (95 mV at 10 mA cm^{-2} and a Tafel slope of 52 mV dec^{-1}), as well as substantial stability in alkaline medium could be achieved [60].

The synthesis of polymorphic $CoSe_2$ with DMF as a solvent was used to increase the effective electrocatalytic active specific surface area (SSA) and reduce ion-transfer resistance. The resulting polymorphic $CoSe_2$ possessed 60 mA cm^{-2} at −250 mV vs. RHE, and a Tafel slope of 31.6 mV dec^{-1}, with a turnover frequency (TOF) of 1.26 s^{-1} [61]. Along with pure polymorphic $CoSe_2$, Kaili Liu et al. [62] generated ternary $CoP_{2x}Se_{2(1-x)}$ nanowires by substituting the Se-atoms for P-atoms in $CoSe_2$. The greatest catalytic performance was achieved by $CoP_{1.37}Se_{0.63}$ in both acidic as well as in alkaline environments, with a current density (10 mA cm^{-2}) at overpotentials of 70 mV and 98 mV, respectively. The remarkable activity was ascribed to the irregular surface with a necklace-like shape that supplied many reactive sites. As per the recently reported work, Dutta et al. [63] produced the best HER activity for all reported cobalt selenides. With an overpotential of 160 mV, the mesoporous CoSSe showed long-term durability of at least 25 h in acidic environments. It has displayed a geometric exchange current density (70 A cm^{-2}), a Tafel slope (52 mV dec^{-1}), and a TOF of 3.34 H_2 s^{-1}. The huge electrochemically active surface area of 38.25 cm^{-2} was the main origin identified for the mechanism analysis. To determine the role of each catalyst constituent, the conjugate histidine was employed to prevent cobalt species while $HAuCl_{43}H_2O$ was employed to inhabit sulfur. According to the results, it was found that the HER activity of the cobalt inhabited electrode (Hist@meso-CoSSe-12 h) was higher than that of the sulfur blocked electrode (Au@mesoCoSSe-12 h), and that of the sulfur as well as the cobalt blocked electrode (Au&Hist@meso-CoSSe-12 h) was even lower. Sulfur, selenium, and SSe play a crucial part in the HER process.

10.5.2.4 Mixed-Metal Selenides

As per the literature, multi-metal nanomaterials exhibit more superior catalytic activity than their components [64]. The improved electrocatalytic H_2 generation productivity is attributable to improved electrical conductivity with architectural stability, and hydrogen absorption Gibbs free energy (ΔG_{H*}) [25]. So, this section will evaluate several reports on multi-metal selenides.

Nowadays, the most widely employed metal elements are Group VIII elements such as iron, cobalt, nickel, and first-row molybdenum. Using a carbon fiber paper skeleton, Zhang et al. [65] fabricated 3D hierarchical $MoSe_2/NiSe_2$ composite nanowires (NWs). With an overpotential of 249 mV, the nanowires delivered 100 mA cm^{-2} current-density, lower than $MoSe_2$ and $NiSe_2$ by 56 and 96 mV, respectively. A hybrid electrode made of topological insulators Bi_2Se_3 and $MoSe_2$ have better electrical conductivity and electronic interaction, according to Yang et al. [66]. XPS revealed that in $MoSe_2/Bi_2Se_3$ hybrids, the Mo $3d_{5/2}$ and Mo $3d_{3/2}$ peaks shifted to the lower-energy area, whereas the Bi-$4f_{7/2}$ and Bi-$4f_{5/2}$ signals relocated to the higher-energy zone. The unique surface of the Bi_2Se_3 substrate transmitted electrons to the $MoSe_2$ matrix in hybrids. UV photoelectron spectroscopy (UPS) indicated that Bi_2Se_3 has a higher Fermi level with a smaller work function than $MoSe_2$. An electrochemical impedance investigation revealed that $MoSe_2/Bi_2Se_3$ had the lowest charge-transfer resistance (R_{ct}) of all the catalysts tested, indicating that the electron transition enhanced the catalyst's electrical conductivity. Se-atoms in the inactive basal plane exhibiting enhanced electrical properties of pure $MoSe_2$ were stimulated by introducing 3% Zn to $MoSe_2$-NSs [29].

Xu et al. [67] demonstrated increased electrical conductivity in $MoSe_2$–Pt hybrid nanoflowers owing to the synergistic interaction of the $MoSe_2$ substrate and Pt-nanoparticles. Further, Zhu et al. [68] recently disclosed the competent HER catalyst of cobalt-doped VSe_2-NSs. ΔG_{H*} dropped to 0.393 eV at a Co-doping concentration of 6.25%. ΔG_{H*} increased when the Co-doping concentration increased. Among all Co-doped VSe_2, $V_{0.86}Co_{0.14}Se_2$ had the superior catalytic abilities, exhibiting 4.28 ion-transfer resistance, 1.3 mF cm^{-2} double-layer capacitance, as well as a Tafel slope of 63.4 mV dec^{-1} in acidic environments.

10.6 Transition Metal Phosphides

In recent years, transition metal phosphides (TMPs), including Ni_xP_y, MoxP, and CoxPy, etc., have been considered as potential HER catalytically active materials [41, 69]. Various levels of phosphorization in TMPs contribute to different HER characteristics and stabilities [70]. According to certain studies, negatively-charged P-atoms can grab positive protons, facilitating their discharge and thereby improving HER's catalytic performance [2]. It should be emphasized that the larger phosphorization degrees in TMPs do not necessarily imply larger HER efficacy, as the conductivity of TMPs is correspondingly regulated by the concentration of P. As previously reported [71, 72], the excessively higher phosphorous content might induce TMP to switch from electrically conducting metals to semi-conducting, entirely non-conducting metal-phosphides, resulting in slow HER catalysis. However, approximately metal-rich phosphides possess metallic characteristics or even superconductivity owing to the excellent coordination of electronegativity with the atomic ratio of metal and phosphorous. Although higher phosphorization creates a higher number of active sites yet lower conductivity, studies have endeavored to counter this by lowering P size by developing a conductive matrix as support, thus improving HER catalytic activity [70].

10.6.1 Various Synthesis Approaches of Metal Phosphides

Transition-metal phosphides preparation methodology is consistently being investigated and improved, and each approach possesses its unique characteristics. Transition-metal phosphide electrocatalysts have a wide range of characteristics. There are a variety of transition metals available, notably Ni_2P, $Ni_{12}P_5$, Ni_5P_4, Cu_3P, Fe_2P, Co_2P, InP, and MoP, etc. [73–75]. To synthesize metal phosphides amid adverse situations, highly combustible phosphorus or very toxic phosphine were required as the phosphorizing reagents. This hampered the development and practical implementation of TMPs. Several techniques are available to synthesize metal phosphides in various respects, comprising metal-organic precursor decomposition techniques [75], solid-phase reaction, electrodeposition [76], solvothermal [77], and many others.

10.6.1.1 A Metal-Organic Precursor Decomposition Approach

A metal-organic precursor decomposition is a prominent approach that avoids challenges associated with conventional phosphide synthesis, including handling and storing highly pyrophoric phosphine gas (PH_3) or white phosphorus. Phosphorus is associated with metal to form metal-organic complexes, which are then thermally decomposed to produce transition-metal phosphides as the end product. In recent years, trioctylphosphine (TOP) has become increasingly popular as a phosphorus source for the thermal-decomposition technique. Additionally, transition-metal phosphides having varied M/P atomic ratios were generated by altering metal-organic and TOP content. Read et al. proposed a flexible synthesis method based on converting commercial metal foils to phosphides via interacting with several organophosphine chemicals [75]. This process may produce metal phosphides at temperatures as low as 291 and 240°C while achieving complete metal to phosphide conversions. There are several drawbacks to this approach, notably the necessity for inert-atmospheric conditions or vacuum environment during the reaction, as well as the highly poisonous gas PH_3 released throughout the reaction mechanism, which results in high proficient and technical demands for the laboratory professional.

10.6.1.2 Solvothermal

The solvothermal synthesis approach is a strategy for the preparation of metal-phosphides nanoparticles in which organic solvents, such as 1-octadecene and oleylamine, are employed as the reaction media. The morphology and crystal architecture of TMP nanostructures can be regulated by modifying the nucleation/growth mechanism with varying temperatures, molecular ratios, or introducing other coordinating solvents [78]. The diameter and topology of NixPy nanoparticles could be modified by altering the P:N molecular ratios throughout the pyrolysis precursor (Ni-acetylacetonate and TOP) [79]. A solvothermal technique employing N-dimethylformamide (DMF)/EDTA was also reported [80]. These solvents were utilized as chelating as well as capping agents in TMP porous microsphere synthesis. However, the utilization of organic solvents in this synthesis generates a combustible and caustic environment. So, this process should be performed in an oxygen-free environment.

10.6.1.3 The Solid-Phase Synthesis Approach

The solid-phase technique is an approach that involves combining solid metal and phosphorus sources, subsequently heating in an inert atmosphere or vacuum. Until recently, TMPs were synthesized by blending metal and red phosphorus in stoichiometric proportions, sealing the mixture in a silica tube, and then heating it to higher temperatures (e.g., 900°C) over extended periods of time (e.g., 8 days). Another solid-phase reaction that reduces reaction temperature and duration is phosphidization of M_XO_Y (metal-oxides), hydroxides, or other solid metallic compounds. This methodology optimizes TMP performance by adjusting temperature and reactant molar ratio. Li et al. [81] synthesized Ni_xP_y catalysts by reacting $Ni(OH)_2$ with $NaH_2PO_2H_2O$ at a 1:5 molar ratio in Ar. As with the metal-organic precursor-decomposition technique, this approach is also temperature sensitive. By adjusting the reaction temperature, the produced Ni_xP_y catalysts can have tunable phases. Pure Ni_2P phase was achieved at a lower reaction temperature of 275°C. As the temperature rises, Ni_5P_4 and NiP_2 acquire the dominant phases. Solid-phase reactions are more cost-effective as well as convenient since they do not necessitate the use of solvents, and their reaction parameters can be easily regulated. Employing this technique, P-rich phosphides with a very high phosphorous content can become a synthesized facial structure, which is preferable since a high degree of phosphorization promotes electrochemical performance.

10.6.1.4 Other Synthetic Methods

There are a variety of different ways for synthesizing TMPs in addition to the techniques described above. Zhou et al. synthesized Sn_4P_3 powders, by employing a high-energy mechanical milling technique [82]. The technique of pulsed laser deposition (PLD) can be utilized to synthesize Sn_4P_3 thin films [83]. Sn:P = 1:3 element molar ratio was used to make the targets, which were made of mixed Sn as well as P powders. According to Barry et al. [4], a solid-vapor reaction approach was designed wherein metal chlorides were utilized to effectively react with P_4 vapors to synthesize the equivalent TMPs. Whenever different kinds of phosphorus are employed, such as white, red, or yellow, the resulting particle sizes of the products differ.

10.6.2 Types of Transition Metal Phosphides

10.6.2.1 Nickel-Phosphides

Nickel phosphide was widely studied because of its resemblance to [NiFe] hydrogenase, a bio-enhanced catalyst [72]. The nickel phosphides crystal morphologies in various phases with varying P:N ratios (i.e., NiP_x) have been reported in the literature. The electronic structures of these compounds, as well as their physicochemical characteristics, are influenced by their composition and architectural variation. To date, P-rich nickel phosphides (NiP_2) [84] and metal-rich nickel phosphides Ni_5P_4 [85], Ni_2P [77], $Ni_{12}P_5$ [86], and Ni_3P [87], have been found to display higher electrocatalytic activity toward the HER. Several analyses were conducted to better comprehend the electrocatalytic activity of NiP_x nanocatalysts. Jiang et al. [84] effectively fabricated nano-sheet arrays of NiP_2 architectures over carbon-cloth (NiP_2/NS) by following a hypophosphite reaction pathway. In H_2SO_4 (0.5 M), the overpotential essential to accomplish cathodic current density of 10 mA cm^2 with 100 mA cm^{-2} is 75 mV and 204 mV, respectively. Given the remarkable conductivity of Ni_5P_4, Laursen et al. [85] constructed Ni_5P_4 microparticles for producing H_2 gas in acidic as well as basic environments, which exhibited a low overpotential of $\eta 10 = 23$ mV with a lower Tafel slope of 33 mV dec^{-1} in H_2SO_4 (1.0 M), comparable to Pt/C. Meanwhile, its HER efficiency in basic medium (1.0 M NaOH) exhibiting a tensile strength of $\eta 10 = 49$ mV surpassed that of the corresponding Ni_2P nanocrystals. The HER effectiveness of NiP_x decreased as the aspect ratio of P ($x < 1$) atoms in Ni-phosphides is lowered. For illustration, Popczun et al. [88] employed organic-phosphine pyrolysis. A control and prominence technique were used to reveal the Ni_2P (001) crystal plane, which theoretical predictions indicated to exhibit excellent electrocatalytic performance.

Similarly, the size of the $Ni_{12}P_5$ nanoparticles produced by Huang et al. [86] was 14 ± 2 nm. These nanoparticles serve as efficient electrocatalysts, particularly HER in an acidic environment, and the overpotential required to accomplish a current density (10 mA cm^{-2}) is approximately 107 mV. Pan et al. [89] extensively compared the HER activity on NiP_x using various ratios. The larger positively charged Ni possessing stronger ensemble impact of P in Ni_5P_4 crystals led to improved electrochemical properties for HER in H_2SO_4 (0.5 M). Further, Vesborg et al. [7] evaluated the electrocatalytic performances of biphasic Ni_5P_4-Ni_2P nanosheet arrays in addition to the previously reported single-phase nickel phosphides. Despite its lower onset overpotential, the biphasic Ni_5P_4-Ni_2P nanosheet displays exceptional long-term stability in acid.

Das et al. [90] have synthesized a core shell-based nanostructure, namely Cuf@Ni_5P_4, where Cuf (copper foam) represented the core, whereas single-phase Ni_5P_4 displayed the shell [90] (Figure 10.2). Moreover, the porous morphology was confirmed via SEM and TEM characterization techniques. Further, the material was utilized for electrochemical applications. Interestingly, a better performance was delivered for HER in acidic media (90 mV, 10 mA cm^{-2}). While, in continuation to this study, we have also reported another work in which the desired synthesis material has demonstrated full water-splitting performance in alkaline medium (1.66 V, 10 mA cm^{-2}) with commendable activity [91]. In addition, the interfacial study has also been explored and concluded that the interface of the Cu [92] plane in Cuf@Ni_5P_4 has efficiently played a significant role in accumulating the charges and becoming depleted through the Ni_5P_4 shell. The study showed the Cuf@Ni_5P_4 was compatible even when hybridized with an external device like a solar cell. Henceforth, the electrodeposited Cuf displayed versatility in the field of energy as fuel generation.

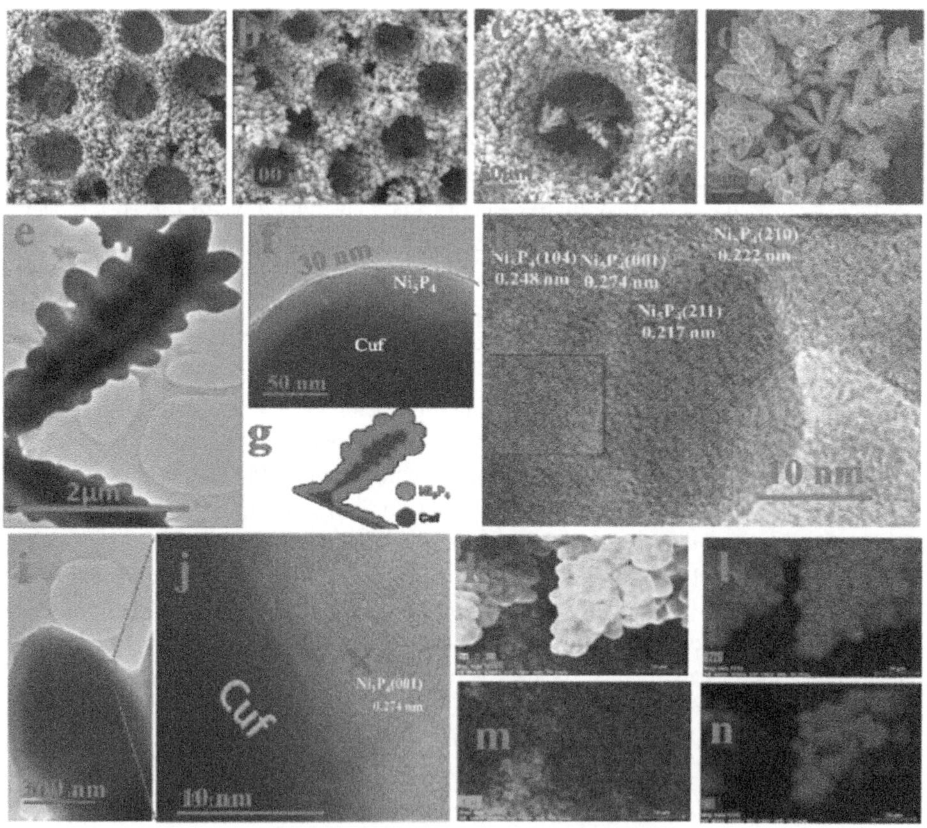

Figure 10.2 (a) SEM image of bare Cuf; (b–d) SEM images of Cuf@Ni_5P_4; (e, f) TEM images of Cuf@Ni_5 P_4; (g) core-shell nano-structure of Cuf@Ni_5P_4; (h) high resolution TEM image of Cuf@ Ni_5P_4; (i–j) HRTEM images of the interface of Cuf@Ni_5P_4 showing (001) plane of Ni_5P_4; and (k–n) elemental mapping of Cuf@Ni_5 P_4 [4], Das et al. (2019), Reproduced with permission from Royal Society of Chemistry.

10.6.2.2 Cobalt Phosphides

Theoretical studies of metallic-cobalt (M-Co) have been projected to exhibit lower energy obstacles for H-atoms, and consequently, Co-based materials are regarded as potential catalysts for the HER mechanism [93]. Another significant component of the TMP (transition metal phosphides) family was demonstrated to be a very reactive as well as acid-resistant electrocatalytic material in several studies [94, 95]. Recently, CoP crystals were explored as electrochemical hydrogen evolution catalysts. In this architecture, the P-atom is surrounded by 6 Co-atoms, resulting in a highly deformed triangular prism architecture, which is characteristic of the MnP type.

Many distinct CoP nanocrystals, comprising nanoparticles [96] and sheets [97], have been constructed and synthesized to date. Figures 10.3a–b [102] illustrate that initially produced homogenous CoP NPs possessing average particle dimensions of 13–2 nm via the organic P-approach. In 0.5 M H_2SO_4, these CoP NPs coated on Ti foil display great HER efficacy with an overpotential of 85 mV. The increased electrocatalytic efficacy was then seen on CoP-NWs (Figures 10.3c–d) [96] in an electrolyte displayed pH ranging from 0 to 14. Through phosphorization of Co_3O_4 precursors [97], developed and produced porous CoP/NS (Figure 10.3e–f) [97] of two-unit cells exhibited a thickness of 1.1 nm. The materials possess a considerable amount of CoP (200) on the crystal plane, known to have remarkable efficiency toward HER [98]. As predicted, the porous CoP/NSs possess excellent electrocatalytic properties, having minimal overpotentials of $\eta 10 = 56$ mV and $100 = 131$ v in H_2SO_4, and long-term sustainability of over 24 h underneath time-dependent current-density measurements. Similarly, Co_2P [99], CoP_2 [100], and CoP_3 nanoparticles have also been illustrated as HER electrocatalysts. Wu et al. [101] generated porous CoP_3 concave

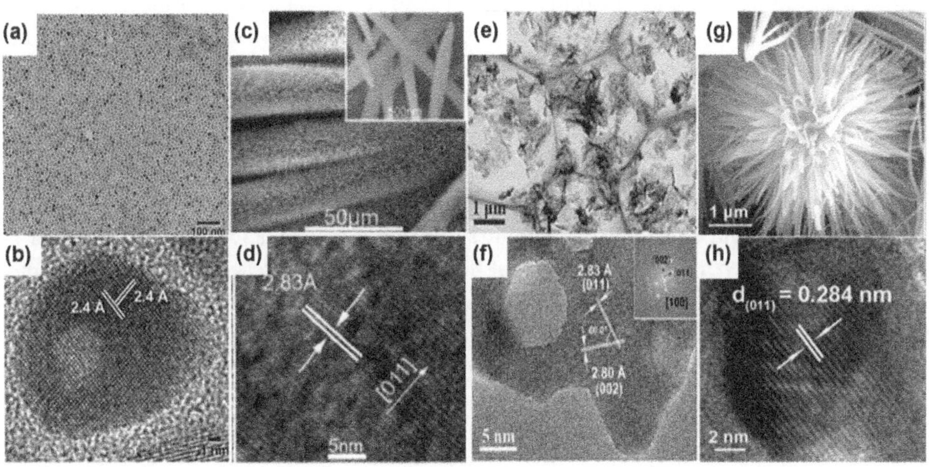

Figure 10.3 (a) TEM; (b) HRTEM images of CoP nano-particles [2], Eric et al. (2014), Reproduced with permission from John Wiley & Sons; (c) SEM image of CoP-NW arrays on carbon-cloth; (d) HRTEM image of CoP-NWs [6], Deng et al. (2016), Reproduced with permission from Royal Society of Chemistry; (e) TEM; and (f) HRTEM images of ultrathin porous CoP-NSs. Reproduced with permission from [97], Royal Society of Chemistry (2017).

polyhedrons by converting Co-MOFs to porous Co_3O_4 and subsequently employing Co_3O_4 having a minor overpotential of $\eta 10 = 78$ mV in H_2SO_4 (0.5 M) exhibiting long-term stabilization in all ranges of pH solutions were achieved by the resulting material. Whereas, Deli Jiang et al. fabricated a novel CoP_3/CoMoP heterogeneous nanosheet supported over nickel foam delivering remarkable HER activity [103], on the other hand, Hongchao et al. developed a facile method to synthesized urchin-like CoP nanocrystals possessing desirable bifunctional activity (HER/OER) with long-term durability [104].

10.6.2.3 Copper Phosphides

Regarding copper phosphide, the Cu_3P has been proposed as a catalyst for electrocatalytic HER [92]. In order to generate the distorted triangular prism, a Cu_3P unit cell is consecutively arranged with four different sites of Cu with one kind of P. According to theoretical predictions, the active site for HER is located on its [101] surface. It should be emphasized that this finding is likewise comparable along with the experimental results [105]. In this respect, Tian et al. [74] developed a hypophosphite-based reaction pathway for directly growing self-supported Cu_3P nanowires on copper foam (Cu_3P/NW). In 0.5 M H_2SO_4, the Cu_3P electrode demonstrated moderate HER performance, having an overpotential of $\eta 10 = 143$ mV as well as a Tafel slope of 67 mV dec^{-1}. Cu_3P nanosheet (Cu_3P/NS) was recently synthesized on nickel-foam by Han et al. [105]. Highly efficient electrocatalytic activity toward HER was demonstrated by the Cu_3P/NS with an overpotential of 105 mV with a Tafel slope of 42 mV dec^{-1} in 1.0 M KOH, respectively. The overpotential, on the other hand, was only marginally reduced after 24 h of operation with a fixed catalytic current density of 10 mA cm^{-2}, displaying its stability for H_2 evolution over a long time period. The performed electrochemical experiment delivered 1200 mol H_2 within an hour.

Note that although Cu_3P was investigated as an HER electrocatalyst, it fell short of the well-studied TMP electrocatalysts based on Ni, Co, [106], or Fe [107]. Accordingly, Wang et al. [108] developed hierarchical Cu_3P nanoparticles treated with an N, P-codoped carbon shell (Cu_3P@NPPC) (Figures 10.4a–b), to further boost HER activity. These nanoparticles demonstrated remarkable HER activity when covered with an N, P-codoped carbon shell (Figures 10.4a–b). This hierarchical Cu_3P@NPPC architecture had a significant SSA (1004.7 m^2 g^{-1}) (Figure 10.4c) and longer-term stability, subsequently 3000 cycles in H_2SO_4 (0.5 M). Cu_3P@NPPC's porosity could efficiently facilitate the transfer of reactants with products, and provide extra active sites for the mechanism of the electrochemical reaction. As an additional approach to enhance HER performance, another hetero-architecture (i.e., NiCoP@Cu_3P) has been developed by taking advantage of the synergistic effect of Cu_3P and NiCo.

10.6.2.4 Molybdenum Phosphides

Because of its superior electrochemical potency for the hydrodesulfurization (HDS) catalytic reaction, molybdenum-phosphide (MoP_x) has been explored extensively as a competitive HER electrocatalytic material [109]. The Pt-like electronic configuration indicates that it may be an effective HER candidate. An electrocatalytic performance of MoP compound toward HER was experimentally ascertained [110] and phosphate-based reaction [111] strategies were employed to prepare the MoP electrocatalyst. They revealed that phosphorization might alter the characteristics of Mo-metal and lead to diverse HER performance under varied phosphorization degrees. Density functional theory simulations

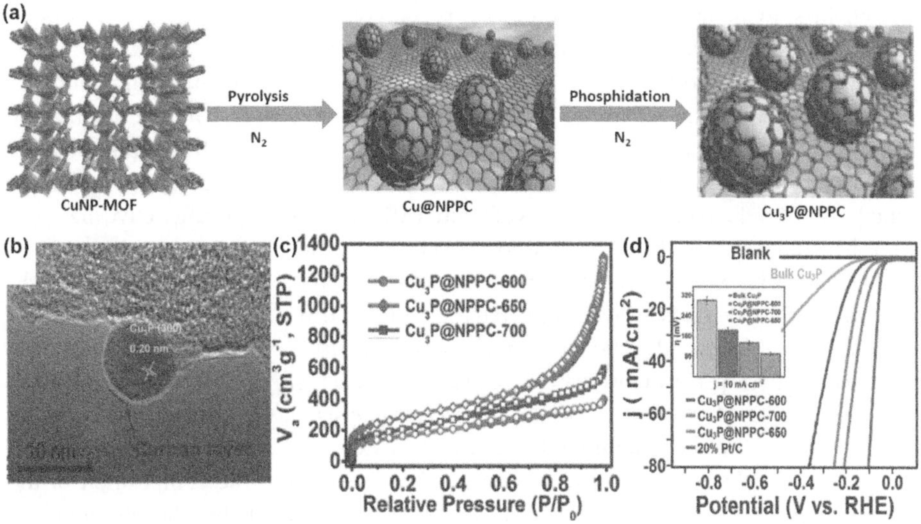

Figure 10.4 (a) The generation of the Cu_3P nano-particles coated with N, P-codoped porous carbon matrix (Cu_3P@NPPC) is depicted schematically; (b) HRTEM analysis of Cu_3P@NPPC [3], Wang et al. (2018), Reproduced with permission from John Wiley & Sons; (c) at 77 k, the plot of Cu_3P@NPPC displaying N_2-sorption isotherms is illustrated; and (d) polarization curves depicting Cu_3P@NPPC for HER in 0.5 M H_2SO_4. Reproduced with permission from [108], Wiley (2018).

revealed that an excellent "H delivery" system implemented on MoP displayed practically zero percent binding toward H-atoms, demonstrating that the system was effective. The P atoms worked as "hydrogen deliverers," adsorbing and desorbing hydrogen at low and high coverage. Further, in acidic and alkaline media, this MoP catalyst demonstrated significant HER activity. Another 54 mV dec^{-1} Tafel slope with an 180 mV overpotential reported using 0.5 M H_2SO_4. Conversely, high-temperature phosphorization of MoP causes agglomeration as well as sintering of MoP nanostructures [112]. However, Wang et al. [41] employed a hypophosphite-based reaction approach to successfully synthesize nanoscale, well-crystallized MoP@nanoparticles on carbon nanotubes (CNTs). Interestingly, well-crystallized MoP nanostructures appeared uniformly bound to CNT sidewalls, preventing MoP nanoparticle accumulation. Superior electrocatalytic performance and high stability in pH-independent environments were achieved by synthesizing the MoP/CNT hybrid material. MoP nanoparticles (MoP@PC) were synthesized by Yang et al. [113] to further enhance the electrochemical HER performance and morphological integrity of MoP. High-temperature heating confined commercial MoO_3 into ultrafine pores of UIO-66 MOF, where it was progressively synthesized to move through *in-situ* carbonization, etching along with phosphorization processes to generate the MoP@PC nano-octahedrons composite. MoP@PC composite, relative to other catalysts like MoN@PC, MoO_2@PC, and Mo_2C@PC, showed exceptional HER activity as well as superior stability resulting from the protection of the carbon shell, as demonstrated by the observations.

10.6.2.5 Mixed-Metal Phosphides

Tuning the composition of metal phosphides could boost catalytic activity significantly. In this regard, cation as well as anion substitution are commonly utilized [114]. Jaramillo et al. used experimentally as well as simulated methodologies to examine the relationship between

HER activities, ΔG_{H^*}, and TMPs. Tang et al. [115] demonstrated that $Fe_{0.5}Co_{0.5}P$ possessed Pt-like electrocatalytic efficacy by producing a series of bimetallic $Fe_xCo_{1-x}P$-NWs ($0 < x < 1$) displayed over carbon-cloth, which delivered superior long-term electrochemical stability for approximately 100 h. The activity of $Fe_xCo_{1-x}P$ nanowires to produce hydrogen is likewise proportional to the Fe substitution ratio. Similarly, substituting P-atoms by S-atoms in metal phosphides to obtain (TMPS) is predicted to promote electrocatalytic HER efficacy. This has resulted in the development and synthesis of many TMPS electrocatalytically active materials within this domain, including CoPS and MoPS [116, 117]. Owing to the change in electronegativity between S and P atoms, the S substitution could shift the d-band center. Incorporating S to the surface layer improved the HER performance of MoP or MoP|S films [118], as well as $MoS_{2(1-x)}$ P_x solid solution [119]. To construct a non-metal electrocatalyst with Pt-like catalytic performance for HER, Jin et al. studied highly active nano-architecture pyrite-type Co-phosphosulphides (CoPS) as electrocatalysts [117]. Similarly, Wang et al. reported high activity and stability of Co-phosphosulfate nanoparticles (CoP|S) developed on CNT [37]. Further analysis revealed that adding P can alter the composition of the chemical interaction between Co and S/P. The cubic structure of the phosphosulphides ($CoS_{2-x}P_x$) was more stable when the sulpfur-atoms were substituted by P. Additionally, the P-substitution depletion antibonding; e.g., * led to an increased chemical stability of the catalytic material throughout the hydrogen generating route.

10.7 Transition-Metal Sulfides

In addition to relatively minimal cost and ease of preparation, transition metal sulfides have been shown to be effective in a broad range of applications. Nevertheless, unlike metallic electrocatalysts, their architecture must be precisely fabricated for practical HER performance. Tan et al. [120] have demonstrated that metal sulfides are not resistant to the ubiquitous sulfur arrangement of catalysts, despite widespread assumptions. While implementing this kind of electrocatalyst, a common difficulty arises in that it delivers remarkable performance at lower hydrogen coverage, while the electrical conductance declines dramatically when the H coverage is increased. As a consequence of its minimal cost and tunability, molybdenum sulfides are one of the most commonly considered alternatives for HER electrocatalysts. Ideally, the lattice of MoS_2 would have an octahedral arrangement, comprising each of the six S atoms surrounding one Mo atom. Nevertheless, such a metallic inter-phase is thermodynamically not favorable and is converted to semiconducting MoS_2. Exfoliation of semiconducting MoS_2 bulk crystal architecture generates metallic MoS_2 single sheets. The HER electrocatalytic performance of a metallic MoS_2 was examined by Gen et al., who obtained an overpotential (175 mV) and a Tafel slope (41 mV dec^{-1}) [121]. Irrespective of whether MoS_2 is in the metallic or semiconducting phases, as with comparable 2D HER electrocatalysts, the reactive sites are present at the plane edges, wherein double-bonded S-atoms contribute significantly in the HER catalytic performance, as demonstrated both theoretically (Figure 10.5) [122] and experimentally [123]. Even though amorphous MoS_2 seems to have an ample amount of unsaturated sulfur atoms that can improve HER activity [123], this is implied at the expense of electrochemical non-durability. This is a major task faced by the MoS_2 electrocatalytic materials, which is to strike a balance between catalytic efficacies with electrochemical stability. $Cu_7S_4@MoS_2$ electrocatalysts had an

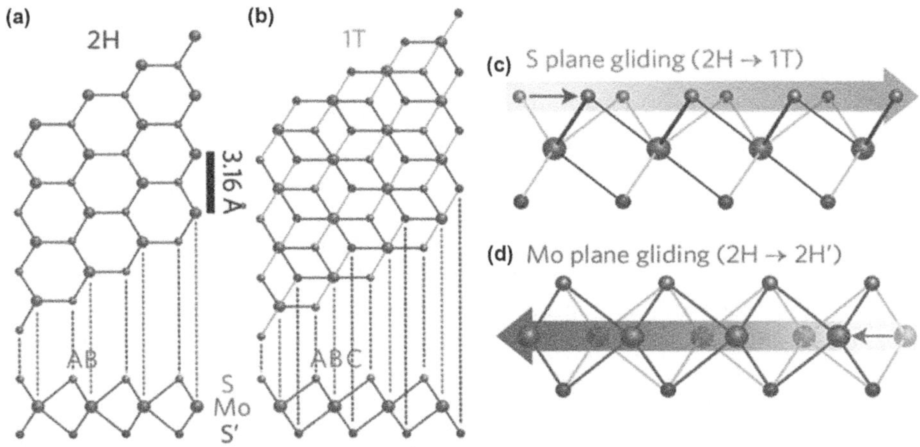

Figure 10.5 (a) Schematic illustration of one-layered MoS_2 with the 2-hydrogen atom; and (b) 1T (b) phases existing within the basal plane along with the cross-sectional views of Mo, top S, and bottom S′. Reproduced from [5] / With permission of Springer Nature.

overpotential of 206 mV with a Tafel slope of 48 mV dec^{-1}, demonstrating the best obtained value for such HER electrocatalytic materials.

Although it is feasible to higher the number of active sites in a 2D layered structure by vertically aligning the layers [66], it does not display any practical significance. A supplementary practical method is to generate active areas within the basal plane, which can be more easily achieved. This can be accomplished through doping, which provides a typically highly effective strategy. A variety of dopants, including N [125], O [126], and Cl [127], have been introduced to enhance the HER performance of MoS_2. However, substituting another transition metal for Mo can alter the charged ion distribution and thus the catalytic performance of potential reactive sites. A geometric transformation in the MoS_2 framework culminates in a much more intricate dopant effect. However, bimetallic systems might exhibit a distinct mechanism in contrast to their counterparts, since the adsorption and desorption can occur at various sites that are associated with surface diffusion.

Similarly, MoS_2 quantum dots demonstrated remarkable performance in the domain of HER [128]. Nevertheless, forming the electroactive material to generate a conducting electrode is a challenge. To boost the electrocatalytic performance significantly, only 10% graphene was introduced as the conducting material. Nevertheless, forming the electroactive material to generate a conducting electrode is also a challenge. To boost the electrocatalytic performance significantly, only 10% graphene was introduced as the conducting material. Nevertheless, it should be highlighted that quantum dots possess reactive sites in their immediate proximity, that can chemically coordinate with the graphene functionalized entities when exposed to light. The purpose of graphene appears to be a conducting catalyst scaffold to promote the homogenous distribution of electrocatalytic active material owing to a large specific surface area (SSA) while permitting electron flow.

Better performance can also be achieved by enhancing the electrocatalytic performance of the basal plane, particularly in which much of the practical emphasis has been focused at the moment. Regarding the electrochemical activity of the basal plane, this circumstance is comparable to those of graphene. However, graphene's curved basal plane has been revealed to have significant electrochemical hydrogen adsorption [129]. Recent computational studies on 2D TM chalcogenides reveal that the basal-plane catalytic performance can be modulated [130, 131]. It has been shown that depositing an MoS_2 monolayer on an electrode other than a glass substrate increases the binding energy of the hydrogen atom toward the basal plane by 0.4 eV, as was demonstrated by Chen et al. MoS_2 has a 2D lattice with a striking resemblance to graphene, although considerably higher elasticity along with Young's modulus [6]. While the thin layer of the electrochemically active material is placed simultaneously onto a flexible architecture, this provides an excellent prospect to govern the electrochemical performance of MoS_2. Besides structural considerations, the charge species distribution of TMs can indeed affect the overall H_{ads} energy, which can impact strain engineering. HER catalytic performance of 1H-NbS_2 could be greatly enhanced by strain engineering [132]. To achieve a successful MoS_2/carbon composite, the carbon chemistry is just as significant as the electrocatalyst themselves. For instance, the degree of oxidation in an MoS_2/graphene oxide nanocomposite is potentially more than twice the overpotential needed for HER [133]. It is feasible for quantum dots to directly interact with the electrocatalyst owing to their minimal sp^2 flat architecture, which provides them with numerous moieties. Several studies have indicated that the interface of C-quantum dots through MoS_2 could decrease the quantity of S^{4+} to promote the generation of $S_2{}^{2-}$ and S^{2-} molecules, that are more favorable in terms of HER performance [40].

Tungsten sulfides have a 2D striking resemblance to that of molybdenum disulfide, and as a consequence, the active sites are concentrated toward the periphery of the crystal. Shifa et al. constructed vertically placed WS_2-NSs by converting tungsten sulfides (WS_2) into a porous tungsten oxide (WO_3) and then deposited graphene sheets on top of the nanosheets to stabilize the architecture [134]. They ascribed remarkable HER effectiveness to the nanoporous WO_3 parental morphological characteristics. While this is an interesting prospect, metal oxides (MO_2) have already been thoroughly explored and are capable of generating diverse nanostructures when subjected to well-defined fabrication techniques. When it related to high current density, the introduction of graphene seemed tremendously beneficial. Like MoS_2 strain engineering, it has shown potential for enhancing WS_2 electrocatalytic activity toward HER [135]. In particular, distorting the 1T phase of tungsten sulfides (WS_2) NSs (nanosheets) generated through the widely used lithium exfoliation approach might trigger the basal plane's electrocatalytic performance, culminating in a significant number of active sites.

Even though cobalt sulfides are a great alternative to OER (oxygen evolution reaction) electrocatalysis, minimal consideration has been focused on its prospective as an electrocatalyst for HER reactions. Its HER activity can be improved to make it a bifunctional water-splitting catalyst. The HER performance of CoS and CoS_2 nanocomposites with CNT has been demonstrated [136].

10.8 Conclusions

A significant role as a future perspective for renewable energy is undoubtedly played by HER; as a consequence, the development of economical and feasible electrocatalysts is a current field of research. Pt is an excellent HER electrocatalyst, yet its high cost renders it unfeasible for numerous applications. Further, the mechanical durability of the nano-composite is a concern that can enhance the operating costs, since platinum needs to be utilized in conjunction with carbon catalytic support. If the HER is being used in portable devices, then the issue of Pt recycling becomes a concern. Under this perspective, shifting to cheaper catalysts sounds imperative rather than optional. Nevertheless, enhancing the conventional Pt electrocatalysts has the potential to dramatically lower operating costs. Even though new possibilities are constantly being identified, the following characteristics of promising electrocatalysts can be summarized as:

1) It is self-evident that a carbon catalyst support with a high surface area is required. However, even with Pt, adequate carbon catalyst support is required to achieve a desirable outcome. Nevertheless, the physical as well as chemical characteristics of carbon have gained a vital role in electrocatalysis.
2) For HER electrocatalysis, the core-shell building units appear to be a feasible methodology to significantly improve the active surface sites while simultaneously rendering the material appropriate for practical advancement. A 2D layered architecture with more active sites on the edges is incredibly effective, but the core architecture is not appropriate for industrial productivity.
3) Doping is an effective strategy of transferring charge to activate catalytic domains. Simultaneous doping might boost a catalyst's catalytic properties. This is not directed toward the electroactive substance, but rather at the carbon catalytic support material. The doping approach can render carbon more electrocatalytically active, allowing it to engage more actively in the charge transfer mechanism while also interacting more efficiently with the catalyst. It is crucial to highlight that the function of electrocatalytic support fluctuates to some extent from the role of catalytic support. In the latter instance, the support liability is to circulate the catalyst worldwide. As an additional purpose, an electrocatalytic provision should aid in the passage of Se- toward the current collector by functioning as a conductor.
4) As with doping, the use of numerous transition metals (TM) during the construction of electrocatalysts can be advantageous in certain applications. This may lead to a distinct HER mechanism owing to the differences in metal adsorption energies. As a consequence, surface diffusion can facilitate the exclusion of intermediates from active sites efficiently.
5) When it comes to monolayer or ultrathin electroactive films, the surface of the substrate plays a critical role in defining their performance. This result might be visible on both flat and particle-coated surfaces. Thus, this stimulus can be geometric or electronic, influencing the higher electrocatalytic characteristics of the upper active sites over the surface. Strain engineering and remodeling of the flat morphology are both aids to trigger the significant electrocatalytic active areas over the basal planes. It is feasible to reduce the charge transfer mechanism by localizing the distribution of charge.

Acknowledgements

T.B. acknowledges the DST SERB for providing a fellowship. This work was financially supported by the DST SERB (CRG/2020/005683) funding agency.

References

1 Heede, R. and Oreskes, N. (2016). Potential emissions of CO_2 and methane from proved reserves of fossil fuels: An alternative analysis. *Global Environmental Change* 36: 12–20.

2 Folkman, S.J. and Finke, R.G. (2017). Electrochemical water oxidation catalysis beginning with Co(II) polyoxometalates: The case of the precatalyst $Co_4V_2W_{18}O_{68}{}^{-10}$. *ACS Catalysis* 7(1): 7–16.

3 You, B., Jiang, N., Liu, X. et al. (2016). Simultaneous H_2 generation and biomass upgrading in water by an efficient noble-metal-free bifunctional electrocatalyst. *Angewandte Chemie International Edition* 55(34): 9913–9917.

4 Barry, B.M. and Gillan E.G. (2009). A general and flexible synthesis of transition-metal polyphosphides via PCl_3 elimination. *Chemistry of Materials* 21(19): 4454–4461.

5 Li, X and Hao, X., Abudula A. et al. (2016). Nanostructured catalysts for electrochemical water splitting: Current state and prospects. *Journal of Materials Chemistry A* 4(31): 11973–2000.

6 Bertolazzi, S., Brivio, J., and Kis A. (2011). Stretching and breaking of ultrathin MoS_2. *ACS Nano* 5(12): 9703–9709.

7 Vesborg, P.C.K. and Jaramillo, T.F. (2012). Addressing the terawatt challenge: Scalability in the supply of chemical elements for renewable energy. *Rsc Advances* 2(21): 7933–7947.

8 Alfredo, U., Gandia Luis, M., and Pablo, S. (2012). Hydrogen production from water electrolysis: Current status and future trends. *Proc IEEE* 100: 410–426.

9 Dalle, K.E., Warnan, J., Leung, J.J. et al. (2019). Electro-and solar-driven fuel synthesis with first row transition metal complexes. *Chemical Reviews* 119(4): 2752–875.

10 Han, L., Dong, S., and Wang, E. (2016). Transition-metal (Co, Ni, and Fe)-based electrocatalysts for the water oxidation reaction. *Advanced Materials* 28(42): 9266–9291.

11 Shalom, M., Ressnig, D., Yang, X. et al. (2015). Nickel nitride as an efficient electrocatalyst for water splitting. *Journal of Materials Chemistry A* 3(15): 8171–8177.

12 Pei, Y., Cheng, Y., Chen, J. et al. (2018). Recent developments of transition metal phosphides as catalysts in the energy conversion field. *Journal of Materials Chemistry A* 6(46): 23220–23243.

13 Zhao, H. and Yuan, Z-Y. (2017). Transition metal–phosphorus-based materials for electrocatalytic energy conversion reactions. *Catalysis Science & Technology* 7(2): 330–347.

14 Indra, A., Menezes, P.W., Zaharieva, I. et al. (2020). Detecting structural transformation of cobalt phosphonate to active bifunctional catalysts for electrochemical water-splitting. *Journal of Materials Chemistry A* 8(5): 2637–2643.

15 Han, N., Yang, K.R., Lu, Z. et al. (2018). Nitrogen-doped tungsten carbide nanoarray as an efficient bifunctional electrocatalyst for water splitting in acid. *Nature Communications* 9(1): 1–10.

16 Chandrasekaran, S., Yao, L., Deng, L. et al. (2019). Recent advances in metal sulfides: From controlled fabrication to electrocatalytic, photocatalytic and photoelectrochemical water splitting and beyond. *Chemical Society Reviews* 48(15): 4178–4280.

17 Tang, C., Cheng, N., Pu, Z. et al. (2015). NiSe nanowire film supported on nickel foam: An efficient and stable 3D bifunctional electrode for full water splitting. *Angewandte Chemie* 127(32): 9483–9487.

18 Zhu, J., Hu, L., Zhao, P. et al. (2019). Recent advances in electrocatalytic hydrogen evolution using nanoparticles. *Chemical Reviews* 120(2): 851–918.

19 Guo, Y., Park, T., Yi, J.W. et al. (2019). Nanoarchitectonics for transition-metal-sulfide-based electrocatalysts for water splitting. *Advanced Materials* 31(17): 1807134.

20 Wei, C., Rao, R.R., Peng, J. et al. (2019). Recommended practices and benchmark activity for hydrogen and oxygen electrocatalysis in water splitting and fuel cells. *Advanced Materials* 31(31): 1806296.

21 Kazantzis, G. (1972). Chromium and nickel. *The Annals of Occupational Hygiene* 15(1): 25–29.

22 Chen, Z., Yang, Y., Ma, Z. et al. (2019). All-solid-state asymmetric supercapacitors with metal selenides electrodes and ionic conductive composites electrolytes. *Advanced Functional Materials* 29(38): 1904182.

23 Burton, M.R., Liu, T., McGettrick, J. et al. (2018). Thin film tin selenide (SnSe) thermoelectric generators exhibiting ultralow thermal conductivity. *Advanced Materials* 30(31): 1801357.

24 Piacenza, E., Presentato, A., Zonaro, E. et al. (2018). Selenium and tellurium nanomaterials. *Physical Sciences Reviews* 3(5): 14.

25 Xu, X., Liang, H., Ming, F. et al. (2017). Prussian blue analogues derived penroseite (Ni, Co) Se_2 nanocages anchored on 3D graphene aerogel for efficient water splitting. *ACS Catalysis* 7(9): 6394–6349.

26 Hou, W., Zheng, B., Qi F. et al. (2018). Self-assembled CNT/Ni_0.85Se-SnO_2 networks as highly efficient and stable electrocatalyst for hydrogen evolution reaction. *Electrochimica Acta* 269: 155–162.

27 Fu, H., Chen, Y., Ren, Z. et al. (2018). Highly dispersed of Ni_0.85Se nanoparticles on nitrogen-doped graphene oxide as efficient and durable electrocatalyst for hydrogen evolution reaction. *Electrochimica Acta* 262: 107–114.

28 Hou, Y., Lohe, M.R., Zhang, J. et al. (2016). Vertically oriented cobalt selenide/NiFe layered-double-hydroxide nanosheets supported on exfoliated graphene foil: An efficient 3D electrode for overall water splitting. *Energy & Environmental Science* 9(2): 478–483.

29 Qian, J., Wang, T., Xia, B. et al. (2019). Zn-doped $MoSe_2$ nanosheets as high-performance electrocatalysts for hydrogen evolution reaction in acid media. *Electrochimica Acta* 296: 701–708.

30 Chenot, C., Robiette, R., and Collin, S. (2019). First evidence of the cysteine and glutathione conjugates of 3-sulfanylpentan-1-ol in hop (*Humulus lupulus* L.). *Journal of Agricultural and Food Chemistry* 67(14): 4002–4010.

31 Samadi, M., Sarikhani, N., Zirak, M. et al. (2018). Group 6 transition metal dichalcogenide nanomaterials: Synthesis, applications and future perspectives. *Nanoscale Horizons* 3(2): 90–204.

32 Gao, D., Xia, B., Wang, Y. et al. (2018). Dual-native vacancy activated basal plane and conductivity of $MoSe_2$ with high-efficiency hydrogen evolution reaction. *Small* 14(14): 1704150.

33 Zhou, H., Wang, Y., He, R. et al. (2016). One-step synthesis of self-supported porous $NiSe_2$/Ni hybrid foam: An efficient 3D electrode for hydrogen evolution reaction. *Nano Energy* 20: 29–36.

34 Lee, C-P., Chen, W-F., Billo, T. et al. (2016). Beaded stream-like $CoSe_2$ nanoneedle array for efficient hydrogen evolution electrocatalysis. *Journal of Materials Chemistry A* 4(12): 4553–4561.

35 Xu, R., Wu, R., Shi, Y. et al. (2016). Ni_3Se_2 nanoforest/Ni foam as a hydrophilic, metallic, and self-supported bifunctional electrocatalyst for both H_2 and O_2 generations. *Nano Energy* 24: 103–110.

36 Zhuang, Z., Peng, Q., Zhuang, J. et al. (2006). Controlled hydrothermal synthesis and structural characterization of a nickel selenide series. *Chemistry – A European Journal* 12(1): 211–217.

37 Wu, X., He, D., Zhang, H. et al. (2016). Ni_0.85Se as an efficient non-noble bifunctional electrocatalyst for full water splitting. *International Journal of Hydrogen Energy* 41(25): 10688–10694.

38 Shi, X., Wang, H., Kannan, P. et al. (2019). Rich-grain-boundary of Ni_3Se_2 nanowire arrays as multifunctional electrode for electrochemical energy storage and conversion applications. *Journal of Materials Chemistry A* 7(7): 3344–3352.

39 Swesi, A.T., Masud, J., Liyanage, W.P.R. et al. (2017). Textured $NiSe_{(2)}$ film: Bifunctional electrocatalyst for full water splitting at remarkably low overpotential with high energy efficiency. *Scientific Reports* 7(1): 2401.

40 Zhou, H., Wang, Y., He, R. et al. (2015). One-step synthesis of self-supported porous $NiSe_2$/Ni hybrid foam: An efficient 3D electrode for hydrogen evolution reaction. *Nano Energy* 20: 26–29.

41 Wang, X., Sun, P., Qin, J. et al. (2016). A three-dimensional porous MoP@C hybrid as a high-capacity, long-cycle life anode material for lithium-ion batteries. *Nanoscale* 8(19): 10330–10338.

42 Cummins, D.R., Martinez, U., Sherehiy, A. et al. (2016). Efficient hydrogen evolution in transition metal dichalcogenides via a simple one-step hydrazine reaction. *Nature Communications* 7(1): 11857.

43 Ravikumar, C.H., Nair, G.V., Muralikrishna, S. et al. (2018). Nanoflower like structures of $MoSe_2$ and MoS_2 as efficient catalysts for hydrogen evolution. *Materials Letters* 220: 133–135.

44 Chia, X. and Pumera, M. (2018). Inverse opal-like porous $MoSe_x$ films for hydrogen evolution catalysis: Overpotential-pore size dependence. *ACS Applied Materials & Interfaces* 10(5): 4937–4945.

45 Nguyen, Q.T., Nguyen, P.D., Nguyen, D.N. et al. (2018). Novel amorphous molybdenum selenide as an efficient catalyst for hydrogen evolution reaction. *ACS Applied Materials & Interfaces* 10(10): 8659–8665.

46 Damien, D., Anil, A., Chatterjee, D. et al. (2017). Direct deposition of $MoSe_2$ nanocrystals onto conducting substrates: Towards ultra-efficient electrocatalysts for hydrogen evolution. *Journal of Materials Chemistry A* 5(26): 13364–13372.

47 Gao, D., Xia, B., Zhu, C. et al. (2018). Activation of the $MoSe_2$ basal plane and Se-edge by B doping for enhanced hydrogen evolution. *Journal of Materials Chemistry A* 6(2): 510–515.

48 Tan, C., Luo, Z., Chaturvedi, A. et al. (2018). Preparation of high-percentage 1T-phase transition metal dichalcogenide nanodots for electrochemical hydrogen evolution. *Advanced Materials* 30(9): 1705509.

49 Mao, B., Bao, T., Yu, J. et al. (2017). One-pot synthesis of $MoSe_2$ hetero-dimensional hybrid self-assembled by nanodots and nanosheets for electrocatalytic hydrogen evolution and photothermal therapy. *Nano Research* 10: 2667–2682.

50 Tang, H., Dou, K., Kaun, C-C. et al. (2014). $MoSe_2$ nanosheets and their graphene hybrids: Synthesis, characterization and hydrogen evolution reaction studies. *Journal of Materials Chemistry A* 2(2): 360–364.

51 Park, S-K., Park, G.D., Ko, D. et al. (2017). Aerosol synthesis of molybdenum diselenide-reduced graphene oxide composite with empty nanovoids and enhanced hydrogen evolution reaction performances. *Chemical Engineering Journal* 315: 355–363.

52 Zhang, L., Sun, L., Huang, Y. et al. (2017). Hydrothermal synthesis of N-doped RGO/ $MoSe_2$ composites and enhanced electro-catalytic hydrogen evolution. *Journal of Materials Science* 52(23): 13561–71351.

53 Lai, F., Yong, D., Ning, X. et al (2017). Bionanofiber assisted decoration of few-layered $MoSe_2$ nanosheets on 3D conductive networks for efficient hydrogen evolution. *Small* 13(7): 1602866.

54 Huang, Y., Lu, H., Gu, H. et al. (2015). A CNT@$MoSe_2$ hybrid catalyst for efficient and stable hydrogen evolution. *Nanoscale* 7(44): 18595–18602.

55 Park, G.D., Kim, J.H., Park, S-K. et al. (2017). $MoSe_2$ embedded CNT-reduced graphene oxide composite microsphere with superior sodium ion storage and electrocatalytic hydrogen evolution performances. *ACS Applied Materials & Interfaces* 9(12): 10673–10683.

56 Deng, S., Zhong, Y., Zeng, Y. et al. (2017). Directional construction of vertical nitrogen-doped 1T-2H $MoSe_2$/graphene shell/core nanoflake arrays for efficient hydrogen evolution reaction. *Advanced Materials* 29(21): 1700748.

57 Kim, J.K., Park, G.D., Kim, J.H. et al. (2017). Rational design and synthesis of extremely efficient macroporous $CoSe_2$–CNT composite microspheres for hydrogen evolution reaction. *Small* 13(27): 1700068.

58 Peng-Zuo, C., Kun, X., Shi, T. et al. (2016). Phase-transformation engineering in cobalt diselenide realizing enhanced catalytic activity for hydrogen evolution in an alkaline medium [J]. *Advanced Materials* 28(34): 7527–7532.

59 Gao, J., Liu, L., Qiu, H-J. et al. (2017). Engineering phase transformation of cobalt selenide in carbon cages and the phases' bifunctional electrocatalytic activity for water splitting. *Nanotechnology* 28(31): 315401.

60 Sun, C., Dong, Q., Yang, J. et al. (2016). Metal–organic framework derived $CoSe_2$ nanoparticles anchored on carbon fibers as bifunctional electrocatalysts for efficient overall water splitting. *Nano Research* 9(8): 2234–2243.

61 Yu, B., Qi, F., Zheng, B. et al. (2018). One-pot synthesis of self-assembled coral-like hierarchical architecture constructed by polymorphic $CoSe_2$ nanocrystals as superior electrocatalyst for hydrogen evolution reaction. *Electrochimica Acta* 277: 161–167.

62 Liu, K., Wang, F., Shifa, T.A. et al. (2017). An efficient ternary $CoP_{2x}Se_{2(1-x)}$ nanowire array for overall water splitting. *Nanoscale* 9: 3995–4001.

63 Dutta, B., Wu, Y., Chen, J. et al. (2018). Partial surface selenization of cobalt sulfide microspheres for enhancing the hydrogen evolution reaction. *ACS Catalysis* 9(1): 456–465.

64 Gao, M-R., Liang, J-X., Zheng, Y-R. et al. (2015). An efficient molybdenum disulfide/ cobalt diselenide hybrid catalyst for electrochemical hydrogen generation. *Nature Communications* 6(1): 1–7.

65 Zhang, L., Wang, T., Sun, L. et al. (2017). Hydrothermal synthesis of 3D hierarchical MoSe /$NiSe_2$ composite nanowires on carbon fiber paper and their enhanced electrocatalytic activity for the hydrogen evolution reaction. *Journal of Materials Chemistry A* 5(37): 19752–19759.

66 Yang, Y., Fei, H., Ruan, G. et al. (2014). Edge-oriented MoS_2 nanoporous films as flexible electrodes for hydrogen evolution reactions and supercapacitor devices. *Advanced Materials* 26(48): 8163–8168.

67 Xu, W., Zhou, X., Qi, Z. et al. (2019). One-pot colloidal synthesis of $MoSe_2$–Pt nanoflowers and their enhanced electrocatalytic hydrogen evolution performance. *Research on Chemical Intermediates* 45(3): 1217–1226.

68 Zhu, Q., Shao, M., Yu, S.H. et al. (2018). One-pot synthesis of Co-doped VSe_2 nanosheets for enhanced hydrogen evolution reaction. *ACS Applied Energy Materials* 2(1): 644–653.

69 Liang, Q., Jin, H., Wang, Z. et al. (2019). Metal-organic frameworks derived reverse-encapsulation Co-NC@Mo_2C complex for efficient overall water splitting. *Nano Energy* 57: 746–752.

70 Pi, C., Huang, C., Yang, Y. et al. (2020). *In situ* formation of N-doped carbon-coated porous MoP nanowires: A highly efficient electrocatalyst for hydrogen evolution reaction in a wide pH range. *Applied Catalysis B: Environmental* 263: 118358.

71 Shi, Y. and Zhang, B. (2016). Recent advances in transition metal phosphide nanomaterials: Synthesis and applications in hydrogen evolution reaction. *Chemical Society Reviews* 45(6): 1529–1541.

72 Liu, P. and Rodriguez, J.A. (2005). Catalysts for hydrogen evolution from the [NiFe] hydrogenase to the Ni_2P (001) surface: The importance of ensemble effect. *Journal of the American Chemical Society* 127(42): 14871–14878.

73 Liang, Y., Liu, Q., Asiri, A.M. et al. (2014). Self-supported FeP nanorod arrays: A cost-effective 3D hydrogen evolution cathode with high catalytic activity. *ACS Catalysis* 4(11): 4065–4069.

74 Tian, J., Liu, Q., Cheng, N. et al. (2014). Self-supported Cu_3P nanowire arrays as an integrated high-performance three-dimensional cathode for generating hydrogen from water. *Angewandte Chemie* 126(36): 9731–9735.

75 Read, C.G., Callejas, J.F., Holder, C.F. et al. (2016). General strategy for the synthesis of transition metal phosphide films for electrocatalytic hydrogen and oxygen evolution. *ACS Applied Materials & Interfaces* 8(20): 12798–12803.

76 Jiang, N., You, B., Sheng, M. et al. (2015). Electrodeposited cobalt-phosphorous-derived films as competent bifunctional catalysts for overall water splitting. *Angewandte Chemie* 127(21): 6349–6352.

77 Tavakkoli, M., Kallio, T., Reynaud, O. et al. (2015). Single-shell carbon-encapsulated iron nanoparticles: Synthesis and high electrocatalytic activity for hydrogen evolution reaction. *Angewandte Chemie International Edition* 54(15): 4535.

78 Tallapally, V., Esteves R.J.A., Nahar, L. et al. (2016). Multivariate synthesis of tin phosphide nanoparticles: Temperature, time, and ligand control of size, shape, and crystal structure. *Chemistry of Materials* 28(15): 5406–5414.

79 Henkes, A.E., Vasquez, Y., and Schaak, R.E. (2007). Converting metals into phosphides: A general strategy for the synthesis of metal phosphide nanocrystals. *Journal of the American Chemical Society* 129(7): 1896–1897.

80 Liu, S., Zhang, H., Xu, L. et al. (2016). Synthesis of hollow spherical tin phosphides (Sn_4P_3) and their high adsorptive and electrochemical performance. *Journal of Crystal Growth* 438: 31–37.

81 Li, J., Li, J., Zhou, X. et al. (2016). Highly efficient and robust nickel phosphides as bifunctional electrocatalysts for overall water-splitting. *ACS Applied Materials & Interfaces* 8(17): 10826–10834.

82 Zhou, W., Yang, H., Shao, S. et al. (2006). Superior high rate capability of tin phosphide used as high capacity anode for aqueous primary batteries. *Electrochemistry Communications* 8(1): 55–59.

83 Wu, J-J. and Fu, Z-W. (2008). Pulsed-laser-deposited Sn_4P_3 electrodes for lithium-ion batteries. *Journal of the Electrochemical Society* 156(1): A22.

84 Jiang, P., Liu, Q., and Sun, X. (2014). NiP_2 nanosheet arrays supported on carbon cloth: An efficient 3D hydrogen evolution cathode in both acidic and alkaline solutions. *Nanoscale* 6(22): 13440–13445.

85 Laursen, A., Patraju, K., Whitaker, M. et al. (2015). Nanocrystalline Ni_5P_4: A hydrogen evolution electrocatalyst of exceptional efficiency in both alkaline and acidic media. *Energy & Environmental Science* 8(3): 1027–1034.

86 Huang, Z., Chen, Z., Chen, Z. et al. (2014). $Ni_{12}P_5$ nanoparticles as an efficient catalyst for hydrogen generation via electrolysis and photoelectrolysis. *ACS Nano* 8(8): 8121–8129.

87 Laursen, A.B., Wexler, R.B., Whitaker, M.J. et al. (2018). Climbing the volcano of electrocatalytic activity while avoiding catalyst corrosion: Ni_3P, a hydrogen evolution electrocatalyst stable in both acid and alkali. *ACS Catalysis* 8(5): 4408–4419.

88 Popczun, E.J., McKone, J.R., Read, C.G. et al. (2013). Nanostructured nickel phosphide as an electrocatalyst for the hydrogen evolution reaction. *Journal of the American Chemical Society* 135(25): 9267–9270.

89 Pan, Y., Liu, Y., Zhao, J. et al. (2015). Monodispersed nickel phosphide nanocrystals with different phases: Synthesis, characterization and electrocatalytic properties for hydrogen evolution. *Journal of Materials Chemistry A* 3(4): 1656–1665.

90 Das, M., Jena, N., Purkait, T. et al. (2019). Single-phase $N_{i5}P_4$–copper foam superhydrophilic and aerophobic core–shell nanostructures for efficient hydrogen evolution reaction. *Journal of Materials Chemistry A* 7(41): 23989–23999.

91 Das, M., Kamboj, N., Purkait, T. et al. (2020). Revealing the structural aspect of ultrastable self-supportive bifunctional electrocatalyst for solar-driven water splitting. *The Journal of Physical Chemistry C* 124(25): 13525–13534.

92 Zhang, Z., Yu, G., Li, H. et al. (2018). Theoretical insights into the effective hydrogen evolution on Cu_3P and its evident improvement by surface-doped Ni atoms. *Physical Chemistry Chemical Physics* 20(15): 10407–17.

93 Zheng, Y., Jiao, Y., Jaroniec, M. et al. (2015). Elektrochemie der wasserstoffentwicklungsreaktion: Optimierung durch korrelation von experiment und theorie. *Angewandte Chemie* 127(1): 52–66.

94 Callejas, J.F., Read, C.G., Popczun, E.J. et al. (2015). Nanostructured Co_2P electrocatalyst for the hydrogen evolution reaction and direct comparison with morphologically equivalent CoP. *Chemistry of Materials* 27(10): 3769–3774.

95 Jiang, P., Liu, Q., Ge, C. et al. (2014). CoP nanostructures with different morphologies: Synthesis, characterization and a study of their electrocatalytic performance toward the hydrogen evolution reaction. *Journal of Materials Chemistry A* 2(35): 14634–14640.

96 Tian, J., Liu, Q., Asiri, A.M. et al. (2014). Self-supported nanoporous cobalt phosphide nanowire arrays: An efficient 3D hydrogen-evolving cathode over the wide range of pH 0–14. *Journal of the American Chemical Society* 136(21): 7587–7590.

97 Zhang, C., Huang, Y., Yu, Y. et al. (2017). Sub-1.1 nm ultrathin porous CoP nanosheets with dominant reactive {200} facets: A high mass activity and efficient electrocatalyst for the hydrogen evolution reaction. *Chemical Science* 8(4): 2769–2775.

98 Kibsgaard, J., Tsai, C., Chan, K. et al. (2015). Designing an improved transition metal phosphide catalyst for hydrogen evolution using experimental and theoretical trends. *Energy & Environmental Science* 8(10): 3022–3029.

99 Ye, C., Wang, M.Q., Chen, G. et al. (2017). One-step CVD synthesis of carbon framework wrapped Co_2P as a flexible electrocatalyst for efficient hydrogen evolution. *Journal of Materials Chemistry A* 5(17): 7791–7795.

100 Wang, J., Yang, W., and Liu, J. (2016). CoP_2 nanoparticles on reduced graphene oxide sheets as a super-efficient bifunctional electrocatalyst for full water splitting. *Journal of Materials Chemistry A* 4(13): 4686–4690.

101 Wu, T., Pi, M., Wang, X. et al. (2017). Three-dimensional metal–organic framework derived porous CoP_3 concave polyhedrons as superior bifunctional electrocatalysts for the evolution of hydrogen and oxygen. *Physical Chemistry Chemical Physics* 19(3): 2104–2110.

102 Popczun, E.J., Read, C.G., Roske, C.W. et al. (2013). Highly active electrocatalysis of the hydrogen evolution reaction by cobalt phosphide nanoparticles. *Angewandte Chemie International Edition* 53(21): 5427–5430.

103 Jiang, D., Xu, Y., Yang, R. et al. (2019). CoP_3/CoMoP heterogeneous nanosheet arrays as robust electrocatalyst for pH-universal hydrogen evolution reaction. *ACS Sustainable Chemistry & Engineering* 7(10): 9309-9317.

104 Yang, H., Zhang, Y., Hu, F. et al. (2015). Urchin-like CoP nanocrystals as hydrogen evolution reaction and oxygen reduction reaction dual-electrocatalyst with superior stability. *Nano Letters* 15(11): 7616–7620.

105 Han, A., Zhang, H., Yuan, R. et al. (2017). Crystalline copper phosphide nanosheets as an efficient janus catalyst for overall water splitting. *ACS Applied Materials & Interfaces* 9(3): 2240–2248.

106 Feng, J-X., Tong, S-Y., Tong, Y-X. et al. (2018). Pt-like hydrogen evolution electrocatalysis on PANI/CoP hybrid nanowires by weakening the shackles of hydrogen ions on the surfaces of catalysts. *Journal of the American Chemical Society* 140(15): 5118–5126.

107 Li, D., Liao, Q., Ren, B. et al. (2017). A 3D-composite structure of FeP nanorods supported by vertically aligned graphene for the high-performance hydrogen evolution reaction. *Journal of Materials Chemistry A* 5(22): 11301–11308.

108 Wang, R., Dong, X-Y., Du, J. et al. (2018). MOF-derived bifunctional Cu_3P nanoparticles coated by a N,P-codoped carbon shell for hydrogen evolution and oxygen reduction. *Advanced Materials* 30(6): 1703711.

109 Clark, P., Wang, X., Deck, P. et al. (2002). Push–pull mechanism of hydrodenitrogenation over silica-supported MoP, WP, and MoS_2 hydroprocessing catalysts. *Journal of Catalysis* 210(1): 116–126.

110 Huang, Z., Chen, Z., Chen, Z. et al. (2014). Cobalt phosphide nanorods as an efficient electrocatalyst for the hydrogen evolution reaction. *Nano Energy* 9: 373–382.

111 Xing, Z., Liu, Q., Asiri, A.M. et al. (2014). Closely interconnected network of molybdenum phosphide nanoparticles: A highly efficient electrocatalyst for generating hydrogen from water. *Advanced Materials* 26(32): 5702–5707.

112 Prins, R. and Bussell, M.E. (2012). Metal phosphides: Preparation, characterization and catalytic reactivity. *Catalysis Letters* 142(12): 1413–1436.

113 Yang, J., Zhang, F., Wang, X. et al. (2016). Porous molybdenum phosphide nano-octahedrons derived from confined phosphorization in UIO-66 for efficient hydrogen evolution. *Angewandte Chemie* 128(41): 13046–13050.

114 Wu, Z., Li, X., Liu, W. et al. (2017). Materials chemistry of iron phosphosulfide nanoparticles: Synthesis, solid state chemistry, surface structure, and electrocatalysis for the hydrogen evolution reaction. *ACS Catalysis* 7(6): 4026–4032.

115 Tang, C., Gan, L., Zhang, R. et al. (2016). Ternary $Fe_xCo_{1-x}P$ nanowire array as a robust hydrogen evolution reaction electrocatalyst with Pt-like activity: Experimental and theoretical insight. *Nano Letters* 16(10): 6617–6621.

116 Liu, W., Hu, E., Jiang, H. et al. (2016). A highly active and stable hydrogen evolution catalyst based on pyrite-structured cobalt phosphosulfide. *Nature Communications* 7(1): 1–9.

117 Cabán-Acevedo, M., Stone, M.L., Schmidt, J.R. et al. (2015). Efficient hydrogen evolution catalysis using ternary pyrite-type cobalt phosphosulphide. *Nature Materials* 14(12): 1245–1251.

118 Kibsgaard, J. and Jaramillo, T.F. (2014). Molybdenum phosphosulfide: An active, acid-stable, Earth-abundant catalyst for the hydrogen evolution reaction. *Angewandte Chemie International Edition* 53(52): 14433–14437.

119 Ye, R., Del Angel-vicente, P., Liu, Y. et al. (2016). High-performance hydrogen evolution from $MoS_{2(1-x)}P_{(x)}$ solid solution. *Advanced Materials* 28(7): 1427–1432.

120 Tan, S.M., Sofer, Z., and Pumera, M. (2015). Sulfur poisoning of emergent and current electrocatalysts: Vulnerability of MoS_2, and direct correlation to Pt hydrogen evolution reaction kinetics. *Nanoscale* 7(19): 8879–8883.

121 Geng, X., Sun, W., Wu, W. et al. (2016). Pure and stable metallic phase molybdenum disulfide nanosheets for hydrogen evolution reaction. *Nature Communications* 7(1): 10672.

122 Hinnemann, B., Moses, P.G., Bonde, J. et al. (2005). Biomimetic hydrogen evolution: MoS_2 nanoparticles as catalyst for hydrogen evolution. *Journal of the American Chemical Society* 127(15): 5308–5309.

123 Merki, D., Fierro, S., Vrubel, H. et al. (2011). Amorphous molybdenum sulfide films as catalysts for electrochemical hydrogen production in water. *Chemical Science* 2(7): 1262–1267.

124 Lin, Y.C., Dumcenco, D.O., Huang, Y.S. et al. (2014). Atomic mechanism of the semiconducting-to-metallic phase transition in single-layered MoS_2. *Nature Nanotechnology* 9(5): 391–396.

125 Zhou, W., Hou, D., Sang, Y. et al. (2014). MoO_2 nanobelts@nitrogen self-doped MoS_2 nanosheets as effective electrocatalysts for hydrogen evolution reaction. *Journal of Materials Chemistry A* 2(29): 11358–11364.

126 Xie, J., Zhang, J., Li, S. et al. (2013). Controllable disorder engineering in oxygen-incorporated MoS_2 ultrathin nanosheets for efficient hydrogen evolution. *Journal of the American Chemical Society* 135(47): 17881–17888.

127 Zhang, X., Meng, F., Mao, S. et al. (2015). Amorphous MoS_xCl_y electrocatalyst supported by vertical graphene for efficient electrochemical and photo-electrochemical hydrogen generation. *Energy & Environmental Science* 8(3): 862–868.

128 Li, F., Li, J., Cao, Z. et al. (2015). MoS_2 quantum dot decorated RGO: A designed electrocatalyst with high active site density for the hydrogen evolution reaction. *Journal of Materials Chemistry A* 3(43): 21772–21778.

129 Eftekhari, A. and Jafarkhani, P. (2013). Curly graphene with specious interlayers displaying superior capacity for hydrogen storage. *The Journal of Physical Chemistry C* 117(48): 25845–25851.

130 Pan, H. (2014). Metal dichalcogenides monolayers: Novel catalysts for electrochemical hydrogen production. *Scientific Reports* 4(1): 5348.

131 Tsai, C., Abild-Pedersen, F., and Nørskov, J.K. (2014). Tuning the MoS_2 edge-site activity for hydrogen evolution via support interactions. *Nano Letters* 14(3): 1381–1387.
132 Chen, X. and Wang, G. (2016). Tuning the hydrogen evolution activity of MS_2 (M = Mo or Nb) monolayers by strain engineering. *Physical Chemistry Chemical Physics* 18(14): 9388–9295.
133 Hu, W-H., Shang, X., Han, G-Q. et al. (2016). MoS_x supported graphene oxides with different degree of oxidation as efficient electrocatalysts for hydrogen evolution. *Carbo* 100: 236–242.
134 Shifa, T.A., Wang, F., Cheng, Z. et al. (2015). A vertical-oriented WS_2 nanosheet sensitized by graphene: An advanced electrocatalyst for hydrogen evolution reaction. *Nanoscale* 7(35): 14760–14765.
135 Putungan, D.B., Lin, S.H., and Kuo, J.L. (2015). A first-principles examination of conducting monolayer 1T'-MX_2 (M = Mo, W; X = S, Se, Te): Promising catalysts for hydrogen evolution reaction and its enhancement by strain. *Phys Chem Chem Phys* 17(33): 21702–21708.
136 Aslan, E., Akin, I., and Patir, I.H. (2016). Highly active cobalt sulfide/carbon nanotube catalyst for hydrogen evolution at soft interfaces. *Chemistry – A European Journal* 22(15): 5342–5349.

11

Devices Development and Deployment Status for Commercial Usage

H_2 Production and CO_2 Utilization

Tulsi Satyavir Dabodiya[1,2], *Twinkle George*[2], *Kapil Dev Singh*[3], *and Arumugam Vadivel Murugan*[2]

[1] *Department of Chemical and Materials Engineering, University of Alberta, Alberta T6G 1H9, Canada*
[2] *Centre for Nanoscience and Technology, Madanjeet School of Green Energy Technologies, Pondicherry University (A Central University), Dr. R. Vankataraman Nagar, Kalapet, Puducherry 605014, India*
[3] *Department of Material Science and Engineering, National Institute of Technology, Hamirpur, Himachal Pradesh 177005, India*

11.1 Introduction

Currently, expanding economies and increasing population raises a soaring demand for global energy supply. Inevitable dependence on fossil fuels for energy generation led to their rapid depletion with a cost to environmental threats and global warming. In the current era of technology and speed, CO_2 remains as a most important global greenhouse gas indirectly causing climate change around the world. The quest for clean and renewable energy technologies is urgently required to sustain the long-term supply without sacrificing ecological deterioration. Molecular hydrogen gas with chemical formula H_2, has emerged as a potential fuel candidate in the current scenario. Hydrogen gas, when subjected to the combustion process, releases water as a combustion by-product with minimum pollution (Figures 11.1a,b). Therefore, production of fuel grade H_2 is the immediate need for the sustainable development of society by green methods.

Carbon will make it possible to reduce the CO_2 content in our atmosphere, as several methods exist to convert CO_2 into useful chemicals or fuels (Figures 11.1c,e). Some of the major challenges to be addressed are: i) source of a carbon-free energy to avoid further emission of CO_2; ii) the immense requirement for catalysts that can help lessen the conversion process of CO_2 activation energy; and iii) the requirement of high temperature or pressure to create the instability of CO_2. Contrarily, the energy needs of the swelling population of the world is majorly met with fossil fuels that directly lead to CO_2 emissions. Figure 11.1d shows the complete CO_2 ecosystem including current generation system and futuristic approach for development techniques with sustainability [1–5]. This chapter consists of two parts: Part I: H_2 production and Part II: CO_2 utilization.

11.1.1 Part I: Devices Development (H_2 Production) and Deployment Status for Commercial Usage

11.1.1.1 Green Methods of Hydrogen Production Process

The abundance of solar energy can resolve the fuel issues of the planet. In this regard, the thermal energy of the solar energy, which is usually wasted, can be made use of efficiently

Green Energy Harvesting: Materials for Hydrogen Generation and Carbon Dioxide Reduction, First Edition. Edited by Pooja Devi.

to produce H_2 by thermal or thermochemical water splitting and biomass gasification [6], thereby making use of both the sun's solar energy and H_2 gas as a fuel.

Solar Biomass Gasification: Hydrogen fuel can be produced by making use of solar energy on biomass such as rice husks and litter waste. This method of water splitting can be considered as sustainable and cost-efficient [8]. The biomass is pyrolyzed using

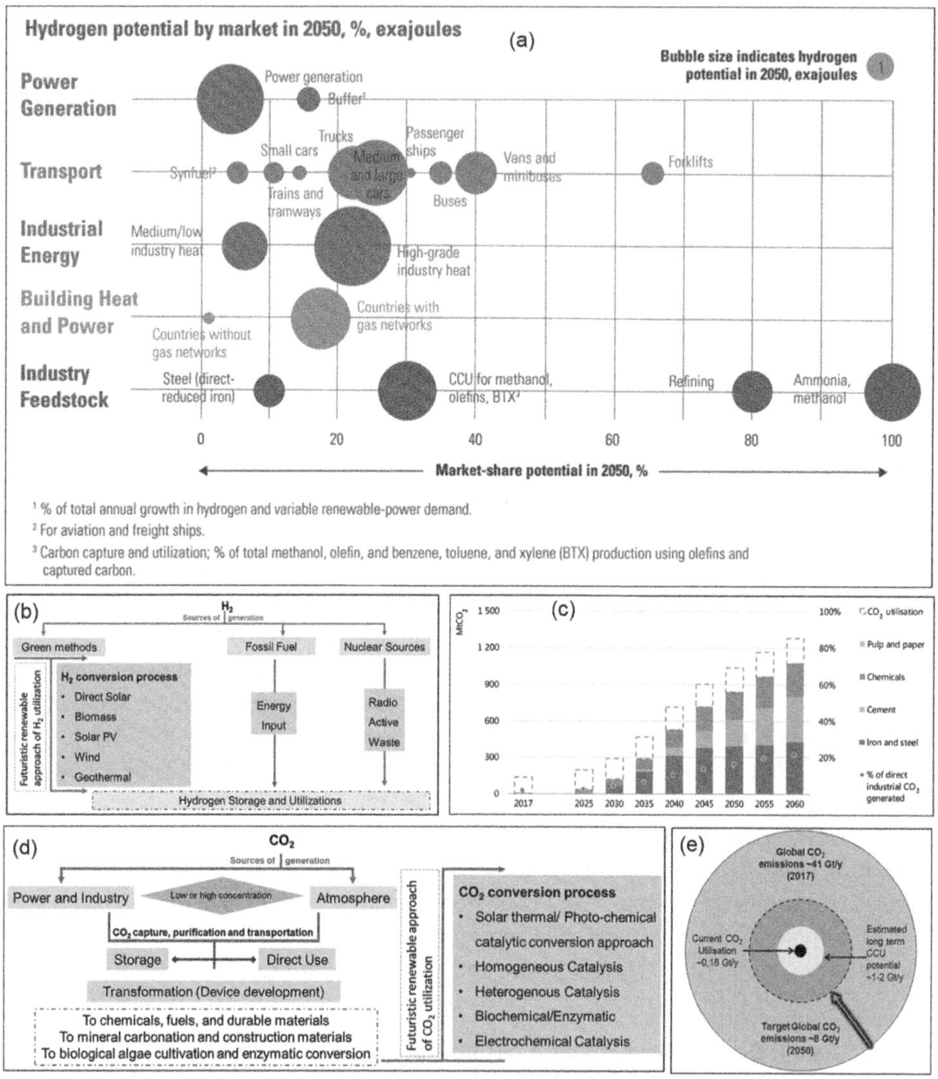

Figure 11.1 (a) Hydrogen market share of world by 2050 in various fields of energy utilizations [6] / United States Department of Energy/ Public Domain; (b) schematic representation of all the sources of hydrogen production for utilization; (c) captured CO_2 for storage by industrial sub-sector and for utilization in the Clean Technology Scenario (CTS) (reproduced by permission [7]); (d) schematic illustration of CO_2 generation and utilization; and (e) status of CO_2 utilization and future goal of CO_2 utilization. Adapted from 135.

solar radiation at a high temperatures, ideally 1600°C. The highest ratio of H_2:CO obtained is 1, making sure that this gas could be used practically for running power plants and engines [9]. In this regard, catalysts could ease up the reaction and may also produce carbon products that could be utilized, such as molten alkali. The production of gas was maintained with an equal ratio for H_2/CO and the thermal efficiency is 40%. The efficiency is given by Equation 11.1.

$$\eta_{reactor} = \frac{\sum_{i=\{CO,CH_4,H_2,\}} \int_{t_i}^{t_2} LHV_i n_{idt}}{\int_{t_i}^{t_2} (Q_{Solar} + Q_{steamgen} + \eta_{feed} \left(LHV_{feed}\right) dt} \tag{11.1}$$

Similarly, using two Frensel lenses to concentrate the heat and irradiation, a solar reactor has been built to make use of the biomass of Jatropha curcas shell. The efficiency of the system was nearly 67.4%. Another such technology that promotes biomass usage for hydrogen evolution is supercritical water gasification (SCWG). A solar plant was developed by Chen et al. to industrialize the SCWG [10]. Similarly, hydrogen is evolved using sewage sludge with the help of an MW solar plant, resulting in a yield of 61.2–67.6 g kg^{-1} (sludge) [11]

Water-Splitting by Solar-Thermal Energy: Further, thermal energy of solar irradiation can be utilized by the thermolysis process for the production of H_2 gas. It has been suggested that low-cost materials capable of H_2 evolution under low temperature cycles are needed for commercialization. Hydrogen gas can be produced by the hydrolysis of the Zn/ZnO thermochemical cycle [12]. The reaction starts at a lower temperature of 250°C and speeds up after 400°C. Here Zn powder is used as a base material for the hydrolysis process in an inert atmosphere. The reaction Equation 11.2 is as follows:

$$Zn + H_2O = ZnO + H_2,\ \Delta H_{500^\circ C} = -110.3\ kJ \tag{11.2}$$

Further, a two-step solar power driven H_2 generation was developed by Abandes et al. [13] using CeO_2/Ce_2O_3, without the use of reducing agents. In a similar fashion, Chambon et al. [14] developed solar reactors that use SnO_2 and ZnO for a similar two-step thermochemical H_2 production. The thermodynamic analysis of the much common SnO_2/SnO hydrogen production was carried out by Bhosale [15]. The reaction equation is given below, wherein H_2 production occurs at the second step (Equations 11.3 and 11.4). The cycle efficiency and SFE is obtained as 41.17 and 49.61%, respectively:

$$SnO_2 \rightarrow SnO + \left(1/2\right) O_2 \tag{11.3}$$

$$SnO + H_2O \rightarrow SnO_2 + H_2 \tag{11.4}$$

However, ZnO-based water-splitting experiments are more attractive owing to the increased SFE and conversion efficiency of 41.1 and 49.5%, which could be further increased with the incorporation of heat recuperations. Also, the water-splitting temperature and the thermal reduction in the case of ZnO is obtained at 1340 and 650 K, respectively [16]. Fe_3O_4/FeO can also be used for water splitting via a two-step cycle [17], wherein magnetite is reduced to wustite (FeO) in an endothermic reaction followed by hydrolysis in an exothermic reaction. The conversion efficiency obtained is 83%.

Mixed oxides of ceria and zirconia are also studied for water splitting, with further developments involving the doping of ceria with gadolinium, lanthanum, yttrium, and praseodymium [18, 19] under inert atmospheric conditions. The reaction equation follows a simple endothermic thermal reduction of ceria with oxygen evolution followed by deoxidization at low temperature to evolve hydrogen gas (as shown in Equations 11.5 and 11.6).

$$M_xCe_{1-x}O_2 + \text{solar energy} \rightarrow M_xCe_{1-x}O_{2-\delta} + \delta/2O_2 \tag{11.5}$$

$$M_xCe_{1-x}O_{2-\delta} + \delta H_2O/CO_2 \rightarrow M_xCe_{1-x}O_2 + \delta H_2/CO \tag{11.6}$$

Ceria-based materials have proven to be highly advantageous when attributed by the second step at lower temperatures while evolving hydrogen gas. However, ZnO-based water-splitting experiments are more attractive owing to the increased SFE, and its conversion has a lower reduction temperature compared to FeO and the energy efficiency is approximately 41.3%. However, KOH or NaOH were used for the production of H_2 that are corrosive in nature, along with the use of costly materials [20].

A novel system involving geothermal and solar energy was investigated by Dincer and Temiz [21], which uses a copper-chlorine thermochemical hydrogen evolution plant. However, it has the disadvantages of being critical and possessing critical requirements such as high-grade heat source and high temperature. The four steps involved in the process are shown in Equations 11.7 to 11.10:

Step 1:

$$2CuCl_2 + H_2O \rightarrow 2Cu_2OCl_2 + 2HCl,\ T = 510\,°C\,(\text{Hydrolysis}) \tag{11.7}$$

Step 2:

$$Cu_2OCl_2 \rightarrow 2CuCl + (1/2)\,O_2,\ T = 390\,°C\,(\text{Thermolysis}) \tag{11.8}$$

Step 3:

$$2CuCl + 2HCl \rightarrow 2CuCl_2 + H_2,\ T = 85°C\,(\text{electrolysis}) \tag{11.9}$$

Step 4:

$$CuCl_2\,(aq) \rightarrow CuCl_2\,(s),\ T = 85°C\,(\text{drying}) \tag{11.10}$$

Further, researchers have analyzed H_2 production, using an MW high temperature solar concentrator [22, 23].

11.1.1.2 Hydrogen Evolution Processes Utilizing Photo Energy

With the help of light, processes such as photocatalytic, PVCs, photobiological, and photoelectrochemical methods can be used to produce H_2 gas.

a) Photocatalytic (PC) and Electrochemical Methods (EC).

Hydrogen evolution reaction (HER), though eco-friendly, requires efficient and economically feasible catalysts that are challenging. In this regard, by using the dip coating and electrochemical reduction-two step methods, commercial carbon is decorated with a palladium catalyst alleviated with MoO_x oxides as MoO_x-Pd [24], which constrains the overall growth of palladium nanoparticles. Mohanty et al. prepared MoS_2 QDs [23], which are low-cost, highly stable, pose lower overpotentials, and are highly active electrochemical catalysts (Figure 11.2). Also, Proton Exchange Membrane (PEMs) could be utilized for their use in acidic electrolytes. When hydrogen evolution is taken into consideration, the most stable catalyst that works, even at low temperatures, is Pt/C [25–27].

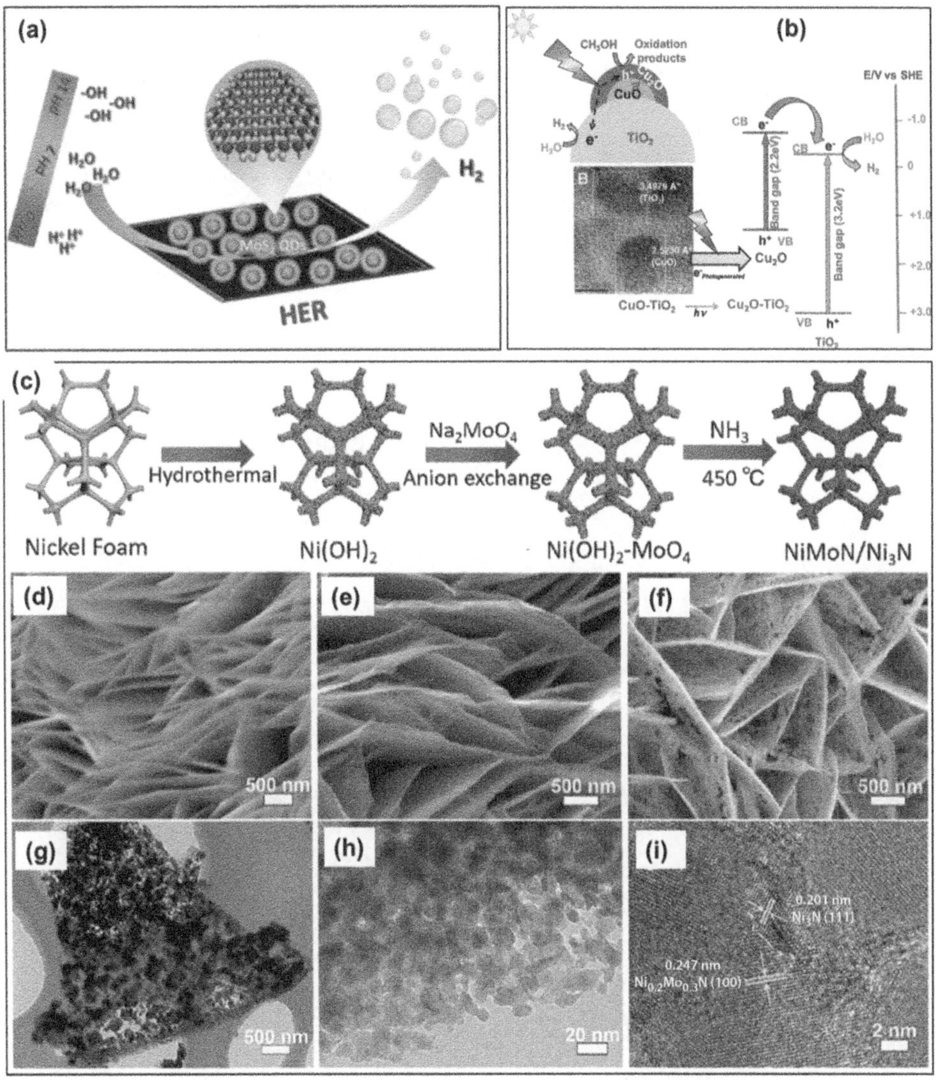

Figure 11.2 (a) MoS_2 QDs, which are low cost, highly stable, pose lower overpotentials, and are highly active electrochemical catalysts (reproduced by permission of [23], American Chemical Society); (b) photocatalytic H_2 production enhanced over Cu/Ti oxide [26] Pai, M. et.al, 2016 / Elsevier; (c) schematic illustration depicting the synthesis procedure followed by the SEM images of samples; (g,h) TEM; and (i) high-resolution TEM images of the NiMoN/Ni_3N heterostructures [32] Hua, W. et. al, 2021 / Elsevier.

In this regard, Pai et al. [27] synthesized $Cu_xTi_{1-x}O_{2-\delta}$ composites with different stoichiometric ratios of x and copper oxide–titanium oxide nanocomposites using the $xCuO-yTiO_2$ via sol–gel method, wherein the photocatalytic H_2 production with the presence of Cu ions (sunlight and UV) has been increased. Here, the substitution of Cu has stabilized the lattice by elongating the Ti–O bonds, also inducing more oxygen vacancies as the substitution causes a decline around Cu ions, which improves the PC activity and H_2 evolution. These are explored with photoreactors to analyze the effect of concentration, formation, role of sacrificial agents, and illumination area on the H_2 production. The reactor promises a hydrogen yield of 1.167 L h m^{-2} under optimal conditions, with 7.5% AQE and SFE of 3.9%. Non-noble metals could be used for commercial applications in the future, such as FeNi-based catalysts widely reported under acidic medium activity [30]

Electrochemical splitting of water consists of HER; i.e., the hydrogen production reaction and the oxygen evolution reaction otherwise commonly known as the OER [31, 32]. Real-time application of electrochemical process requires catalysts that are non-corrosive and stable [28, 33]. In this regard, a durable alkali $NiMoN/Ni_3H$ heterostructure is obtained through interface engineering [28]. The porous nanosheets provide enough active sites,

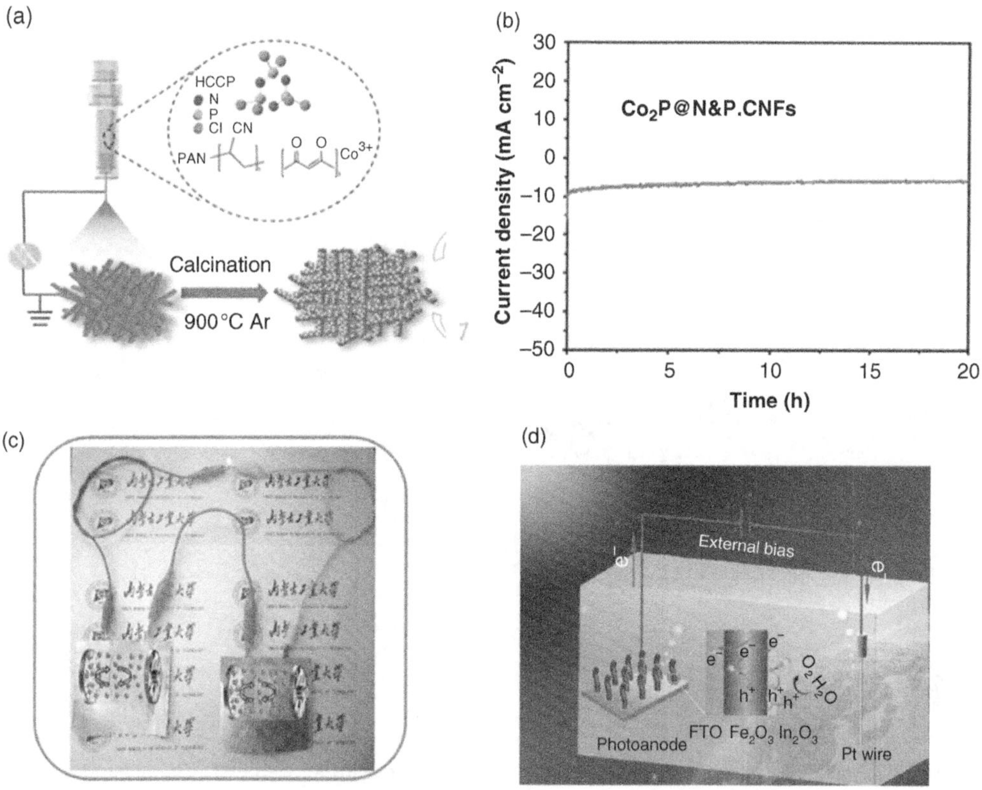

Figure 11.3 (a–c) Co_2P encapsulated N,P dual-doped carbon nanofibers (Co_2P@N,P-CNFs) are developed via electrospinning technology followed by the carbonization process [136] Sun, X. et.al., 2021 / Elsevier; (d) decoration of In_2O_3 nanolayers on hematite nanoarray photoanode [34] Yi, S. S. et.al, 2021 / Elsevier; (e) schematic illustration of W:BiVO$_4$ and an Si solar cell; and (f) energy band diagram of the fabricated photoelectrode device (reproduced by permission of [40], Springer).

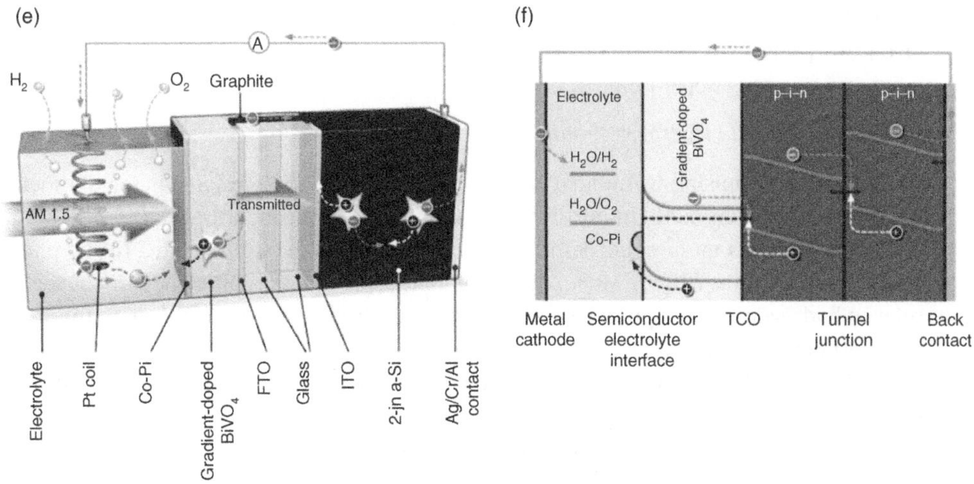

Figure 11.3 (Continued)

thereby increasing the reactivity rate and requires only an overpotential of 28 mV. Further, WO_3 can be utilized along with (Ag, Cu)$InGaSe_2$ as a cathodic electrocatalyst of solar cell modules for hydrogen production Figure 11.3 [34].

Yi et al. [36]. established that hematite-based photoelectrodes for solar water splitting, with worm-like Fe_2O_3 films, with its surface carrying In_2O_3 nanolayers (NLs), remarkably improves the photoelectrochemical water oxidation activity. Co_3S_4 nanoparticle-doping provides superior photoelectrochemical performance and high stability enhanced as improved photogenerated carrier separation and the reduction potential on g-C_3N_4/α-Fe_2O_3 S-type heterojunction photocatalyst. On a similar note, Sun et al. [36] developed a graphite carbon nitride (g-C_3N_4)-based catalyst using a magnetically actuated artificial cilia array that could improve the g-C_3N_4 activity. This improved the photocatalytic activity under visible light and the PC H_2 evolution efficiency increased by 75% magnetic field strength enhanced in the range 0 to 12.50 Hz. For electrochemical (EC) and photoelectrochemical (PEC) hydrogen production, ternary pyrite-type cobalt phosphosulfide (CoPS) [39] nanostructure materials have exceptional operational stability at very low overpotentials such as 48 mV. When these nanoparticles are integrated into Si micropyramids, the photocurrent is enhanced tremendously.

For large-scale production of photocatalytic H_2, a recent study on the use of a tubular kind of reactor with a compound parabolic collector has been conducted. Ren et al. [41] studied the crystal fluid dynamics of catalyst–liquid phase flow reactors. However, achieving this is challenging, as interfacial anomalies create defects serving as recombination centers [40, 42]. Accordingly, a $BiVO_4$ photoanode was fabricated using spray coating and tungsten doping, wherein the efficiency of separation of charge is $\eta = 80\ \%$ [43], by equilibrating the Fermi energy levels of tungsten-doped $BiVO_4$ and pristine $BiVO_4$, creating a depletion layer that enhances the band bending and charge separation.

The CSE (Charge-separation efficiency) is quantified in Equation 11.11, as reported by Dotan et al. [44]:

$$J_{ph} = J_{abs} \times \eta_{sep} \times \eta_{cat} \tag{11.11}$$

where the total photocurrent density is denoted by J_{ph}, the rate of photon absorption is given by J_{abs}, η_{cat} gives the efficiency of the catalytic H_2O oxidation, and η_{sep} depicts the efficiency of carrier-separation.

b) Solar PV Cell and Hydrogen Production

PV+PEC is a promising technology for H_2 production, as the semiconductor photoelectrode can convert the sunlight into chemical fuels capable of generating H_2 [45, 46]. Zhang et al. [47] studied a mesoscopic perovskite tandem solar cell, coupled with a $BiVO_4$ photoelectrohemical cell, with an MoS_x cathode which achieved 3.1% efficiency in converting solar to hydrogen and the photocurrent density was found at 3.0 mA cm^{-2} when there was no external bias (Figure 11.4).

In this regard, $BiVO_4$, the position of VB edge occurred at 2.4 Vs RHE and the bandgap of the material was 2.4 eV (E1/2 = 1.23 V vs. RHE) [44, 49, 50].

IPCE, the incident photon-to-current efficiency is given by Equation 11.12.

$$\mathrm{IPCE} = \frac{1240I}{\lambda J_{light}} \tag{11.12}$$

where I expresses the photocurrent density (mA cm^{-2}). The incident wavelength is given by λ (nm), and J_{light} denotes the irradiance measured (mW cm^{-2}) at a particular λ [51].
A triple junction PV cell consisting of InGaP/GaAs/Ge [52], where the front side is covered by PMMA and the backside is covered by Ni metal foil, acts as a catalyst for the oxygen evolution reaction along with providing protection from corrosion and improvizing the stability. The device was capable of producing an SHE of 14% at 1–15 suns and 12% at 39 suns. The maximum SHE is found using Equation 11.13, while considering 100% Faradaic efficiency (ηF).

$$\mathrm{STH} = \frac{\text{Power from hydrogen produced}}{\text{power of light incident on to PE Ccell}} \tag{11.13}$$

The amount of hydrogen generated is calculated using (Equation 11.14).

$$V_{\text{annual}}(\text{hydrogen}) = (1\text{m}^2)(2.5\frac{L}{m^2}) \times \int_{\text{APE1}}^{\text{APE2}} \text{P(APE)STHd(APE)} \tag{11.14}$$

where the mean photon energy is given as *APE* and the irradiance of the illumination is depicted by *P*. The angle of illumination hugely impacts the production of hydrogen in a PV-EC system [53]. Green et al. developed a hybrid perovskite solar cells [54]. Though triple mesoscopic structures have shown low open circuit voltage, they have higher stability [55]. The effect of energies such as the magnetic field, pulsating electric field, and ultrasonic fields on water splitting has been studied by Burton et al. [56], who proved that they could enhance efficiency of the hydrolysis process, owing to the decrease of bond strength between H_2O. Excess electricity produced by PVs, could be generated for H_2 production [55, 57].

c) Upscaling Challenges and Large-Scale Applications

Upscaling PCE of PSC has been challenging. Several companies have started producing large-scale application of PSC, such as Microquanta Semiconductor (China), with a 17.8 cm^{-2} active area capable of PCE of 17.4%. Solliance (The Netherlands) use laser

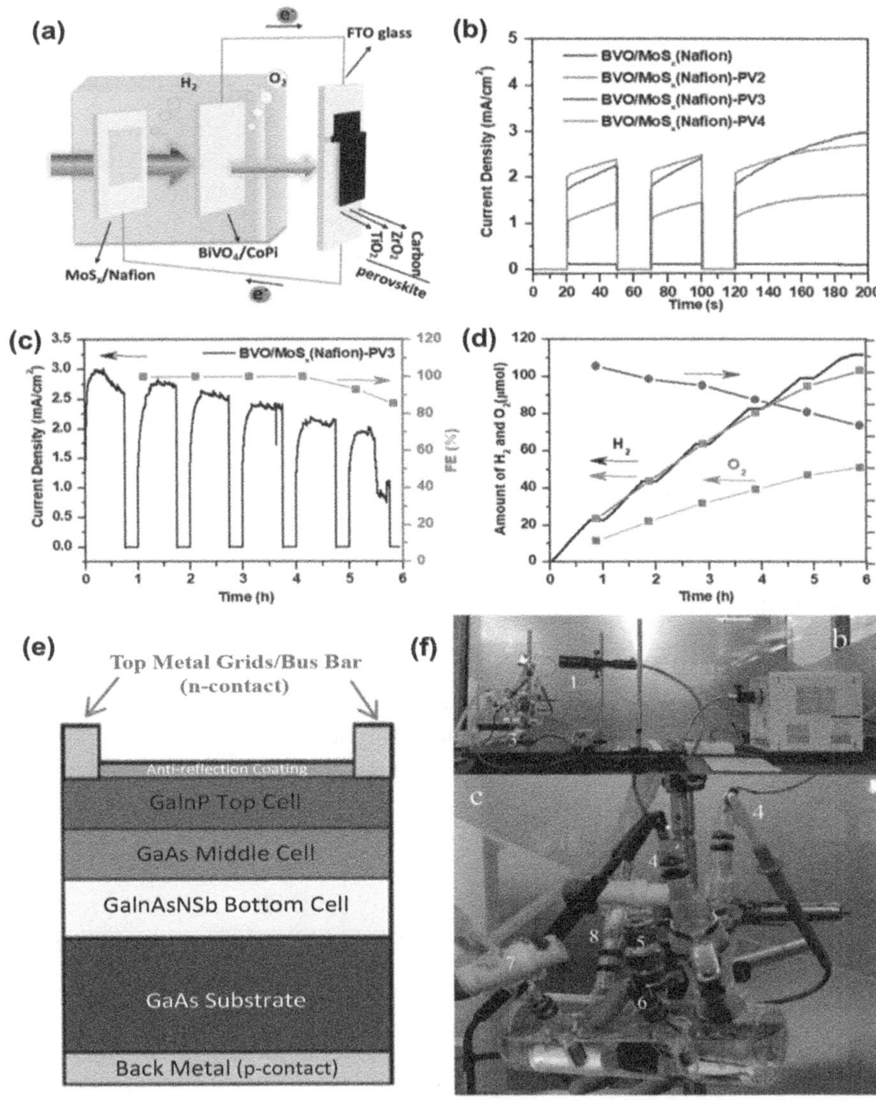

Figure 11.4 (a) Schematic architecture of tandem PV/PEC device; (b) photocurrents of the different systems; (c) current of the BVO/MoS_x (Nafion)–PV_3 tandem device with light on-off cycle and the Faraday efficiency (FE); (d) theoretical amount of hydrogen production (black line), measured amounts of hydrogen (red line), and oxygen (green line), and corresponding solar-to-hydrogen efficiency in each hour (blue line). Reproduced from [42] / American chemical society; (e) schematic view of the triple junction (3-J) PV-cell; and (f) complete PV-electrolysis setup [137] Nadeem, M. A., Idriss, H. et al., 2020 / Elsevier.

ablation and a 169 cm^{-2} area capable of producing 10% [58]. Similarly, Greatcell Solar, Australia have fabricated a 10 cm × 10 cm solar module with a conversion efficiency of 12%. The efficiency of PV cells depends on the design and architecture, thereby requiring more research in the field.

d) Z-Scheme for Hydrogen Evolution.

For large-scale production of H_2, solar energy conversion is an economic choice, and could be done by a Z-scheme technology [59, 60] involving photoexcitation of an O^2 evolution photocatalyst (OEP) and H_2 evolution photocatalyst (HEP). For this, semiconductors capable of water oxidation or reduction are used [61, 62]. Wang et al. developed lanthanum- and Rh-codoped $SrTiO_3$ as an HEP and Mo-doped $BiVO_4$ as an OEP embedded into a gold layer as photocatalysts [63].

The apparent quantum yield (AQY) of the Z-scheme H_2O splitting is represented by Equation 11.5.

$$\mathrm{AQY}(\%) = \left[4 \times n(\mathrm{H_2})\right] / n(\mathrm{photon}) \times 100] \qquad (11.15)$$

In a similar note, a novel all solid state heterojunction photocatalyst $Mn_{0.2}Cd_{0.8}S/CeO_2$, which is a 0D/2D junction, was synthesized by Wang et al. The rate of H_2-generation was nearly 8.7 mmol $h^{-1}g^{-1}$, which is 119 times higher when compared to that of CeO_2 when taken with an aqueous solution of Na_2S/Na_2SO_3 as the electrolyte [63]. Maden and Domen [64] reviewed the oxysulfides, nitrides, and other non-oxides for PC water splitting under visible light. The Z-scheme catalyst $g\text{-}C_3N_4/Bi_4NbO_8Cl$ [65] shows promising PC for H_2 generation using visible-light irradiation ($\lambda > 420$ nm) with a capability of producing H_2 6.9 times greater than pristine g- C_3N_4 and 67.2 times more than unaltered Bi_4NbO_8Cl.

11.1.1.3 Hybrid Processes of Hydrogen Production

Solar thermal energy could be utilized efficiently by using one part for water heating and steam production and one part for electricity production; in this way, 82% SHE can be achieved [66].

Recently, H_2 gas generation from aq solutions of ethanol, methanol, bioethanol, and glycerol in liquid phase is possible by using an optical fiber photoreactor illuminated by UV LEDs and 1% wt% Au/TiO_2 [67]. Photoelectrocatalytic cells have been used to generate H_2 gas by Thongthep et al. [68], by making use of Cu_2O film. Organic dye degradation and H_2 production has been achieved together with a PEC cell using Cu_2O photocathode and a $WO_3/BiVO_4$ photoanode. An Au/TiO_2 photocatalyst for removing organic pollutants and H_2 production has been conducted under direct sunlight [69]. Bismuth-doped titanium dioxide ($BiO_{x-1}TiO_2$) anode and stainless-steel cathode was studied for a PV+EC system [70].

11.1.1.4 Solar Biohydrogen

Biohydrogen is produced using feed from photofermentation with HAU-M1 bacteria [71]. Here, photofermentation involves the H_2 production along with CO_2 with the help of anaerobic bacteria (*Rhodobium, Rhodobacter, Rhodopseudomonas*) for photosynthesis [72]. Though low yield is a challenge and drawback of the photo-assisted biological hydrogen production, it is advantageous when compared to steam autothermal re-forming partial oxidation [73]. Biological hydrogen has been gaining attention, though the treatment of organic waste is essential for improvizing the fermentative process. Xu et al. [74] developed photosynthetic microbial microreactors that are droplet-based and dispersible with dual functionality, which can be exploited for photobiological hydrogen generation in the presence of oxygen. Photobiological hydrogen production explicitly depends on the nature of the organisms and the substrate along with their concentrations.

11.1.1.5 Economic Analysis and Pilot-Scale Study

There is currently a need to consider scaling up photo-assisted H_2 production. Various research has been carried out in this arena, while Das [75] did scale-up studies on fermentative H_2 by using cheap raw materials and methods that are competitive. A need for genetically-engineered robust micro-organisms could help with this issue. In this regard, *Rhodobacter capsulatus* (Hup−) mutant was used for photofermentation in an outdoor pilot style solar tubular photobioreactor. Light exposure affected the H_2 production [76], while the yield determination is done according to the H_2 generated daily compared to acetic acid fed and acetic acid consumed; 95–99% of the gas produced H_2, and the rest is CO_2 by volume. However, the cost of operating outdoor photo-bioreactors is excessive as it requires temperature control and recirculation. A pH of 4.5 and a temperature of 35°C is maintained for the operating conditions of the dark fermentation unit, while in the case of the photofermentation unit, it is 30°C and the pH is 7.0. It is observed that the overall rate of biogas production reached $87.8 \pm 3.8\ m^{-3}day^{-1}$ with nearly 68% of H_2 content [77].

The dark fermentation reaction shown in Equation 11.16 can be given as an example.

$$C_6H_{12}O_6 + 2H_2O \rightarrow 2CH_3COOH + 2CO_2 + 4H_2 \tag{11.16}$$

The products from the above dark fermentation reaction are then converted into hydrogen using the reaction shown in Equation 11.17.

$$2CH_3COOH + 4H_2O + \text{light} \rightarrow 4CO_2 + 8H_2 \tag{11.17}$$

The photo-fermentation strains, by making use of the effluents of the dark fermentation as substrates, could produce H_2 [78]. The reaction in Equation 11.18 occurs in hydrogen-generation reactors.

$$C_6H_{12}O_6 + 6H_2O \rightarrow 6CO_2 + 12H_2 \tag{11.18}$$

Using separate strains for each reactor fed glucose could be converted to hydrogen. If both Equation 11.20 and 11.21 takes place in a single reactor, it is dark fermentation [79]. Research teams of Vatsala and Zhang implemented pilot-scale dark and photo-fermentation for hydrogen generation that investigated major areas of influence such as the quantity and the retention time [79, 80].

11.1.1.6 Economic Analysis, Challenges, and Future Perspectives

It is vital to note that H_2 production depends on cost of feedstock, gas separation, and pretreatment, which could become negligible if the PCE (photochemical-efficiency) is improved [81], which further increases the need for developing biological hydrogen production technology [82]. The economic viability of various production methods has been studied by Sathyaprakasan and Kannan [83], including factors such as operating costs, capital costs, design type, water power, labor, culture, and supplies. On the basis of capital cost, construction cost, setup costs, water cost, electricity, and labor, Sharma and Kaushik [84] compared both dark and photo-fermentation processes. However, the major drawback is that only 33% is converted to hydrogen of inorganic substances, with side products being organic acids. A maximum of 28% STH efficiency is obtainable if the cost of hydrogen (LCOH) is taken to be \$5.9 kg^{-1}, which approximates to c-Si solar farms (\$4.9 kg^{-1}), other factors including operating hours and concentrators [85].

The challenges to realize an hydrogen-based economy can be categorized as:

i) *Production Challenges and Storage*: the major challenge associated with H_2 production is its transportation and storage. Large-scale application of H_2 production could be coupled with carbon capture and utilization technology. Unlike other fuels, H_2 can easily be stored in any state of matter such as compressed gas, liquified, or a solid metal/hybrid complex. Among this, easier storage is capable for the liquid organic hydrogen carrier (LOHC), as it poses advantages over conventional storages. The system conprises of hydrogen-rich and hydrogen-deficient organic compounds that could store hydrogen, thereby requiring the ultimate need for selecting thermodynamic feasible hydrogen and dehydrogen processes which are exothermic and endothermic in nature, respectively [85, 86]. When taking electric vehicles into consideration, nearly 700 bar hydrogen pressure is needed for refueling, which requires multiple compression stages.

ii) *Challenges During Utilization*: the cost of purification can be reduced if methods of using less pure hydrogen are researched. In this regard, less pure H_2 can be used for combustion, which can increase the stability of an ethanol engine during combustion under lean-burn conditions [87]. PVs utilizing electrolysis to produce H_2 and LOHC could evolve H_2; in this regard, Muller et al. studied several methods of utilizing LOHCs and observed that fuel cells can be coupled along with them [88]. However, for practical application of the hydrogen fuel cell, highly pure H_2 is required, thereby requiring to purify the gas several times before it can be used.

iii) *Challenges Related to Safety*: it is known that H_2 leakage can cause a fire hazard, as H_2 is extremely combustible (up to 74% in air and up to 94% in oxygen) and its properties of being odorless and colorless makes it harder to detect by the human senses. This further leads to the need of suitable hydrogen detection sensors in every system producing, storing, or transporting hydrogen gas, as well proper ventilation [87].

iv) *Challenges Related to Commercialization*: When commercialization and practical application is taken into consideration, the catalyst precursors should be abundant and cheap. Additionally, developing and upgrading existing facilities for H_2 production should be entertained. Lastly, the initial investment in changing the fuel cell combustion engine is large, making a fuel cell car costlier than the ion chromatograph (IC) engine car, hence introduction of subsidies for fuel cell and awareness programs for enlightening people should be conducted to break the social barriers that play another important role in developing an H_2-dependent life style.

11.1.1.7 Status of Commercial of H_2 Generation

One of the global initiatives taken by the European Commission and 19 member countries in 2003, was the International Partnership for H_2 and Fuel Cells in the Economy (IPHE), in order to fulfill future needs of fuels by facilitating and improvizing the usage of clean energy technologies such as H_2. They have invested approximately $1 billion annually, with the primary focus of the partnership being the codes, its regulations, the safety and standards, outreach, and proper awareness to establish technologies based on H_2. In the United States, the current production rate of H_2 is nearly 14 M tonnes a year [88–90]. Companies such as Hydrogenics have invested in power-to-gas. In

Toronto (Canada) there exists a 2-MW power-to-gas facility and several bus refueling stations. At the EU, the R&D projects have been funded by Fuel Cells and H_2 Joint Undertaking (FCH JU) since 2008 to technologize H_2 for commercialization. The FCH JU has invested ~€893 million for the development of H_2 production, fuel cells, storage, and its delivery in the arena of transportation and other storage applications. In the current era, Asia can be considered as a place with maximum deployment of resources in the field of H_2 economy. During 2019, in Japan, the Ministry of Economy, Trade, and Industry released an increased budget for developing hydrogen-based technologies. The ENE-FARM (energy and farm), a government-aided small-scale CHP system with <5 kW for residential application is considered as the most successful program, considering that 270,000 units were installed by early 2019 itself [89–93]. Toshiba, deploying H_2, includes a international and national PEM-based electrolyzer, in addition to holding a 220-kW trial in Hokkaido, that would lead to an industry-scale hydrogen export industry by utilizing a low-carbon hydrogen supply chain. Further, Kawasaki Heavy Industries and the Chiyoda Corporation, utilize carbon and stores in addition to obtaining the hydrogen from coal and natural gas. Since 2017, Korea has more than 300 MW of fuel cells in operating condition and holds the largest fuel cell park with more than 50 Mwe, in addition to the rising number of fuel cells-based EVs. Interestingly, Hyundai has developed several numbers of EVs (trucks) for Switzerland and electric-based buses in Korea, while they have promised to increase their production by multiple-fold by 2030. In the case of Africa, DST of South Africa launched Hydrogen South Africa (HySA). Initiatives in Australia include its National H_2 Strategy that plans to: i) export H_2 to Asian countries such as Japan and South Korea; ii) reduced dependence on trade-in fuels; and iii) energy systems that control the electricity grid. Another initiative by the Australian Government was the partnership with Germany, the European Commission, the IPHE, H_2 Council, and others in 2018. Similarly, the United Kingdom has also made remarkable efforts to utilize clean energy such as partnering Siemens with Rutherford Appleton for hydrogen production. This system could be used for energy storage, renewable energy export, and reduction of emissions [88–90].

Numerous refueling stations operate in the USA, Australia, Japan, and the EU to produce hydrogen for vehicle refueling for lightweight and heavyweight public transport. ITM Power, 2013, utilized a PEM electrolyzer, to inject hydrogen into the German gas-distribution network [91, 92, 93]. The energy demand of the world in the near future would make it dependent on global production of hydrogen. Although several potential markets may develop, it is unclear which renewable energy will dominate.

11.1.2 Part II: Devices Development and Deployment Status for Commercial Usage: CO_2 Utilization

11.1.2.1 CO_2 Utilization Procedures

a) *Reforming System of Transformation Into Value-Added Procedure*: This process is generally known as methane reforming due to its main component being methane (CH_4). The three main types of methane reforming processes involved in CO_2 utilization, for production of syngas ($CO+H_2$) are:
   (1) *Dry reforming (DRM)*: DRM is an extremely endothermic reaction process shown in Equation 11.19 [94].

$$CH_4 + CO_2 \rightarrow 2CO + 2H_2,\ \Delta H_{STP} = 248\ kJ.mol^{-1} \quad (11.19)$$

(2) *Bi-reforming (BRM)*: Methane dissociates into major greenhouse gases such as water vapor, CO_2, CH, and ozone (O_3), as in Equation 11.20 [94].

$$3CH_4 + 2H_2O + CO_2 \rightarrow 4CO + 8H_2,\ \Delta H_{STP} = 220\ kJ.mol^{-1} \quad (11.20)$$

(3) *Oxy-CO_2 reforming (ORM)*: The most promising among the three of syngas production is oxy-methane reforming (partial oxidation of methane, as in Equation 11.21 [95].

$$3CH_4 + CO_2 + O_2 \rightarrow 4CO + 6H_2,\ \Delta H_{STP} = 175.1\ kJ.mol^{-1} \quad (11.21)$$

b) *CO_2 Hydrogenation System of Transformation Into Value-Added Procedure*
The addition of H_2 to CO_2 is defined as CO_2 hydrogenation, owing to higher Gibbs free energy (ΔG°) of H_2 than CO_2 and which is more favorable. The excessive energy generation during heat or light as a thermal or photothermal reaction facilitates the CO_2 hydrogenation, as describe in the reaction Equation 11.22.

$$CO_2 + 2H_2 \rightarrow C + 2H_2O,\ \text{-90}\ kJ.mol^{-1} \quad (11.22)$$

c) *CO_2 Process of Transformation*
The carboxylation process involves assigning functional CO_2 molecules to reactants for the production of ureas (RRNCONRR), organic carbonates (ROC(O)OR), carbamates ($R_1R_2NCOOR_3$), and other valuable chemicals [93].

d) *Mineralization Process of Transformation*
Carbonation or CO_2 mineralization is an extensively used procedure in which CO_2 reaction alkaline earth oxides such as CaO and MgO produce metal carbonates. There are two types of mentalization procedures, known as *In-situ* mineralization and *Ex-situ* mineralization [96, 97]

e) *Electrochemical Reduction*
In electrochemical CO_2 reduction, by using electrical energy, the reduction of CO_2 is made possible. The cathodic reaction of the general form of electrochemical reduction is given in Equation 11.23 [98].

$$xCO_2 + nH^+ + ne^- \rightarrow product + yH_2O,\ CO_2 reduction(CO_2R) \quad (11.23)$$

f) *Photochemical Reduction*
PEC CO_2 conversion in water is an artificial photosynthesis technique that follows the reactions through charge transfer for C=O bond breakage, as shown in Equations 11.24–11.27 [99].

$$2H_2O(l) \rightarrow O_2(g) + 4H^+(aq) + 4e^- \quad (11.24)$$

$$CO_2(aq) + e^- \rightarrow CO_2^{-}(aq) \quad (11.25)$$

$$H^+ + e^- \rightarrow H(\text{hydrogen radical formation}) \quad (11.26)$$

$$CO_2 + e^- \rightarrow CO_2^-(aq)\,(\text{anion radical formation}) \quad (11.27)$$

where, $CO_2^{\bullet-}$ is the intermediate state [99].

g) *Plasma Catalysis*

Bypassing slow reactions in thermal catalysts via a reactive gas phase is known as plasma catalysis. Plasma-catalytic CO_2 hydrogenation into CH_3OH has recently attracted great interest.

11.1.2.2 Process of CO_2 Conversion

Some of the CO_2 reduction products are given in Equations 11.28–11.32 [100].

$$CO_2 + 2H^+ + 2e^- \rightarrow CO + H_2O, \quad (11.28)$$

$$CO_2 + 2H^+ + 2e^- \rightarrow HCO_2H, \quad (11.29)$$

$$CO_2 + 6H^+ + 6e^- \rightarrow CH_3OH + H_2O, \quad (11.30)$$

$$CO_2 + 8H^+ + 8e^- \rightarrow CH_4 + 2H_2O, \quad (11.31)$$

$$CO_2 + e^- \rightarrow CO^{\bullet-}_2, \quad (11.32)$$

a) *Molecular Catalyst*

The transition metals complexes host multiple binding modes of CO_2 with the help of electron transfer-rich agents (e^-) for further CO_2 transformation. Different metal catalysts and ligand structures are highly tunable to achieve fast kinetics and long-term stability [100–102].

b) *Photochemical (PC) and Photoelectrochemical (PEC) System*

When one or both electrochemical electrodes are semiconductor photoelectrodes, they can be considered under the PC and PEC systems. Herein, charge carriers generated by photons are driven in one or both half reactions, with one or more catalysts covering the solar cell and a protecting layer, wholly forming the photoelectrode. PC and PEC systems are classify into a homogeneous CO_2 photoreduction system and a heterogeneous CO_2 photoreduction system.

11.1.2.3 CO_2 Conversion Approach

The significant approach of CO_2 reduction using solar energy are classified below:

- Bio-photosynthethic
- Photoelectrochemical (PEC)
- Combined Photo-Voltaic+Electrochemical (PV+EC)
- Photocatalytic/photosynthetic (PS/PC)
- Photothermal

Among these approaches, CO_2 reduction based on bio-photosynthesis involves bio mimicking of photosynthesis with enzyme-reduced redox reactions, with energy molecules being artificial microbes or photocatalysts for photosynthesis [103]. However, in the case of

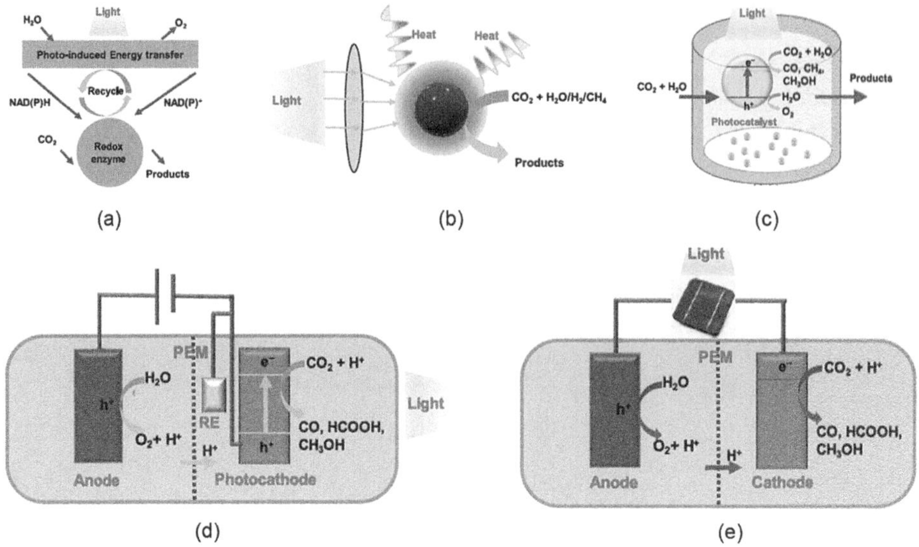

Figure 11.5 Schematic of various solar-assisted CO_2 conversion methods. Reproduced from [107] / American chemical society.

photothermal reduction, concentrated solar radiations are utilized to split CO_2, while biocatalyst and new devices could be utilized for converting CO_2 into fuels or chemicals in microbial PEC CO_2 reduction [104]. The PC, PS, PEC, and PV+EC approaches are widespread and much sort after, as illustrated in Figures 11.5a–e [105]. Though the functions appear almost identical, the difference lies in the thermodynamics of the Gibbs free energy $\Delta G < 0$ for PC, as it is accelerated by the catalysts and is thermodynamically favorable, but in the case of PS, $\Delta G > 0$ makes it thermodynamically unfavorable.

To determine the effectiveness and clarity of the reduction process, the fundamentals should be kept in check, such as:

1) The light to fuel conversion efficiency should be accurate.
2) The oxidation of donors must be considered, as the donors can help in CO_2 reduction either by direct oxidation of electron donors to produce carbon products or form intermediates that help in CO_2 reduction.

Hence, it is necessary to evaluate the fate of the donors while considering the overall CO_2 reduction [106, 107].

From a device perspective, photoelectrode-based PEC can make use of both PS and electrocatalysis for reduction of CO_2. The electrodes arrangement in a device can be classified into three configurations, such as Photoanode-dark cathode, Photocathode-dark anode, and Photocathode-photoanode and Photoanode-photocathode. Among these, a solar cell can be coated with catalysts to act as a photoelectrode. The PV cell can be coupled with an electrolyzer to make the PV+EC system. This separation allows easy selection of materials and easy processability relating to computability and stability. The method have successfully been used in experiments for splitting of water for H_2 production.

a) Performance Metrics

Quantitative measurement of electrocatalysis can be done using metrics like the onset potential, the Faradaic efficiency (FE), Current density, and the Energy efficiency (EF) [108]. Here, the onset potential is practical voltage and the difference between equilibrium potential and onset potential is the overpotential. FE denotes the selectivity of products defined by Equation 11.33.

$$\varepsilon = \frac{anF}{Q} \tag{11.33}$$

However, the Energy efficiency (EF) reflects the utilization of overall energy in obtaining the final product and is explained by Equation 11.34.

$$E_{enery} = \frac{E_{Eq}}{E_{Eq} + \eta} X\varepsilon_{Faradaic} \tag{11.34}$$

where *a* is the number of transferred electrons, *n* denotes the amount of specific product in mols, *F* is the Faraday's constant ($F = 96485\ C\ mol^{-1}$), and *Q* measures the quantity of charge consumed in reduction in coulombs (C), and E_{Eq} and η imply equilibrium potential and overpotential [109].

The main aspects of performance evaluation of efficient CO_2 reduction devices are selectivity, durability, and activity. In the case of PS/PC, the formation rate of products is considered which is normalized over the catalyst weight or the geometric area if substrates are taken into consideration. Another method to measure the light to fuel efficiency includes the Quantum efficiency (QE). In the case of PCE, the Faradaic efficiency is widely used. Additionally, the product detection potential is measured along with the j, which as the current per unit area is another important metric. Solar-to-Fuel efficiency (SFE) is another important key metric for measurement. However, SFE, is applicable only in cases where there is no voltage supplied. Then again, IPCE and absorbed light-current conversion efficiency also is an important metric in terms of device performance measurements. In the case of PV+EC, voltage and current densities similar to PEC are taken into consideration along with SFE [101, 110, 111]. Some of the other important metrics are: Overpotential, Faradaic efficiency (FE), Catalytic selectivity (CS), Turnover number, and Turnover frequency (TOF) = catalytic turnovers per unit time.

b) PS/PC System of Utilization

In the past decade, the activity, selectivity, and durability of PS/PC CO_2 reduction has been investigated. The surface activated cobalt oxide NPs (nanoparticles) as catalyst has shown a high formation rate reaction involving the CO_2 reduction to CH_4 and improved selectivity. For enhancement of catalytic efficiency, N-bromosuccinimide treatment is involved, resulting in the formation of Co_3O_4 with coordinated Br on the surface (Figure 11.6a). Further enhancement of CH_4 formation rate was increased by utilizing surface modification with sulfur elevating to 10 mmol $g^{-1}h^{-1}$ from 80 mmol $e^{-1}g^{-1}h^{-1}$, holding a QE of 2.3 5 and selectivity of 98% at 405 nm [101]. In the case of determining selectivity, the formation of a stable Cu-C-O-In intermediate at the Cu-In dual site in the case of sulfur-deficient $CuIn_5S_8$ was the key feature, thereby achieving nearly 100% selectivity for light driven

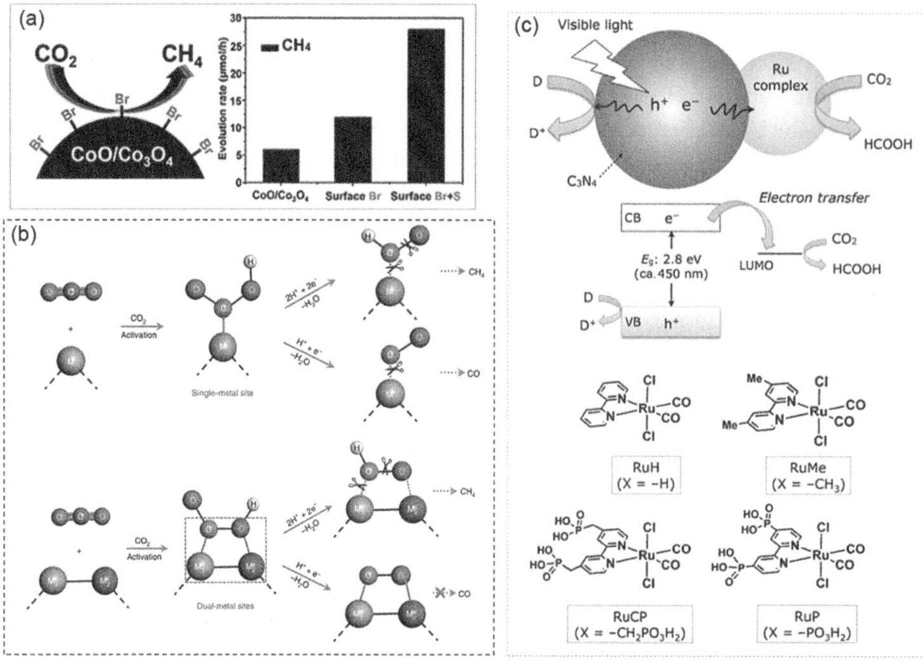

Figure 11.6 For PS/PC system (a) cobalt oxide nanoparticles, by surface activation for CO_2 reduction (reproduced with permission of [115], American Chemical Society); (b) catalytic systems (reproduced with permission of [113], Springer, Nature); and (c) CO_2 reduction using a Ru complex/C_3N_4 hybrid photocatalyst, along with its structure (reproduced with permission of [112], John Wiley & Sons).

CO_2 reduction to CH_4 (inset of Figure 11.6b) [112]. Similarly, in Figure 11.6c, a hybrid system of Ru complex and C_3N_4 carbon nitride selectively reduced Co_2 [111]. Terpyridine modification of iridium PC with 2-phenylpyridine (ppy) and 2,2′:6′,2″-terpyridine (tpy) ligand resulted in a turnover number (TON) of 265, involving a QE of 0.10, showing high durability of over 10 days [114].

c) PEC System of Utilization

A selective CO_2 reduction to $HCOO^-$, accompanied with high current density and operational stability, has been observed when mesoporous bismuth ns are transformed by cathode reactions of ultrathin atomic-sized bismuth oxycarbonate (−1.1 V vs. RHE, ~18 mA cm^{-2}, F ≈ 100%, OS = 12 h). Further, 1.5% efficiency was observed in solar to formate conversion when Bi NS was combined with an Ir/C anode in fuel cells, the performance justified by the enlarged surface of the mesoporous 2D NS morphology, abundant under coordinated Bi sites and the stability of the structure [116]. Figures 11.7d–g show the complete cell configuration, including the measured parameters using PEC configuration. Recently, a new photocathode was prepared out of an $MAPbI_3$ solar cell, by coating it with an alloy layer of $In_{0.4}Bi_{0.6}$, which successfully achieved an appreciable current density of 5.5 mA cm^{-2} at −0.6 V vs. RHE [117]. The device showed good operational stability of 1.5 h, with HCOOH producing FE of 100%. Similarly, a TiO_2-coated Cu_2O photoanode was coupled with a rhenium bipyridyl catalyst, which was able to obtain current density of ~1.5 mA cm^{-2} at −1.73 V vs. Fc^+/Fc, producing an FE of 100% [118] (inset of Figures 11.7a–c.).

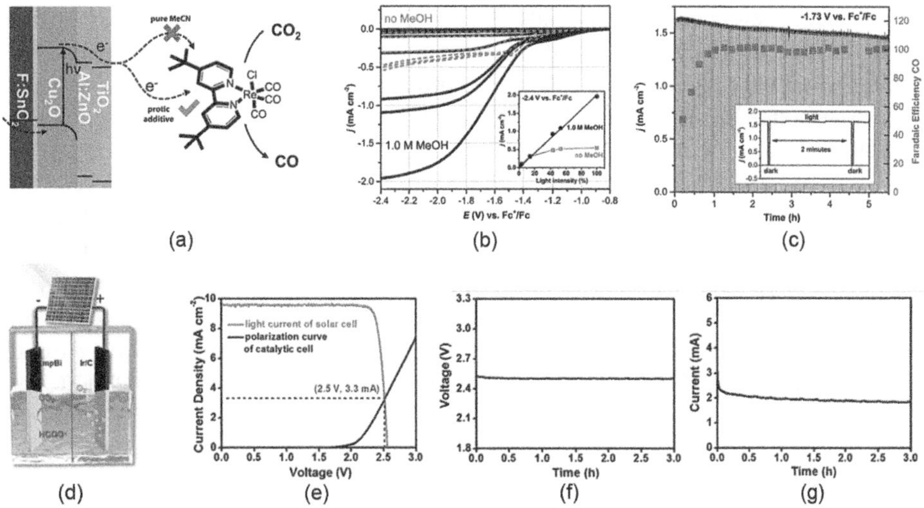

Figure 11.7 (a) Illustration of a photoelectrochemical cell with Cu_2O photocathodes and an Re-based molecular catalyst; (b) LSVs in the absence (red) and presence (black) of 1.0 M MeOH under several intensities; (c) cathodic current density and CO evolution efficiency of the Cu_2O photocathode. Both current density (black line) and Faradaic efficiency toward carbon monoxide (red dots) are reported (adapted from of [118]). PEC system of CO_2 generation via solar-driven CO_2 RR/OER full-cell electrolysis; (d) schematic showing the experimental setup for the solar-driven full-cell electrolysis; (e) I–V curve of the GaInP/GaInAs/Ge solar cell under illumination; (f) voltage changes; and (g) current evolution during the solar electrolysis (reproduced with permission of [122], John Wiley & Sons).

PEC CO_2 reduction to alcohols are also been studied; for example, methanol was produced by an Li-doped $CuFeO_2$ photoanode capable of an FE of almost 96.7% at −0.63 V [114]. Another challenge and objective of a PEC cell rescuing CO_2 is its durability. In this regard, a cell of the same photoanode containing $CuFeO_2$/CuO and Pt anode produced HCOOH for more than a week with a solar to formate efficiency of nearly 1% sans and external bias (FE >90%) [120] The photoanode-dark cathode configuration has also been investigated; for example, by considering a tandem GaAS/InGaP/TiO_2/Ni photoanode, a Ti mesh cathode coated with Pd/C nanoparticles has been constructed for solar driven CO_2 reduction [121]. In this configuration, a separate bipolar membrane was used along with separate catholyte and anolyte to allow problem-free operation. The device was capable of achieving SFE as appreciable as 10% with CO_2 to formate FE as >94% without any bias at a current density of 8.5 mA cm^{-2}. The surface and electrolyte interface is affected by the electrolyte material, such as FE values are obtained for highly reduced products such as CH_3OH, CH_4, and C_2H_5OH.

d) PV+EC CO_2 System of Utilization

PV+EC configuration that has produced much success in H_2 evolution has also turned its attention toward CO_2 reduction. Here, the two configurations are: i) coupled systems incorporated within the same unit two functions, and ii) conventional PV panels connected to EC cells. The bipolar membrane used to separate the anode and cathode, and the solar cell utilizing CO_2 reduction into CO and O, was done by using a GaInP/GaInAs/Ge PV [123]. It was capable of achieving a solar-to-CO efficiency of 13.4% with an SFE of 14.4%, while achieving an operational stability of 5 h. In an integrated solar-to-CO_2 reduction system, the unit achieved an efficiency of 13.9% and an SFE of 15.6% with an appreciable operational stability of 19 h [124]. By the incorporation of a Zn/Zn(II) redox mediator, two-step reduction medium-assisted CO_2 reduction was developed. Here, the medium acts as an electron carrier during PS, and this has achieved an efficiency of 15.6% [125]. Similarly, by using a gas diffusion electrode, 19% solar-to-CO efficiency has been achieved [95].

Commercial catalysts such as Ag and Au, onto which CO_2 can be selectively reduced to CO, are mainly used in PV+EC systems [126, 127]. In the case of PC/PS, CO and CH_4 are the major products, as both oxidation and reduction take place on the same particle. In a similar notation, HCOOH produced mainly by PEC and PV+EC becomes oxidized by the holes generated by the incidence of light or by the intermediate oxidizing agents in the case of PV+EC followed by PEC and then PS/PC. In the case of durability, PEC and PV+EC systems maintain a high operational stability of less than 1 h. In the case of PS/PC, a duration of 20 h is widely seen, with instability the reason for the corrosion from redox reactions of the active material.

Recently, $CsPbBr_3$ nanocubes rapidly facilitated the easy charge carriers' extraction [128]. In the case of PV+EC, the charge carriers are extracted on a nanosecond or picosecond scale. Conclusively, it can be said that however similar conditions and catalysts are utilized, the products formed are different in each approach. The summary of PS/PC, PEC, and PV+EC systes is given in Table 11.1.

11.1.2.4 Status of Commercial of CO_2 Generation

Industrial attention given toward utilizing CO_2 by producing several chemicals and fuels has been tremendously increasing in the past years. Bayers Chempark, a pilot plant in Germany, was opened in 2011 to produce polyurethanes by utilizing CO_2. DNV GL, another pilot unit, developed reduction of CO_2 to organic products of industrial interest. Similarly, a

Table 11.1 Challenges and summary of PS/PC, PEC, and PV+EC systems.

	PS/PC	PEC	PV+EC
Economic Feasibility	Catalyst remains costly Device setup cheap	Expensive setup as well catalyst requirements	Expensive setup as well as catalysts (as electrode size and membranes are considered)
Stability	Solid–Liquid interface causes stability concerns	Electrode–electrolyte interface causes stability issues	Relatively stable
Light Absorption	One or more light sources required	Photoanode/ photocathode/both	Single junction or multijunction can be utilized
Charge Collection	Not required		Rapid collection in PV cell
Charge Transfer	Both reactions take place on the same particle	Slow transfer between substrate/medium and electrode	Electro-chemical interface initiates reaction
Other Aspects	High surface area required	High surface area required, which initiates surface recombination	Stable electrocatalysts needed with minimum surface recombination

catalyst to make acrylic acid from ethylene and CO_2 was developed by BASF (Baden Aniline and Soda Factory) as an effective alternative for the bio-based route initially utilized by BASF and Arkema. Similarly, succinic acid production by Bioamber, Mitsubishi Chemicals, and acetone synthesis by Evonik or Roquette, etc. are other examples of the biotechnology route of utilizing CO_2. A plant in Svartsengi, Iceland synthesizes 5 Mton per year of methanol from CO_2, using H_2 produced from geothermal sources. In Germany, methane from CO_2 along with the production and storage of wind energy is for ultimate automotive use.

In the case of utilizing CO_2 to produce advance functional materials such as polymers and fine chemicals, developing alternate synthesis routes is more applicable than reducing GHG emission. Initiatives taken by the industrial sector include Bayer focused on production of polyoils for polyurethane. BASF's initiative for the production of polypropylene carbonate and Novomer Inc. for initiating the production of the polypropylene carbonate commercial sector includes Asahi Kasei's for their phosgene-free production of aromatic polycarbonate starting from ethylene epoxide, bisphenol-A, and CO_2. Several commercial plants have been developed from this, such as Bioamber, a starting facility in Pomacle, France, which uses an *Escherichia coli* strain and wheat-derived glucose as a substrate and has an annual capacity of 2000 tonnes. DSM (now Sabic) has been investing in the production of bio-succinic acid. BASF, with its Purac subsidiary, has invested in the production of bio-succinic acid using a BASF-developed bacterial strain (*Basfi succiniproducens*) and glycerine or glucose as a feedstock [129–131].

Production of longer chain alcohols is another route for utilizing CO_2; in this regard, the Mitsubishi Chemical Holdings Corporation has been investigating, at the University of California, the successful production of butanol and hexanol by applying genetically-modified algae. Similarly, at ULCA (University of California, Los Angeles), the bacteria, *E. coli* has been engineered to convert CO_2 into alcohols, and cyanobacterium has been used to produce isobutanol. Commercialization of the fermentation of syngas to ethanol by *C. ljungdahlii* has been developed, by combing biomass gasification, syngas fermentation, and distillation of ethanol from the reactor effluent [130, 131].

The European Chemical Industry, through the Sustainable Process Industry Public–Private Partnership initiative, aims for 30% reduction in fossil fuel energy by 2030, by utilizing CO_2 that could derive from CO_2 valorization technologies. In collaboration with Lotus Engineering (UK), FSK Technology (UK), PARC (Palo Alto, CA), and GT-Systems (UK), Edwards et al. presented the concept of obtaining methanol by proper utilization of CO_2 and combining it with renewable H_2. In this regard, GE showed that methanol, when implemented as a fuel, can be burned without needing any modifications that are major to the combustor ~20% lower emission of NO_x [130]. Therefore, it is evident that there is a huge potential for utilizing CO_2 for the production of syngas [128–133].

11.2 Conclusions and Future Prospects

H_2 produced from solar energy is a vital sustainable and environment-friendly route for energy.

In comparison to thermal processes, the photo-assisted processes of H_2 production are facile as a low temperature is only required. The economic analysis announces the competitiveness of H_2 production being on par with other conventional fuels. Hence, novel synthesis techniques that are facile and cost-effective are required for obtaining materials capable of high energy conversion. Hydrogen biomass is low cost and also not a nonrenewable process, which could utilize sunlight to convert biomass and even pollutants in waste water to H_2, which would eventually have a positive effect on pollution and renewable energy production. This present chapter also focuses on the available approaches of the solar-driven CO_2 reduction systems, with their respective measurement metrics. The selectivity, durability, and activity of the PS/PC, PEC, and PV+EC systems have been discussed in detail.

In summary, it could be observed that:

i) Product distribution remains different for different approaches
ii) PV+EC could be improvized for industrial applications in the near future.

Further, Gas Diffusion Electrodes are a new trend that could be exploited properly for eliminating the existing limitations, such as sluggish transport with minimum capacity and less turnover at the activation sites. Selectivity remains as a major metric if the nature of active sites is understood. In order to design photoanode materials with high selectivity, understanding of atomistic active sites and conversion mechanism are required. Hybrid photoelectrode systems provide a great platform for solar-CO_2 reduction and utilization. Some of the important criteria for achieving a good hybrid photoelectrode system are: Light absorption, Charge transport, Charge transfer kinetics and Stability, and robustness of the material and device.

It is no wonder that not many materials exist that can fulfill all the criteria. However, they can be taken care of by integrating different materials for a common benefit such as incorporating PS/PC and PEC materials together. Also, it is observed that insertion of a highly conductive material into the photoelectrode could improve its charge carrier transport properties as well the photocurrent generation properties. Organic photoelectrodes have also been utilized for exploiting the attractive features of each material. Similarly, the surface-to-bulk ratio effects on the surface functional groups remains unexplored. Plasmonic catalysis is a promising as well emerging area in CO_2 utilization methods, as these particles can harness light with their hot carriers and use it for photocatalytic CO_2 reduction.

A buried junction, if designed properly, can incorporate PV and EC technologies together for advantageous usage, as they can overcome every issue specifically dealing with stability, thereby increasing the durability. Value-added anode processes, by adding other anodic oxidation reactions to the existing anodic oxygen evolution and cathodic CO_2 reduction, which consumes less power, can achieve high-value products which can provide openings for the renewable next-generation PC+PEC systems for a better future.

In PV+EC systems, the majority of the studies focus on the catalysts, while the electrodes, membrane, and the membrane assembly are ignored. Replacement of a liquid electrolyte with a solid electrolyte can be a better alternative, as liquid electrolytes require energy-intensive downstream separation as the salt is dissolved in the liquid.

References

1 Jacobson, T.A., Kler, J.S., Hernke, M.T. et al. (2019). Direct human health risks of increased atmospheric carbon dioxide. *Nature Sustainability* 2(8): 691–701.

2 Jarvis, S.M. and Samsatli, S. (2018). Technologies and infrastructures underpinning future CO_2 value chains: A comprehensive review and comparative analysis. *Renewable & Sustainable Energy Reviews* 85: 46–68.

3 Rogelj, J., Den Elzen, M., Höhne, N. et al. (2016). Paris agreement climate proposals need a boost to keep warming well below 2C. *Nature* 534(7609): 631–639.

4 De Luna, P., Hahn, C., Higgins, D. et al. (2019). What would it take for renewably powered electrosynthesis to displace petrochemical processes? *Science* 364(6438): 3506.

5 Villadsen, S N., Fosbøl, P.L., Angelidaki, I. et al. (2019). The potential of biogas: The solution to energy storage. *ChemSusChem* 12(10): 2147–2153.

6 DOE, US (2020). Hydrogen strategy, enabling a low-carbon economy. *Office of Fossil Energy* 24: 3–10.

7 IEA (2019). *The role of CO_2 storage. Paris*: IEA. https://www.iea.org/reports/the-role-of-co2-storage

8 Strezov, V., Anawar, H.M., and Strezov, V. eds. (2019). *Renewable Energy Systems From Biomass.* CRC Press, Taylor and Francis group.

9 Weldekidan, H., Strezov, V., Li, R. et al. (2020). Distribution of solar pyrolysis products and product gas composition produced from agricultural residues and animal wastes at different operating parameters. *Renewable Energy* 151: 1102–1109.

10 Chen, J., Xu, W., Zuo, H. et al. (2019). System development and environmental performance analysis of a solar-driven supercritical water gasification pilot plant for hydrogen production using life cycle assessment approach. *Energy Conversion and Management* 184: 60–73.

11 Li, X., Shen, Y., Wei, L. et al. (2020). Hydrogen production of solar-driven steam gasification of sewage sludge in an indirectly irradiated fluidized-bed reactor. *Applied Energy* 261: 114229.

12 Vishnevetsky, I. and Epstein, M. (2007). Production of hydrogen from solar zinc in steam atmosphere. *International Journal of Hydrogen Energy* 32(14): 2791–2802.

13 Abanades, S. and Flamant, G. (2006). Thermochemical hydrogen production from a two-step solar-driven water-splitting cycle based on cerium oxides. *Solar Energy* 80(12): 1611–1623.

14 Chambon, M., Abanades, S., Flamant, G. et al. (2011). Thermal dissociation of compressed ZnO and SnO_2 powders in a moving-front solar thermochemical reactor. *AIChE Journal* 57(8): 2264–2273.

15 Bhosale, R.R. (2017). Thermodynamic analysis of solar driven SnO_2/SnO based thermochemical water splitting cycle. *Energy Conversion and Management* 135: 226–235.

16 Bhosale, R.R. (2018). Thermodynamic efficiency analysis of zinc oxide based solar driven thermochemical H_2O splitting cycle: Effect of partial pressure of O_2, thermal reduction and H_2O splitting temperatures. *International Journal of Hydrogen Energy* 43(32): 14915–14924.

17 Charvin, P., Abanades, S., Flamant, G. et al. (2007). Two-step water splitting thermochemical cycle based on iron oxide redox pair for solar hydrogen production. *Energy* 32(7): 1124–1133.

18 Le Gal, A., Abanades, S., Bion, N. et al. (2013). Reactivity of doped ceria-based mixed oxides for solar thermochemical hydrogen generation via two-step water-splitting cycles. *Energy & Fuels* 27(10): 6068–6078.

19 Krenzke, P.T. and Davidson, J.H. (2014). Thermodynamic analysis of syngas production via the solar thermochemical cerium oxide redox cycle with methane-driven reduction. *Energy & Fuels* 28(6): 4088–4095.

20 Charvin, P., Abanades, S., Lemort, F. et al. (2007). Hydrogen production by three-step solar thermochemical cycles using hydroxides and metal oxide systems. *Energy & Fuels* 21(5): 2919–2928.

21 Temiz, M. and Dincer, I. (2021). Concentrated solar driven thermochemical hydrogen production plant with thermal energy storage and geothermal systems. *Energy* 219: 119554.
22 Brendelberger, S., Rosenstiel, A., Lopez-Roman, A. et al. (2020). Performance analysis of operational strategies for monolithic receiver-reactor arrays in solar thermochemical hydrogen production plants. *International Journal of Hydrogen Energy* 45(49): 26104–26116.
23 Mohanty, B., Mitra, A., Jena, B. et al. (2020). MoS_2 quantum dots as efficient electrocatalyst for hydrogen evolution reaction over a wide pH range. *Energy & Fuels* 34(8): 10268–10275.
24 Xu, J., Zhang, C., Liu, H. (2020). Amorphous MoO_X-stabilized single platinum atoms with ultrahigh mass activity for acidic hydrogen evolution. *Nano Energy* 70: 104529.
25 Smiljanic, M., Rakocevic, Z., Maksic, A. et al. (2014). Hydrogen evolution reaction on platinum catalyzed by palladium and rhodium nano-islands. *Electrochimica Acta* 117: 336–343.
26 Vasić, D.D., Pašti, I.A., Mentus, S.V. et al. (2013). DFT study of platinum and palladium overlayers on tungsten carbide: Structure and electrocatalytic activity toward hydrogen oxidation/evolution reaction. *International Journal of Hydrogen Energy* 38(12): 5009–5018.
27 Pai, M.R., Banerjee, A.M., Rawool, S.A. et al. (2016). A comprehensive study on sunlight driven photocatalytic hydrogen generation using low cost nanocrystalline Cu-Ti oxides. *Solar Energy Materials and Solar Cells* 154: 104–120.
28 Hua, W., Sun, H., Liu, H. et al. (2021). Interface engineered NiMoN/Ni_3N heterostructures for enhanced alkaline hydrogen evolution reaction. *Applied Surface Science* 540: 148407.
29 Li, D., Liu, H., Feng, L. et al. (2020). A review on advanced FeNi-based catalysts for water splitting reaction. *Energy & Fuels* 34(11): 13491–13522.
30 Wang, D., Li, Q., Han, C. et al. (2019). Atomic and electronic modulation of self-supported nickel-vanadium layered double hydroxide to accelerate water splitting kinetics. *Nature Communications* 10(1): 1–12.
31 Qian, G., Yu, G., Lu, J. et al. (2020). Ultra-thin N-doped-graphene encapsulated Ni nanoparticles coupled with MoO_2 nanosheets for highly efficient water splitting at large current density. *Journal of Materials Chemistry A* 8(29): 14545–14554.
32 Kibsgaard, J. and Chorkendorff, I. (2019). Considerations for the scaling-up of water splitting catalysts. *Nature Energy* 4(6): 430–433.
33 Xiao, X. and Huang, D. (2018). Engineering NiS/Ni_2P heterostructures for efficient electrocatalytic water splitting. *ACS Applied Materials & Interfaces* 10(5): 4689–4696.
34 Pehlivan, I., Atak, G., Niklasson, G. et al. (2021). Electrochromic solar water splitting using a cathodic WO_3 electrocatalyst. *Nano Energy* 81: 105620.
35 Yi, S., Wang, Z., Li, H. et al. (2021). Coupling effects of indium oxide layer on hematite enabling efficient photo-electrochemical water splitting. *Applied Catalysis B: Environmental* 283: 119649.
36 Sun, M., Wang, Q., Dai, B. et al. (2020). Construction of a facile recyclable graphene-like C_3N_4 cilia array for effective visible-light-responsive photocatalytic hydrogen production. *Energy & Fuels* 34(8): 10290–10298.
37 Cabán-Acevedo, M., Stone, M., Schmidt, J. et al. (2015). Efficient hydrogen evolution catalysis using ternary pyrite-type cobalt phosphosulphide. *Nature Materials* 14(12): 1245–1251.
38 Ren, Y., Zhao, L., Jing, D. et al. (2016). Investigation and modeling of CPC based tubular photocatalytic reactor for scaled-up hydrogen production. *International Journal of Hydrogen Energy* 41(36): 16019–16031.

39 Sun, X., Liu, H., Xu, G. et al. (2021). Embedding Co_2P nanoparticles into N&P co-doped carbon fibers for hydrogen evolution reaction and supercapacitor. *International Journal of Hydrogen Energy* 46(2): 1560–1568.

40 Abdi, F., Han, L., Smets, A. et al. (2013). Efficient solar water splitting by enhanced charge separation in a bismuth vanadate-silicon tandem photoelectrode. *Nature Communications* 4(1): 1–7.

41 Eron, M. and Rothwarf, A. (1985). Interface charging and solar-cell characteristics: $CuInSe_2/CdS$. *Journal of Applied Physics* 57(6): 2275–2279.

42 Saad, M.and Kassis, A. (2003). Effect of interface recombination on solar cell parameters. *Solar Energy Materials & Solar Cells* 79(4): 507–517.

43 Dotan, H., Sivula, K., Grätzel, M. et al. (2011). Probing the photo-electrochemical properties of hematite (α-Fe_2O_3) electrodes using hydrogen peroxide as a hole scavenger. *Energy & Environmental Science* 4(3): 958–964.

44 Zhang, S., Shen, L., Ye, T. et al. (2020). Noble-metal-free perovskite–$BiVO_4$ tandem device with simple preparation method for unassisted solar water splitting. *Energy & Fuels* 34(4): 5016–5023.

45 Feng, J., Huang, H., Fang, T. (2019). Defect engineering in semiconductors: Manipulating nonstoichiometric defects and understanding their impact in oxynitrides for solar energy conversion. *Advanced Functional Materials* 29(11): 1808389.

46 Moniz, J.A., Shevlin, A.S., Martin, J.D. (2015). Visible-light driven heterojunction photocatalysts for water splitting: A critical review. *Savio Energy & Environmental Science* 8(3): 731–759.

47 Zhang, K., Ma, M., Li, P. et al. (2016). Water splitting progress in tandem devices: Moving photolysis beyond electrolysis. *Advanced Energy Materials* 6(15): 1600602.

48 Nadeem, M.A. and Idriss, H. (2020). Effect of pH, temperature, and low light flux on the performance (16% STH) of coupled triple junction solar cell to water electrolysis. *Journal of Power Sources* 459: 228074.

49 Hu, Y., Wu, Y., Feng, J. et al. (2018). Rational design of electrocatalysts for simultaneously promoting bulk charge separation and surface charge transfer in solar water splitting photoelectrodes. *Journal of Materials Chemistry A* 6(6): 2568–2576.

50 Suarez, C. M., Hernández, S., Russo, N. et al. (2015). $BiVO_4$ as photocatalyst for solar fuels production through water splitting: A short review. *Applied Catalysis A: General* 504: 158–170.

51 Ziani, A., Al-Shankiti, I., Khan, M.A. et al. (2020). Integrated photo-electrocatalytic (PEC) systems for water splitting to hydrogen and oxygen under concentrated sunlight: Effect of internal parameters on performance. *Energy & Fuels* 34(10): 13179–13185.

52 Rong, Y., Hu, Y., Mei, A. et al. (2018). Challenges for commercializing perovskite solar cells. *Science* 361(6408): 8235.

53 Sonker, R.K., Shastri, R., Johari, R. et al. (2021). Superficial synthesis of CdS quantum dots for an efficient perovskite-sensitized solar cell. *Energy & Fuels* 35(9): 8430–8435.

54 Hassani, H., Zaouche, F., Rekioua, D. et al. (2020). Feasibility of a standalone photovoltaic/battery system with hydrogen production. *Journal of Energy Storage* 31: 101644.

55 Burton, N.A., Padilla, R.V., Rose, A.H. et al. (2021). Increasing the efficiency of hydrogen production from solar powered water electrolysis. *Renewable & Sustainable Energy Reviews* 135: 110255.

56 Allouhi, A. (2020). Management of photovoltaic excess electricity generation via the power to hydrogen concept: A year-round dynamic assessment using Artificial Neural Networks. *International Journal of Hydrogen Energy* 45(41): 21024–21039.

57 Ganjehsarabi, H. (2019). Performance assessment of solar-powered high pressure proton exchange membrane electrolyzer: A case study for Erzincan. *International Journal of Hydrogen Energy* 44(20): 9701–9707.

58 Caldwell, R. (2018). Quarterly report & Appendix 4C, quarter 3 – 2018; *Great Cell Solar* Dyesol Ltd., Global Leaders in Perovskite Solar Cell Technology

59 Wang, Q., Li, Y., Hisatomi, T. et al. (2015). Z-scheme water splitting using particulate semiconductors immobilized onto metal layers for efficient electron relax author links open overlay panel. *Journal of Catalysis* 328: 308–315.

60 Hisatomi, T., Kubota, J., Domen, K. et al. (2014). Recent advances in semiconductors for photocatalytic and photo-electrochemical water splitting. *Chemical Society Reviews* 43(22): 7520–7535.

61 Yuan, Y.P., Ruan, L.W., Barber, J. et al. (2014). Hetero-nanostructured suspended photocatalysts for solar-to-fuel conversion. *Energy & Environmental Science* 7(12): 3934–3951.

62 Wang, Q., Hisatomi, T., Ma, S.S.K. et al. (2014). Core/shell structured La-and Rh-codoped $SrTiO_3$ as a hydrogen evolution photocatalyst in Z-scheme overall water splitting under visible light irradiation. *Chemistry of Materials* 26(14): 4144–4150.

63 Wang, Y., Hao, X., Zhang, L. et al. (2020). Rational design of all-solid-state 0D/2D $Mn_{0.2}Cd_{0.8}S/CeO_2$ direct Z-scheme for photocatalytic hydrogen evolution. *Energy & Fuels* 34(2): 2599–2611.

64 Maeda, K. and Domen, K. (2007). New non-oxide photocatalysts designed for overall water splitting under visible light. *The Journal of Physical Chemistry C* 111(22): 7851–7861.

65 You, Y., Wang, S., Xiao, K. et al. (2018). Z-scheme g-C_3N_4/Bi_4NbO_8Cl heterojunction for enhanced photocatalytic hydrogen production. *ACS Sustainable Chemistry & Engineering* 6(12): 16219–16227.

66 Rabady, R.I. (2014). Solar spectrum management for effective hydrogen production by hybrid thermo-photovoltaic water electrolysis. *International Journal of Hydrogen Energy* 39(13): 6827–6836.

67 Taboada, E., Angurell, I., Llorca, J. et al. (2014). Hydrogen photoproduction from bio-derived alcohols in an optical fiber honeycomb reactor loaded with Au/TiO_2. *Journal of Photochemistry and Photobiology A: Chemistry* 281: 35–39.

68 Thongthep, P., Moonmangmee, S., Ponchio, C. et al. (2021). Solar/photo-electrocatalytic cell development for H_2 production and simultaneous organic dye degradation. *Materials Science in Semiconductor Processing* 124: 105597.

69 Salgado, S.Y.A., Zamora, R.M.R., Zanella, R. et al. (2016). Photocatalytic hydrogen production in a solar pilot plant using a Au/TiO_2 photo catalyst. *International Journal of Hydrogen Energy* 41(28): 11933–11940.

70 Liu, J., Zhou, W., Fan, S. et al. (2019). Coproduction of hydrogen and butanol by *Clostridium acetobutylicum* with the biofilm immobilized on porous particulate carriers. *International Journal of Hydrogen Energy* 44(23): 11617–11624.

71 Jiang, D., Ge, X., Zhang, T. et al. (2020). Effect of alkaline pretreatment on photo-fermentative hydrogen production from giant reed: Comparison of NaOH and $Ca(OH)_2$. *Bioresource Technology* 304: 123001.

72 Reaño, R.L. (2020). Assessment of environmental impact and energy performance of rice husk utilization in various biohydrogen production pathways. *Bioresource Technology* 299: 122590.

73 Mona, S., Kumar, S.S., Kumar, V. et al. (2020). Green technology for sustainable biohydrogen production (waste to energy): A review. *Science of the Total Environment* 728: 138481.

74 Xu, Z., Wang, S., Zhao, C. et al. (2020). Photosynthetic hydrogen production by droplet-based microbial micro-reactors under aerobic conditions. *Nature Communications* 11(1): 1–10.

75 Das, D. (2009). Advances in biohydrogen production processes: An approach towards commercialization. *International Journal of Hydrogen Energy* 34(17): 7349–7357.

76 Boran, E., Özgür, E., Yücel, M. et al. (2012). Biohydrogen production by *Rhodobacter capsulatus* Hup−mutant in pilot solar tubular photobioreactor. *International Journal of Hydrogen Energy* 37(21): 16437–16445.

77 Zhang, Q., Zhang, Z., Wang, Y. et al. (2018). Sequential dark and photo fermentation hydrogen production from hydrolyzed corn stover: A pilot test using 11 m^3 reactor. *Bioresource Technology* 253: 382–386.

78 Zhang, Q., Wang, Y., Zhang, Z. et al. (2017). Photo-fermentative hydrogen production from crop residue: A mini review. *Bioresource Technology* 229: 222–230.

79 Zhang, Z., Li, Y., Zhang, H., He, C. et al. (2017). Potential use and the energy conversion efficiency analysis of fermentation effluents from photo and dark fermentative bio-hydrogen production. *Bioresource Technology* 245: 884–889.

80 Vatsala, T.M., Raj, S.M., Manimaran, A. et al. (2008). A pilot-scale study of biohydrogen production from distillery effluent using defined bacterial co-culture. *International Journal of Hydrogen Energy* 33(20): 5404–5415.

81 Zhang, Q., Lu, C., Lee, D.J.T. et al. (2017). Photo-fermentative hydrogen production in a 4 m^3 baffled reactor: Effects of hydraulic retention time. *Bioresource Technology* 239: 533–537.

82 James, B.D., Baum, G.N., Perez, J., and Baum, K.N. (2009). Technoeconomic boundary analysis of biological pathways to hydrogen production, subcontract Report No. NREL/SR-560-46674; *National Renewable Energy Laboratory (NREL)*, Golden, CO, USA.

83 Sharma, S., Basu, S., Shetti, N.P. et al. (2020). Waste-to-energy nexus for circular economy and environmental protection: Recent trends in hydrogen energy. *Science of the Total Environment* 713: 136633.

84 Sathyaprakasan, P. and Kannan, G. (2015). Economics of bio-hydrogen production. *International Journal of Environmental Science and Development* 6(5): 352–356.

85 Sharma, M., Kaushik, A., and Rathore, D. (2017). Biohydrogen economy: Challenges and prospects for commercialization. In: *Biohydrogen Production: Sustainability of Current Technology and Future Perspective* (eds Singh, A. and Rathore, D.). New Delhi: Springer.

86 Khan, M.A., Al-Shankiti, I., Ziani, A. et al. (2021). Demonstration of green hydrogen production using solar energy at 28% efficiency and evaluation of its economic viability. *Sustainable Energy & Fuels* 5(4): 1085–1094.

87 Modisha, P.M., Ouma, C.N., Garidzirai, R. et al. (2019). The prospect of hydrogen storage using liquid organic hydrogen carriers. *Energy & Fuels* 33(4): 2778–2796.

88 Müller, K., Brooks, K., Autrey, T. et al. (2018). Releasing hydrogen at high pressures from liquid carriers: Aspects for the H_2 delivery to fueling stations. *Energy & Fuels* 32(9): 10008–10015.

89 Du Toit, M.H., Avdeenkov, A.V. et al. (2018). Reviewing H_2 combustion: A case study for non-fuel-cell power systems and safety in passive autocatalytic recombiners. *Energy & Fuels* 32(6): 6401–6422.

90 Badwal, S.P., Giddey, S., Munnings, C. et al. (2018). Emerging technologies, markets and commercialization of solid-electrolytic hydrogen production. *Wiley Interdisciplinary Reviews: Energy & Environment* 7(3): e286.

91 Schoenung, S.M. and Keller, J.O. (2017). Commercial potential for renewable hydrogen in California. *International Journal of Hydrogen Energy* 42(19): 13321–13328.

92 Badwal, S.P.S., Giddey, S., Kulkarni, A. et al. (2015). Direct ethanol fuel cells for transport and stationary applications: A comprehensive review. *Applied Energy* 145: 80–103.

93 Giddey, S., Badwal, S.P.S., Munnings, C. et al. (2017). Ammonia as a renewable energy transportation media. *ACS Sustainable Chemistry & Engineering* 5(11): 10231–10239.

94 Kamkeng, A.D., Wang, M., Hu, J. et al. (2021). Transformation technologies for CO_2 utilisation: Current status, challenges and future prospects. *Chemical Engineering Journal* 409: 128138.

95 Mohanty, U.S., Ali, M., Azhar, M.R. et al. (2021). Current advances in syngas ($CO + H_2$) production through bi-reforming of methane using various catalysts: A review. *International Journal of Hydrogen Energy* 46(65): 32809–32845.

96 Choudhary, T.V. and Choudhary, V.R. (2008). Energy-efficient syngas production through catalytic oxy-methane reforming reactions. *Angewandte Chemie International Edition* 47(10): 1828–1847.

97 Zhao, Y., Liu, Z., and Meyers, R.A. (2019). Transformation of CO_2 into valuable chemicals. In: *Green Chemistry and Chemical Engineering. Encyclopedia of Sustainability Science and Technology Series* (eds Han, B. and Wu, T.). New York: Springer.

98 Hills, C.D., Tripathi, N., Carey, P.J. et al. (2020). Mineralization technology for carbon capture, utilization, and storage. *Frontiers in Energy Research* 8: 142.

99 Nitopi, S., Bertheussen, E., Scott, S.B. et al. (2019). Progress and perspectives of electrochemical CO_2 reduction on copper in aqueous electrolyte. *Chemical Reviews* 119(12): 7610–7672.

100 Kumaravel, V., Bartlett, J., Pillai, S.C. et al. (2020). Photo-electrochemical conversion of carbon dioxide (CO_2) into fuels and value-added products. *ACS Energy Letters* 5(2): 486–519.

101 Kumar, B., Llorente, M., Froehlich, J. et al. (2012). Photochemical and photo-electrochemical reduction of CO_2. *Annual Review of Physical Chemistry* 63: 541–569.

102 Benson, E.E., Kubiak, C.P., Sathrum, A.J. et al. (2009). Electrocatalytic and homogeneous approaches to conversion of CO_2 to liquid fuels. *Chemical Society Reviews* 38(1): 89–99.

103 Savéant, J.M. (2008). Molecular catalysis of electrochemical reactions. Mechanistic aspects. *Chemical Reviews* 108(7): 2348–2378.

104 Yano, J., Haber, J.A., Gregoire, J.M. et al. (2014). JCAP research on solar fuel production at light sources. *Synchrotron Radiation News* 27(5): 14–17.

105 Jia, J., O'Brien, P.G., He, L. et al. (2016). Visible and near-infrared photothermal catalyzed hydrogenation of gaseous CO_2 over nanostructured Pd@Nb_2O_5. *Advanced Science* 3(10): 1600189.

106 He, J. and Janáky, C. (2020). Recent advances in solar-driven carbon dioxide conversion: Expectations versus reality. *ACS Energy Letters* 5(6): 1996–2014.

107 Kamat, P.V. and Jin, S. (2018). Semiconductor photocatalysis: "tell us the complete story!" *ACS Energy Letters* 3(3): 622–623.

108 Kramm, U.I., Marschall, R., Rose, M. et al. (2019). Pitfalls in heterogeneous thermal, electro-and photocatalysis. *ChemCatChem* 11(11): 2563–2574.

109 Ma, S., Kenis, P.J., and Ru-Huei, J. (2013). Electrochemical conversion of CO_2 to useful chemicals: Current status, remaining challenges, and future opportunities. *Current Opinion in Chemical Engineering* 2(2): 191–199.

110 Pei, Y., Zhong, H., Jin, F. et al. (2021). A brief review of electrocatalytic reduction of CO_2-Materials, reaction conditions, and devices. *Energy Science & Engineering* 9(7): 1012–1032.

111 Kuriki, R., Sekizawa, K., Ishitani, O. et al. (2015). Visible-light-driven CO_2 reduction with carbon nitride: Enhancing the activity of ruthenium catalysts. *Angewandte Chemie International Edition* 54(8): 2406–2409.

112 Schreier, M., Gao, P., Mayer, M.T. et al. (2015). Efficient and selective carbon dioxide reduction on low cost protected Cu_2O photocathodes using a molecular catalyst. *Energy & Environmental Science* 8(3): 855–861.

113 Yuan, J., Yang, L., Hao, C. et al. (2019). Communication: Lithium-doped $CuFeO_2$ thin film electrodes for photoelectrochemical reduction of carbon dioxide to methanol. *Journal of the Electrochemical Society* 166(14): H718.

114 Choi, J.Y., Lim, C.K., Park, B. et al. (2019). Surface activation of cobalt oxide nanoparticles for photocatalytic carbon dioxide reduction to methane. *Journal of Materials Chemistry A* 7(25): 15068–15072.

115 Kang, U., Choi, S.K., Ham, D.J. et al. (2015). Photosynthesis of formate from CO_2 and water at 1% energy efficiency via copper iron oxide catalysis. *Energy & Environmental Science* 8(9): 2638–2643.

116 Wang, Y., Liu, J., Wang, Y. et al. (2018). Efficient solar-driven electrocatalytic CO_2 reduction in a redox-medium-assisted system. *Nature Communications* 9(1): 1–8.

117 Cheng, W.H., Richter, M.H., Sullivan, I. et al. (2020). CO_2 reduction to CO with 19% efficiency in a solar-driven gas diffusion electrode flow cell under outdoor solar illumination. *ACS Energy Letters* 5(2): 470–476.

118 Ma, M., Liu, K., Shen, J. et al. (2018). *In situ* fabrication and reactivation of highly selective andstable Ag catalysts for electrochemical CO_2 conversion. *ACS Energy Letters* 3(6): 1301–1306.

119 Zhao, S., Jin, R., Jin, R. et al. (2018). Opportunities and challenges in CO_2 reduction by gold-and silver-based electrocatalysts: From bulk metals to nanoparticles and atomically precise nanoclusters. *ACS Energy Letters* 3(2): 452–462.

120 Samu, G.F., Balog, Á., De Angelis, F.C. et al. (2019). Electrochemical hole injection selectively expels iodide from mixed halide perovskite films. *Journal of the American Chemical Society* 141(27): 10812–10820.

121 Yang, H., Han, N., Deng, J. et al. (2018). Selective CO_2 reduction on 2D mesoporous Bi nanosheets. *Advanced Energy Materials* 8(35): 1801536.

122 Beauprez, J.J., De Mey, M., Soetaert, W.K. et al. (2010). Microbial succinic acid production: Natural versus metabolic engineered producers. *Process Biochemistry* 45(7): 1103–1114.

123 Cheng, K.K., Zhao, X.B., Zeng, J. et al. (2012). Biotechnological production of succinic acid: Current state and perspectives. *Biofuels, Bioproducts and Biorefining* 6(3): 302–318.

124 Zhang, K., Sawaya, M.R., Eisenberg, D.S. et al. (2008). Expanding metabolism for biosynthesis of nonnatural alcohols. *Proceedings of the National Academy of Sciences* 105(52): 20653–20658.

125 Gao, D., Wu, X., Wang, P. et al. (2021). Selenium-enriched amorphous $NiSe^{1+}_x$ nanoclusters as a highly efficient cocatalyst for photocatalytic H_2 evolution. *Chemical Engineering Journal* 408: 127230.

126 Lai, G.J., Lyu, L.M., Huang, Y.S. et al. (2021). Few-layer WS_2–MoS_2 in-plane heterostructures for efficient photocatalytic hydrogen evolution. *Nano Energy* 81: 105608.

127 Wang, Y., Meng, X., Hu, Q. et al. (2021). Visible-light driven $ZnIn_2S_4/TiO_{2-x}$ heterostructure for boosting photocatalytic H_2 evolution. *International Journal of Hydrogen Energy* 46(9): 6262–6271.

128 Banerjee, R., Pal, A., Ghosh, D. et al. (2021). Improved photocurrent response, photostability and photocatalytic hydrogen generation ability of CdS nanoparticles in presence of mesoporous carbon. *Materials Research Bulletin* 134: 111085.

129 Samaniego-Benitez, J.E., Lartundo-Rojas, L., García-García, A. et al. (2021). One-step synthesis and photocatalytic behavior for H_2 production from water of ZnS/MoS_2 composite material. *Catalysis Today* 360: 99–105.

130 Magnin, J.P. and Deseure, J. (2019). Hydrogen generation in a pressurized photobioreactor: Unexpected enhancement of biohydrogen production by the phototrophic bacterium *Rhodobacter capsulatus*. *Applied Energy* 239: 635–643.

131 Mirza, S. S., Qazi, J.I., Liang, Y. (2019). Growth characteristics and photofermentative biohydrogen production potential of purple non sulfur bacteria from sugar cane bagasse. *Fuel* 255: 115805.

132 Li, X., Sun, Y., Xu, J. et al. (2019). Selective visible-light-driven photocatalytic CO_2 reduction to CH_4 mediated by atomically thin $CuIn_5S_8$ layers. *Nature Energy* 4(8): 690–699.

Index

Note: Page numbers for figures are in *italics*, tables are in **bold**

Green Energy Harvesting: Materials for Hydrogen Generation and Carbon Dioxide Reduction, First Edition. Edited by Pooja Devi.

d

e

i

k

l

m

n

q

r

s

t

u

v

w

x

y

z